Mathematics in Ancient Iraq

Mathematics in Ancient Iraq

A Social History

ELEANOR ROBSON

PRINCETON UNIVERSITY PRESS
PRINCETON AND OXFORD

LIBRARY OF CONGRESS CATALOGING-IN-PUBLICATION DATA
Robson, Eleanor.
    Mathematics in ancient Iraq : a social history / Eleanor Robson.
        p.   cm.
    Includes bibliographical references and index.
    ISBN 978-0-691-09182-2 (cloth : alk. paper)   1. Mathematics, Ancient—Iraq—History.
2. Mathematics—Iraq—History.   I. Title.
    QA22.R629 2008
    510.935—dc22   2007041758

British Library Cataloging-in-Publication Data is available

This book has been composed in Sabon and Swiss 721

Printed on acid-free paper. ∞

press.princeton.edu

Printed in the United States of America

10 9 8 7 6 5 4 3 2 1

For Luke, my love

CONTENTS

FIGURES

TABLES

PREFACE

Old Babylonian mathematics has a unique place in the interests and affections of mathematicians as the world's first 'pure' mathematics. This special status rests on the abstraction and sophistication of the sexagesimal place value system, a highly accurate approximation for the square root of 2, and the apparently Pythagorean complexities of the famous cuneiform tablet Plimpton 322. Throughout the twentieth century Old Babylonian mathematics was compared favourably to the mathematics of its contemporary and neighbour, Middle Kingdom Egypt, which was considered to consist of little more than utilitarian rules of thumb, and positioned as the forerunner to the 'miracle' of classical Greek mathematics. Assyriologists, on the whole, have been less enamoured of the subject, seeing it as overly complex and marginal to mainstream concerns within the intellectual and socio-economic history of ancient Iraq. Mathematics barely features in the many textbooks, handbooks, and encyclopedias of ancient Mesopotamia, despite the fact that the overwhelming majority of cuneiform tablets are primarily records of quantitative data.

These two attitudes, admittedly caricatured here, are perhaps a natural reaction to the way in which modern experts on the subject have tended to present their work. But over the last few decades a new picture has developed, from the perspective of the social and intellectual history of ancient Iraq. That is not to say that the historians of the mid-twentieth century were somehow ignorant or negligent: they made enormous strides in the internal analysis of Babylonian mathematics. Rather, they were working within a philosophical paradigm that, however mathematically attractive, is not very productive for historians of the subject.

There are good philosophical grounds for believing that abstract mathematical objects such as numbers and sets in some sense exist, independent of our beliefs about them. And historians and popularisers of Babylonian mathematics have primarily been mathematicians, with predominantly realist leanings that have informed their historiographical stance. On this view, mathematical ideas and techniques are *found* or *discovered* by individuals or groups. Thus at is simplest the realist historical enterprise consists of identifying Platonic mathematical objects in the historical record and equating the terminology used to describe and manipulate them with their modern-day technical counterparts. The emphasis is on tracing mathematical *sameness* across time and space; for instance looking at historical instances of 'Pythagoras' Theorem'. But for the mathematics of ancient Iraq, realism ground to a halt as a productive historical methodology in the mid-twentieth century. Once the decipherments were made and the ancient sources rewritten in modern symbolic notation there was nothing left to say. The field stagnated for many decades.

In the 1970s, however, new philosophical and historical movements began to stir. Henk Bos and Hubert Mehrtens articulated a programme for considering the historical relationships between mathematicians and society, while David Bloor argued forcefully that mathematics itself was socially constructed.[1] Constructivism argues that mathematics is not discovered but *created* by social groups, just as spoken languages are—albeit that society is often professionally restricted and its methods for reaching agreement are highly formalised. Within the contemporary global community of professional mathematicians it is probably fair to say that social constructivism is a minority view, even amongst anti-realists, but it is increasingly recognised as a powerful means of understanding mathematics in history. The realist historical enterprise is of necessity a descriptive one: it is simply a matter of uncovering what, and perhaps how, the ancients correctly knew about mathematics. But, as this book aims to show, internal mathematical developments are of course a necessary part of the history of mathematics but they are not sufficient to account for the particular form mathematical expression took. On the constructivist historical view, the emphasis is on difference, localism, and choice: why did societies and individuals *choose* to describe and understand a particular mathematical idea or technique one particular way as opposed to any other? How did the social and material world in which they lived affect their mathematical ideas and praxis?

Rather than emphasising the periods and places from which a lot of traditional evidence survives, namely Old Babylonian scribal schools, I have chosen to take a wider view of mathematics and numeracy, seeking mathematical thought and practices in all periods of Mesopotamian history over a span of roughly three millennia. I have tried to even-handedly assign a chapter of roughly equal length to each five hundred-year period, even though that occasionally means stretching the conventional periodisation that is based on the political history of kings and empires. There is of course a great deal of fluidity and continuity, as well as disjunction, between these artificial divisions: intellectual, conceptual, and social history do not always fall as neatly into blocks of time as dynastic history does. Nevertheless, it seemed to me that a chronological framework should be the overarching structure of the book, thereby avoiding the somewhat prevalent tendency to treat all 'Babylonian' mathematics as an essentialising whole, disregarding whether individual documents are from the first centuries of the second millennium BCE or the last centuries of the first. Framed by a methodological and explanatory introduction (much of which can be skipped by Assyriological readers) and a historiographical epilogue, the books falls into two equal halves. Chapters 2–4 cover early Mesopotamia, the age of city states and short-lived territorial empires, while chapers 6–8 are devoted to later Mesopotamia, the time of great empires. The focus is necessarily almost exclusively on the south of the region, but chapter 5 tackles Assyria, Babylonia's supposedly innumerate northern neighbour.

This book does not pretend to be an exhaustive survey of the field. It does not aim to replace existing works, such as Neugebauer's *The exact sciences in antiquity* or more recent specialist studies like Jens Høyrup's *Lengths, widths, surfaces* or my own *Mesopotamian mathematics*.[2] In an attempt to compensate for that, each chapter begins with a survey of the background history and the pertinent published mathematical sources. The rest of each chapter is split into three sections, each typically focussing on a different topic in the world of school, institutional administration (particularly issues of land and labour), and social culture. Translations of primary sources are integrated into the narrative throughout, but many more can be found in my contribution to Princeton's *The mathematics of Egypt, Mesopotamia, China, India and Islam: a sourcebook*.[3] The entire published corpus—over 950 tablets as of December 2006—is also online at <http://cdl.museum.upenn.edu/dccmt>.

Philadelphia
December 2006

ACKNOWLEDGMENTS

It is a true pleasure to record my gratitude to all those who have inspired, corrected, or distracted me since this book's inception. I am very lucky to have such intelligent, generous, and supportive colleagues, students, and friends—not all of whom, despite my best efforts of recall, are listed here. My apologies if I have overlooked you.

The idea grew out of a series of lectures for the Mathematical Association of America's Summer Institute in the History of Mathematics and Its Uses in Teaching in the summer of 1999, organised by Victor Katz, Fred Rickey, and the late Karen Dee Michalowicz (although Niek Veldhuis tells me that I was already contemplating it in early 1996). Since then I have given many papers in many places as a means of driving the project forward: my sincere thanks to all who invited, hosted, listened to, and responded to me.

Most of the research and much of the writing was carried out during a postdoctoral research fellowship at All Souls College, Oxford, in 2000–3. The project underwent a year-long hiatus in 2004–5, during which I got to grips with the pleasures and rigours of a lectureship in the Department of History and Philosophy of Science at the University of Cambridge, and dealt with the deaths of three irreplaceable mentors. My indebtedness to David Fowler, Jeremy Black, and Roger Moorey should be apparent throughout this book. David bestowed on me a cheerfully iconoclastic approach to the historiography of ancient mathematics, Jeremy a profound pleasure in the complexities and breadth of cuneiform textuality, and Roger a serious regard for the material culture of tablets and their archaeological contexts. Jens Høyrup, who showed that explanatory history of Babylonian mathematics can indeed be written, is, I am happy to say, still very much alive and thriving. His influence also runs deep throughout what follows. Even though I do not invariably agree with him, it is always with great respect and real gratitude that he has provided such a strong model to kick against. If my arguments break down in the process that is my fault, not his. No blame should be laid at the doors of any of my role models, living or dead, for any failure of mine to live up to their standards.

Most recently my marvellous HPS colleagues have led me by example to new ways of thinking about the social history of science and ancient scholarship, and provided the best possible environment in which to explore them.

And while I have been finishing and tidying up this manuscript over the final months of 2006, on sabbatical leave in the Babylonian Section of the University of Pennsylvania Museum of Archaeology and Anthropology, Sumerologists extraordinaire Steve Tinney and (for one week only!) Niek Veldhuis have provided their usual irresistible mix of encouragement and diversion—not to mention the run of Steve's office and personal library.

Many more people are implicated somehow in the writing of this book (but absolved from all responsibility). In Cambridge: Annette Imhausen, Geoffrey Lloyd, Karin Tybjerg; Peter Lipton, Tamara Hug, Liba Taub; Nicholas Postgate, Augusta McMahon, Martin Worthington. In Oxford over the years: Graham Cunningham, Fran Reynolds, Jon Taylor, Gábor Zólyomi; John Bennet, Chris Gosden, Andrew Wilson; the Warden and Fellows of All Souls. In Philadelphia at various times: Paul Delnero, Phil Jones, Fumi Karahashi, Erle Leichty; Barry Eichler, Grant Frame, Ann Guinan, and the rest of the Tablet Room crew. In other Assyriological locales around the world: Kathryn Slanski, John Steele, and Cornelia Wunsch read chapters and provided online answering services; Heather Baker, Philippe Clancier, Ben Foster (who knew there was so much Sargonic maths out there? well, you did), Eckart Frahm, Michael Jursa, Lee Payne, Seth Richardson, Ignacio Marquez Rowe, and Michel Tanret gave help and amusement at key points. Many colleagues in history of mathematics have been equally crucial: Jackie Stedall and June Barrow-Green especially; Jeremy Gray and the late John Fauvel; the Oxford history of maths colloquium and the British Society for the History of Mathematics; Serafina Cuomo, Duncan Melville, Reviel Netz, and Gary Urton.

This book would have been impossible with access to tablets. I warmly thank Nawala al-Mutawalli and Donny George (Iraq Museum), Tony Brinkman (University of Chicago Oriental Institute), Annie Caubet and Béatrice André-Salvini (The Louvre), Ulla Kasten (Yale Babylonian Collection), Mohammed Reza Karegar (Iran Bastan Museum), Joachim Marzahn (Vorderasiatisches Museum, Berlin), the late Michael O'Connor (Catholic University of America Semitics Collection), Jane Siegel (Columbia University Rare Book and Manuscript Library), and especially Christopher Walker (British Museum) and Helen Whitehouse (Ashmolean Museum). Most of my travels were funded by the British Academy or All Souls College. I am also grateful to the British School of Archaeology in Iraq for funding the project Measure for Measure: Old Babylonian Metrology and Pedagogy, some outcomes of which are presented in chapter 4.

David Ireland commissioned this book for Princeton; Vickie Kearn saw it patiently to completion. Production editor Mark Bellis and copy editor Will Hively oversaw the transition from manuscript to book with good humour, courtesy, and prompt and exacting attention to detail. Thank you all for everything. Thank you too to the book's readers, who are collectively responsible for many useful improvements.

In between times, Christine Shimmings has kept me fit and sane. Fran, Jackie, Jaimie, June, Niek, Rowan, and Steve all deserve more than a meagre thank-you for being the most fantastic friends a girl could ask for. But those who have put up with most during the writing of this book, and probably have least to gain from it being finished, are Tom and Bo Treadwell, and most of all my lovely Luke. This dedication is hardly recompense, my darling, for all the love and support you've given me over the last ten years; here's to another ten, and more.

Mathematics in Ancient Iraq

CHAPTER ONE

Scope, Methods, Sources

The mathematics of ancient Iraq, attested from the last three millennia BCE, was written on clay tablets in the Sumerian and Akkadian languages using the cuneiform script, often with numbers in the sexagesimal place value system (§1.2). There have been many styles of interpretation since the discovery and decipherment of that mathematics in the late nineteenth and early twentieth centuries CE (§1.1), but this book advocates a combination of close attention to textual and linguistic detail, as well as material and archaeological evidence, to situate ancient mathematics within the socio-intellectual worlds of the individuals and communities who produced and consumed it (§1.3).

1.1 THE SUBJECT: ANCIENT IRAQ AND ITS MATHEMATICS

Iraq—Sumer—Babylonia—Mesopotamia: under any or all of these names almost every general textbook on the history of mathematics assigns the origins of 'pure' mathematics to the distant past of the land between the Tigris and Euphrates rivers. Here, over five thousand years ago, the first systematic accounting techniques were developed, using clay counters to represent fixed quantities of traded and stored goods in the world's earliest cities (§2.2). Here too, in the early second millennium BCE, the world's first positional system of numerical notation—the famous sexagesimal place value system—was widely used (§4.2). The earliest widespread evidence for 'pure' mathematics comes from the same place and time, including a very accurate approximation to the square root of 2, an early form of abstract algebra, and the knowledge, if not proof, of 'Pythagoras' theorem' defining the relationship between the sides of a right-angled triangle (§4.3). The best-known mathematical artefact from this time, the cuneiform tablet Plimpton 322, has been widely discussed and admired, and claims have been made for its function that range from number theory to trigonometry to astronomy. Most of the evidence for mathematical astronomy, however, comes from the later first millennium BCE (§8.2), from which it is clear that Babylonian astronomical observations, calculational models, and the sexagesimal place value system all had a deep impact on the later development of Old World astronomy, in particular through the person and works of Ptolemy. It is hardly surprising, then, that ever since its discovery a century ago the mathematics of ancient Iraq has claimed an important role in the history of early mathematics. Seen as the first flowering of 'proper' mathematics, it has been hailed as the cradle from which classical Greek mathematics, and therefore the Western tradition, grew. But, as laid out over the course of this book, the mathematical culture of ancient Iraq was much richer, more complex, more diverse, and more human than the standard narratives allow.

The mathematical culture of ancient Iraq was by no means confined to the borders of the nation state as it is constructed today. The name al-'Iraq (Arabic 'the river shore') is first attested about a century after the Muslim conquests of the early seventh century CE,[1] while the lines on modern maps which delimit the territory of Iraq are the outcome of the division of the collapsing Ottoman empire amongst European powers at the end of the First World War. The mathematics of pre-Islamic Iraq, as it has been preserved, was written on small clay tablets in cuneiform writing. Because, as argued here, mathematics was an integral and powerful component of cuneiform culture, for present purposes it will be a useful first approximation to say that cuneiform culture and mathematical culture were more or less co-extensive. The core heartland of the cuneiform world was the very flat alluvial plain between Baghdad and the Gulf coast through which the Tigris and Euphrates flow (figure 1.1). It was known variously in antiquity as Sumer and Akkad, Babylonia, Karduniaš, or simply The Land. The Land's natural resources were primarily organic: reeds, small riverine trees, and other plant matter, but most importantly the earth itself. Alluvial clay was the all-purpose raw material par excellence, from which almost anything from sickle blades to monumental buildings could be manufactured. Equally, when judiciously managed the soil was prodigiously fertile, producing an abundance of arable crops (primarily barley), as well as grazing lands for herds (sheep and goats but also cattle). Even the wildest of marshlands were home to a rich variety of birds and fish and the all-purpose reeds, second only to clay in their utility. What the south lacked, however, were the trappings of luxury: no structural timber but only date-palms and tamarisks, no stone for building or ornamentation other than small outcrops of soft, dull limestone, and no precious or semi-precious stones at all, let alone any metals, base or precious. All these had to be imported from the mountains to the north, east, and west, in exchange for arable and animal products.

The centre of power shifted north at times, to northern Iraq and Syria east of the Euphrates, known in ancient times as Assyria, Subartu, Mitanni, or the land of Aššur. Life here was very different: rainfall could be counted on for wheat-based agriculture, building stone was abundant, and mountainous sources of timber and metal ores relatively close to hand. Conversely, the dates, tamarisks, and reeds of the south were absent here, as were the marshes with their rich flora and fauna. Overland trade routes ran in all directions, linking northern Iraq with the wider world.[2]

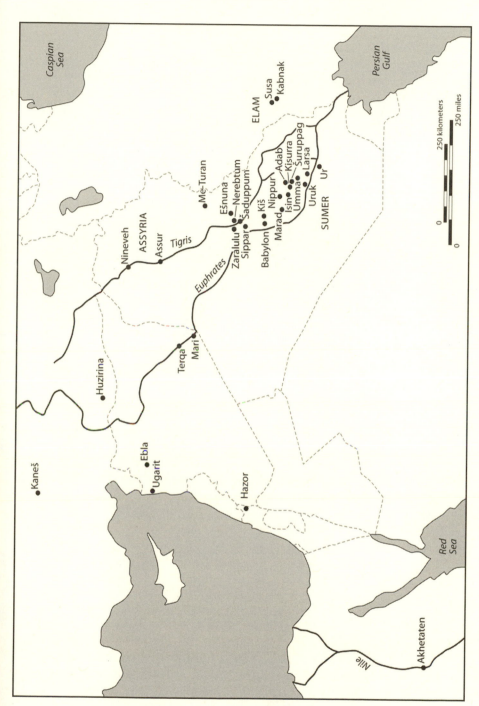

Figure 1.1 Map of the ancient Middle East, showing the major findspots of mathematical cuneiform tablets.

The fluid peripheries over which these territories had at times direct political control or more often cultural influence expanded and contracted greatly over time. At its maximum extent cuneiform culture encompassed most of what we today call the Middle East: the modern-day states of Turkey, Lebanon, Syria, Israel and the Palestinian areas, Jordan, Egypt, and Iran. Chronologically, cuneiform spans over three thousand years, from the emergence of cities, states, and bureaucracies in the late fourth millennium BCE to the gradual decline of indigenous ways of thought under the Persian, Seleucid, and Parthian empires at around the beginning of the common era. The history of mathematics in cuneiform covers this same long stretch and a similarly wide spread (table 1.1).

The lost world of the ancient Middle East was rediscovered by Europeans in the mid-nineteenth century (table 1.2). Decades before the advent of controlled, stratigraphic archaeology, the great cities of Assyria and Babylonia, previously known only through garbled references in classical literature and the Bible, were excavated with more enthusiasm than skill, yielding vast quantities of cuneiform tablets and *objets d'art* for Western museums.[3] The complexities of cuneiform writing were unravelled during the course of the century too, leading to the decipherment of the two main languages of ancient Iraq: Akkadian, a Semitic precursor of Hebrew and Arabic; and Sumerian, which appeared to have no surviving relatives at all.

In the years before the First World War, as scholars became more confident in their interpretational abilities, the first mathematical cuneiform texts were published.[4] Written in highly abbreviated and technical language, and using the base 60 place value system, they were at first almost impossible to interpret. Over the succeeding decades François Thureau-Dangin and Otto Neugebauer led the race for decipherment, culminating in the publication of their rival monumental editions, *Textes mathématiques babyloniens* and *Mathematische Keilschrifttexte*, in the late 1930s.[5] By necessity, scholarly work was at that time confined to interpreting the mathematical techniques found in the tablets, for there was very little cultural or historical context into which to place them. For the most part the tablets themselves had no archaeological context at all, or at best could be attributed to a named city and a time-span of few centuries in the early second millennium BCE. The final reports of the huge and well-documented excavations of those decades were years away from publication and nor, yet, were there any reliable dictionaries of Akkadian or Sumerian.

After the hiatus of the Second World War, it was business as usual for the historians of cuneiform mathematics. Otto Neugebauer and Abraham Sachs's *Mathematical cuneiform texts* of 1945 followed the paradigm of the pre-war publications, as did Evert Bruins and Marguerite Rutten's *Textes mathématiques de Suse* of 1961.[6] Neugebauer had become such a towering figure that his methodology was followed by his successors in the

TABLE 1.1
Overview of Mathematics in Ancient Iraq

Dates	Political History	Mathematical Developments
Fourth millennium BCE	Urbanisation Uruk period, c. 3200–3000	Pre-literate accounting Commodity-specific number systems
Third millennium BCE	Early Dynastic period (city states), c. 3000–2350 First territorial empires of Akkad and Ur, c. 2350–2000	Literate numeracy Sophisticated balanced accounting Sexagesimal place value system
Second millennium BCE	Babylonian kingdom, c. 1850–1600 Amarna age, c. 1400	Old Babylonian (OB) pedagogical mathematics: geometry, algebra, quantity surveying Spread of cuneiform culture and numeracy across the Middle East
First millennium BCE	Assyrian empire, c. 900–600 Neo-Babylonian empire, c. 600–540 Persian empire, c. 540–330 Seleucid empire, c. 330–125	Beginnings of systematic observational astronomy Reformulation of cuneiform mathematics Mathematical astronomy
First millennium CE	Parthian empire, c. 125 BCE–225 CE Sasanian empire, c. 225–640 Abbasid empire, c. 750–1100	Last dated cuneiform tablet, 75 CE Astronomical activity continues Baghdadi 'House of Wisdom': al-Khwarizmi, decimal numbers and algebra
Second millennium CE	Mongol invasions, c. 1250 Ottoman empire, c. 1535–1920 Modern Iraq, c. 1920–	(Iraq a political and intellectual backwater) Rediscovery of ancient Iraq and cuneiform culture

TABLE 1.2
The Rediscovery of Cuneiform Mathematics

Date	Event
1534–1918	Iraq under Ottoman rule
Before 1800	Travellers' tales of ancient Babylonia and Assyria
1819	Tiny display case of undeciphered cuneiform at British Museum
1842	Anglo-French rediscovery of ancient Assyria; priority disputes
1857	Assyrian cuneiform officially deciphered at Royal Asiatic Society in London
1871	Discovery of Babylonian flood story in British Museum
1877	Discovery of Sumerian language and civilisation; no mention in Bible
1880–	Mass recovery of cuneiform tablets in Babylonia
1889–	First decipherments of cuneiform astronomy and sexagesimal place value system
1900	First Old Babylonian (OB) mathematical problems published
1903–	Progress in understanding sexagesimal numeration and tables
1916	First decipherment of OB mathematical problem
1920	Formation of modern Iraqi state
1927	Neugebauer's first publication on OB mathematics
1927–39	Neugebauer and Thureau-Dangin's 'golden age' of decipherment
1945	Neugebauer and Sachs's final major publication on OB mathematics
1955	Neugebauer and Sachs publish mathematical astronomy
1956	First volume of *Chicago Assyrian Dictionary* published (finished in 2008)
1968	Ba'athist coup in Iraq
1976–	Increased interest in third-millennium mathematics
1984	First volume of *Pennsylvania Sumerian Dictionary* published (now online)
1990	Høyrup's discourse analysis of OB maths; war stops excavation
1996–	Developing web technologies for decipherment tools and primary publication
2003–	Iraq War and aftermath result in major archaeological looting and the virtual collapse of the State Board of Antiquities and Heritage

discipline, though often without his linguistic abilities. Cuneiformists put mathematical tablets aside as 'something for Neugebauer' even though he had stopped working on Babylonian mathematics in the late 1940s. Since there was almost no further output from the cuneiformists, historians of mathematics treated the corpus as complete. In the early 1950s the great Iraqi Assyriologist Taha Baqir published a dozen mathematical tablets from his excavations of small settlements near Baghdad, but virtually the only other editor of new material was Bruins, who tended to place short articles in the small-circulation Iraqi journal *Sumer* (as did Baqir) or in *Janus*, which he himself owned and edited.[7] All attempts at review or criticism met with such vitriolic attacks that he effectively created a monopoly on the subject.

Meanwhile, since Neugebauer's heyday, other aspects of the study of ancient Iraq had moved on apace. The massive excavations of the pre-war period, and the more targeted digs of the 1950s onwards, were being published and synthesised. The monumental *Chicago Assyrian dictionary* gradually worked its way through the lexicon of Akkadian, volume by volume. The chronology, political history, socio-economic conditions, and literary, cultural, religious, and intellectual environments of Mesopotamia were the subjects of rigorous, if not always accessible, scholarship. In the course of the 1970s and '80s attention turned to much earlier mathematical practices, as scholars led by Marvin Powell and Jöran Friberg found and analysed the numeration, metrology, and arithmetic of the third millennium BCE, from sites as far apart as Ebla in eastern Syria and Susa in southwestern Iran.[8] Denise Schmandt-Besserat began to formulate her mould-breaking theories of the origins of numeracy and literacy in the tiny calculi of unbaked clay that she had identified in prehistoric archaeological assemblages all over the Middle East.[9]

Nevertheless, it would be no exaggeration to say that between them, Neugebauer's renown for scholarly excellence and Bruins's reputation for personal venom seriously stifled the field of Babylonian mathematics until their deaths in 1990. It is perhaps no coincidence that 'Algebra and naïve geometry', Jens Høyrup's seminal work on the language of Old Babylonian algebra, was also published in that year, signalling a paradigm shift away from the history of Mesopotamian mathematics as the study of calculational techniques and their 'domestication' into modern symbolic algebra. Høyrup's work was in effect a discourse analysis of Mesopotamian mathematics: a close scrutiny of the actual Akkadian words used, and their relationship to each other. In this way he completely revolutionised our understanding of ancient 'algebra', showing it to be based on a very concrete conception of number as measured line and area.[10] An interdisciplinary project based in Berlin developed further important new methodologies in the early 1990s, leading to the computer-aided decipherment of the complex metrologies in the proto-literate temple accounts from late fourth-millennium Uruk which had resisted satisfactory interpretation for over eighty years.[11] Uruk also provided new sources from the other end of the chronological spectrum, as Friberg published mathematical tablets from the latter half of the first millennium BCE.[12]

In the past decade, large numbers of new mathematical tablets have come to light, both from excavation and from renewed study of old publications and large museum collections, and are now attested from almost every period of cuneiform culture. The published corpus now comprises over 950 tablets (table B.22). Still the largest body of evidence, though, is the pedagogical mathematics—exercises set and solved, metrological and mathematical tables copied and recopied—from the early second millennium BCE or Old Babylonian period. This currently accounts for over 80 percent of the published sources, not far short of 780 tablets. There are fewer than sixty known mathematical tablets from the whole of the third millennium, on the other hand (about 6 percent), and just over twice that number from the millennium and a half after 1500 BCE (some 13 percent). Thus the main focus of attention is still therefore on the large body of Old Babylonian material.

With some exceptions, the new generation of scholarship has taken a long time to filter through to the wider historical community. Cuneiformists have been put off by technical mathematics, historians of mathematics by technical Assyriology. Thus mathematics tends to be ignored in general histories of the ancient Near East, and even though it has an inviolable place at the beginning of every maths history textbook, the examples found there are for the most part still derived from a few out-of-date general works. Neugebauer's *The exact sciences in antiquity*, first published in 1951, was justly influential, but Van Der Waerden's derivative *Science awakening* (first English edition 1954) and later *Geometry and algebra in ancient civilizations* (1983) are both deeply Eurocentric and diffusionist. All in all it is time for a new look, from a new perspective—which is what this book sets out to do.

1.2 THE ARTEFACTS: ASSYRIOLOGICAL AND MATHEMATICAL ANALYSIS

Perhaps the most important methodological thread running through this book is that although mathematics is most immediately the product of individuals, those individuals are shaped and constrained by the society in which they live, think, and write. In order to understand the mathematics of a particular people as richly as possible, historians need to contextualise it. This approach is especially important for comprehending the mathematics of ancient Iraq, where anonymous tradition was prized over named authorship and we are more often than not completely unable to identify the work or influence of individuals within the written tradition. But context, crucial though it is, has to be paired with scrupulous attention to the mathematical, linguistic, and artefactual details of the tablets themselves. In order to demonstrate this, on the following pages a typical example, in the standard style of primary publication, is used to explain the basics of the media, script, numeration, and language of the sources, and to exemplify the usual methods of decipherment, interpretation, and publication. The final section demonstrates some of the different ways in which contextualisation can add layers of meaning to the interpretation of individual objects.

The primary publication of a cuneiform tablet should normally comprise at least a hand-copy (scale drawing) and transliteration, and often a photograph and translation as well. The sample tablet, 2N-T 30 (figure 1.2), has been partially published twice before: once as a rather blurry photograph, and once as a transliteration and translation based on that photograph. The hand-copy presented here is also based on that photograph, and on personal inspection of the tablet in Baghdad in March 2001.[13]

1 45	0;00 01 45
1 45	0;00 01 45
⅓ kuš₃ ½ šu-si-ta-am₃	A square is ⅓ cubit, ½ finger on each
ib₂-[si₈]	side.
a-šag₄-bi en-nam	What is its area?
a-šag₄-bi	Its area
9 še igi-5-ǧal₂' še-kam	is 9 grains and a 5th of a grain.

As the photograph shows, the text is not on a flat writing surface like paper or papyrus, but on a small cushioned-shaped tablet of levigated clay (that is, clay that has been cleaned of all foreign particles so that it is pure and smooth), measuring about 7.5 cm square by 2.5 cm thick at its maximum extent. Clay tablets varied in size and shape according to place and time of manufacture, and according to what was to be written on them; they could be as small as a postage stamp or as large as a laptop computer, but more usually were about the size of a pocket calculator or a mobile phone (though often rather thicker). Scribes were adept at fashioning tablets to the right size for their texts, making the front side, or *obverse*, much flatter than the back, or *reverse*. Some specialised genres of document traditionally required particular types of tablet, as in the case of 2N-T 30: it is square, with text only in the bottom right and top left corners; the rest of the tablet (including the reverse) is blank. Tablets were more usually rectangular, with the writing parallel to the short side and covering the whole surface of the clay. When the scribe reached the bottom of the obverse, instead of turning the tablet through its vertical axis (as we would turn the pages of a book),

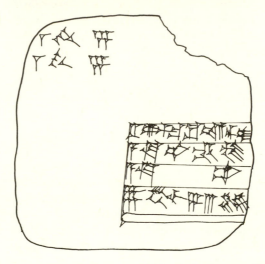

Figure 1.2 A mathematical exercise on an Old Babylonian cuneiform tablet.
(2N-T 30 = IM 57828. Photograph: F. R. Steele 1951, pl. 7, courtesy of the
University of Pennsylvania Museum of Archaeology and Anthropology; drawing
by the author.)

he would flip it horizontally, so that the text continued uninterrupted over
the lower edge and onto the reverse.[14] The text would thus finish at the top
edge of the tablet, next to where it had started. Scribes could also use the
left edge of the tablet to add a few extra lines or a summary of the document's contents. Larger tablets were ruled into columns, and here the same

conventions applied: on the obverse the columns were ordered left to right and then the final column ran directly down the tablet and onto the other side, becoming the first column on the reverse of the tablet. The remaining columns were then ordered right to left, ending up in the bottom left corner of the reverse, right next to the top of the obverse.

Tablets were rarely baked in antiquity, unless there was some special reason for doing so, or it happened by accident in a fire. Tablets in museum collections tend to be baked as part of the conservation process, where they are also catalogued and mended. Tablets were often broken in antiquity— this one has lost its top right corner—and where possible have to be pieced together like a three-dimensional jigsaw puzzle or a broken cup, some-times from fragments that have ended up in different collections scattered around the world. Often, though, missing pieces are gone forever, crum-bled into dust. This tablet has two catalogue numbers, one from excava-tion and one from the museum that now houses it. The excavation num-ber, '2N-T 30', signifies that it was the thirtieth tablet (T 30) to be found in the second archaeological season at Nippur (2N) in 1948, run jointly by the Universities of Chicago and Pennsylvania. This is the designation used in the archaeological field notes, enabling its original findspot and context to be traced; we shall return to this later. The museum number, 'IM 57828', indicates that it is now in the Iraq Museum in Baghdad (IM) and was the 57,828th artefact to be registered there. Like almost all museum numbers, it gives no information at all about the tablet's origins.

The writing on the tablet is a script composed of wedge-shaped impres-sions now known as cuneiform (figure 1.3). It runs horizontally from left to right (whatever the direction of the columns it is in) and where there are line rulings (as on the bottom right of this tablet) the signs hang down from the lines rather than sitting on them. As can be seen from the top left of the tablet, though, the lines were not always ruled.

The appearance and structure of cuneiform changed significantly over the course of its history, but some common key features remained through-out. In the earliest written documents, from the late fourth millennium BCE, there were two distinct sign types. Numbers were *impressed* into the surface of the clay, while other signs were *incised* into stylistic representa-tions. Most of the non-numerical signs were *logograms* and *ideograms*; that is, they represented whole words or idea. A small subset acted as *de-terminatives*, classifying the words they were attached to as belonging to a particular category, such as vessels, wooden objects, or place-names. De-terminatives did not represent parts of speech but were aids to reading the words with which they were associated. Many of the non-numerical signs were *pictographic*; that is, they looked like the objects they represented, but others were more abstract shapes. Each sense unit was grouped into *cases*, rectangular areas ruled onto the surface of the tablet.

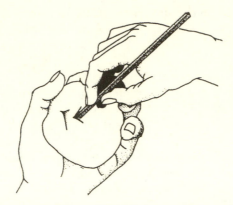

Figure 1.3 Writing cuneiform on a clay tablet with a stylus. (Drawing by the author.)

The early writing system was very limited in scope and function, but over the first half of the third millennium BCE it gradually acquired more power and flexibility by using the same signs to represent not just an idea or a thing, but the sound of that word in the Sumerian, and later Akkadian, language. Thus, as writing became language-specific and acquired the ability to record *syllables* it became increasingly important to arrange the signs into lines on the surface of the tablet, following the order of the spoken language. At the same time, the signs themselves lost their curviform and pictographic visual qualities, becoming increasingly cuneiform and visually abstract. However, signs retained their ideographic significance even as they acquired new syllabic meanings. In other words, they became *multivalent*; that is, a single cuneiform sign could have as many as twenty different meanings or *values* depending on the context in which it was used. For instance (table 1.3), the sign DUG functioned as a determinative in front of the names of pottery vessels; as a logogram with the meaning 'pot' (Sumerian *dug* = Akkadian *karpatum*) or 'cup' (Sumerian *lud* = Akkadian *luṭṭum*); and as an Akkadian syllabogram with values *dug, duk, duq, tuk, tuq* and *lud, lut, luṭ*. Conversely, the writing system also encompassed *homovalency*; that is, the potential for different signs to represent the same syllables; some of these, however, were at least partially contextually determined. For instance, there are three other signs with the value 'dug': one is just a syllable, while the others are logograms for the adjective 'good' and the verb 'to speak'. The total repertoire of cuneiform signs is around six hundred, but not all of those signs, or all possible values of a sign, were in use at any one time or in any one genre of text.[15]

In modern transliteration, determinatives are shown in superscript (there are none in 2N-T 30), and Sumerian syllables in normal font. Akkadian

TABLE 1.3
Four Different Cuneiform Signs that Take the Value *Dug*

Late IV mill.	III mill.	II mill.	I mill.	Determinative Values	Logographic Values	Main Syllabic Values
(sign)	(sign)	(sign)	(sign)	dug 'pottery vessel'	dug = *karpatum* 'pot'; lud = *luṭṭum* 'cup'	*dug, duk, duq, tuk, tuq* and *lud, lut, luṭ*
—	(sign)	(sign)	(sign)	—	du₁₂ = *zamārum* 'to play (a musical instrument)'; tuku = *rašûm* 'to acquire'	*tug, tuk, tuq, duk, raš*
(sign)	(sign)	(sign)	(sign)	—	dug₃ = *ṭâbum* 'to be good, sweet'; ḫi = *balālum* 'to mix'	*ḫi, ḫe, ṭa*
(sign)	(sign)	(sign)	(sign)	—	dug₄ = *qabûm* 'to speak'; gu₃ = *rigmum* 'noise'; inim = *awatum* 'word'; ka = *pûm* 'mouth'; kir₄ = *būṣum* 'hyena'; kiri₃ = *appum* 'nose'; zu₂ = *šinnum* 'tooth'; zuḫ = *šarāqum* 'to steal'	*ka, qa*

syllables are written in italics, and logograms in small capitals. (But in trans-lations, this book shows all untranslated words—such as metrological units—in italics, whether they are in Sumerian or Akkadian.) Homovalent signs are distinguished by subscript numbers, as for instance kuš₃ or si₈. Signs that are missing because the tablet has broken away are restored in square brackets: [si₈] or in half brackets ⌈ĝeš⌉ if the surface is damaged and the sign not clearly legible; uncertain signs can also be signalled by a super-script question mark: še? and erroneously formed signs indicated with an exclamation mark: ĝal₂!. Sometimes scribes omitted signs; editors restore them inside angle brackets: <ti>. When a scribe has erroneously written an extra sign it is marked in double angle brackets: «na». Modern editor's glosses are given in parentheses: (it is), with comments in italics: (*sic*).

Cuneiform was used to record many different languages of the ancient Middle East, just as the Roman alphabet today is not reserved for any one language or group of languages. Only two were used extensively for mathematics, however: Sumerian and Akkadian. Sumerian was probably the first language in the world to be written down and maybe because of that it appears to be a language isolate: that is, it is genetically related to no other known language, living or dead. It seems to have died out as a mother tongue during the course of the late third and early second millennium BCE, but continued to be used as a literary and scholarly language (with a status similar to Latin's in the Middle Ages) until the turn of the common era. It does, of course, have features in common with many languages of the world. In the writing system four vowels can be distinguished—a, e, i, u—and fifteen consonants b, d, g, ğ, ḫ, k, l, m, n, p, r, s, š, t, and z (where ḫ is approximately equivalent to kh or Scottish ch in 'loch', ğ to ng in 'ring', and š to sh in 'shoe'). Words are formed by *agglutinating* strings of grammatical prefixes and suffixes to a lexical stem, as can happen to some extent in English, for instance with the stem 'do': un-do-ing. The main feature of the Sumerian case system is *ergativity*: grammatical distinction between the subjects of transitive and intransitive verbs. Compare in English the intransitive 'the man walked' with transitive 'the man walked the dog'; in Sumerian the second 'man' is in the ergative case, 'the man(-erg.) walked the dog'. Its gender system is a simple dichotomy, people/others, and its word order is verb-final. Instead of tense, Sumerian uses aspect, distinguishing between completed and incomplete action. Thus to completely Sumerianise 'the man walked the dog', we would have 'man(-erg.-pers.) dog(-other) (he-it-)walk(-compl.). It is perhaps not surprising that there is still much argument about the details of how Sumerian works and exactly how it should be translated.[16]

Akkadian, by contrast, is very well understood. It is a Semitic language, indirectly related to Hebrew and Arabic. Like Sumerian it was written with the four vowels a, e, i, u and all the consonants of Sumerian (except ğ) as well as q, ṣ (emphatic s, like ts in 'its'), ṭ (emphatic t), and a glottal stop, transliterated: '. Like other Semitic languages, Akkadian works on the principle of *roots* composed of three consonants, which carry the lexical meaning of words. For instance, the root *mḫr* carries the sense of equality and opposition, while *kpp* signifies curvature. Words are constructed by *inflection*, that is, by the addition of standard patterns of prefixes, suffixes, and infixes in and around the root which bear the grammatical meaning (tense, aspect, person, case, etc.). Thus *maḫārum* and *kapāpum* are both verbal infinitives 'to oppose, to be equal', and 'to curve'. The first-person present tense is *amaḫḫar* 'I oppose', the simple past *amḫur* 'I opposed'. Nouns, adjectives, and adverbs are all derivable from these same roots, so that the reflexive noun *mithartum* 'thing that is equal and opposite to itself' means a square and *kippatum* 'curving thing' is a circle.[17]

The mixed cuneiform writing system enabled polysyllabic Akkadian words to be represented with a logogram, usually the sign for the equivalent Sumerian stem; indeed some signs, especially weights and measures, were almost never written any other way. So while it *appears* that 2N-T 30 is written in Sumerian, it may equally be in highly logographic Akkadian: it is impossible to tell. And just as it is sometimes difficult to judge the language a scribe was *writing* in, we cannot even begin to determine what language he might have been speaking or thinking in. The death of Sumerian as a written and/or spoken language is still a hotly debated issue, but it must have taken place over the last third of the third millennium and/or the first half of the second.[18]

The tablet 2N-T 30 uses two different systems of numeration. The numbers on their own in the top left corner are written in the *sexagesimal place value system* (SPVS), while those in the main text belong to various *absolute value systems* that were used for counting, weighing, and measuring. The earliest known written records, the temple account documents from late fourth-millennium Uruk, used a dozen or so very concrete numeration systems that were not only absolute in value but were also determined by the commodity being counted or measured (table A.1 and see chapter 2). Over the course of the third millennium this bewildering variety of metrologies was gradually rationalised to five: length; area, volume, and bricks; liquid capacity; weight; and the cardinal (counting) numbers (table A.3 and see chapter 3). Although the number sixty became increasingly prominent, all systems except the weights retained a variety of bases (comparable to more recent pre-decimal Imperial systems) and all used signs for metrological units and/or different notations for different places. For instance, the length of the square in 2N-T 30 is ⅓ cubit and ½ finger. A cubit was approximately 0.5 m long, and comprised 30 fingers—so the square is 10½ fingers long, c. 17.5 cm.

But the interface between the different systems was never perfect: for instance, there was an area unit equal to 1 square rod, but none equal to 1 square finger or 1 square cubit. In other words, the exercise on 2N-T 30 is not a simple matter of squaring the length in the units given: it is first necessary to express that length in terms of rods, where the two systems meet nicely. Then the length can be squared, to give an area expressed in (very small) fractions of a square rod (c. 36 m²) or, more appropriately, integer numbers of smaller area units. The area system borrowed the sexagesimal divisions of the mina weight for its small units, so that a sixtieth part of a square rod is called a shekel (c. 0.6 m², or 2.4 square cubits), and a 180th part of a shekel is known as a grain (c. 33 cm², or 12 square fingers).

The area in 2 N-T 30 is almost too small for the area system to handle, so the scribe has approximated it as 9⅓ grains.

The function of the sexagesimal place value system (SPVS), which came into being during the last centuries of the third millennium (§3.4), was thus to ease movement between one metrological system and another. Lengths, for instance, that were expressed in sexagesimal fractions of the rod instead of a combination of rods, cubits, and fingers, could be much more easily multiplied into areas expressed first in terms of square rods, and then converted to more appropriate units if necessary. The SPVS, in other words, was only a calculational device: it was *never* used to record measurements or counts. That is why it remained a purely positional system, never developing any means of marking exactly how large or small any number was, such as the positional zero or some sort of boundary marker between integers and fractions. In other words, these deficiencies were not the outcome of an unfortunate failure to grasp the concept of zero, but rather because neither zeros nor sexagesimal places were necessary within the body of calculations. For the duration of the calculation, then, the absolute value of the numbers being manipulated is irrelevant; only their relative value matters— as long as the final result can be correctly given in absolute terms.

This is apparent on the top left corner of 2N-T 30, where the scribe has expressed 10½ fingers as a sexagesimal fraction of a rod, namely as 1 45. In modern transliteration sexagesimal places are separated by a space or comma, rather like the reading on a digital clock. But in translation it is often useful to show the absolute value of a sexagesimal number if known, so we can write 0;00 01 45, where the semicolon is a 'sexagesimal point' marking the boundary between whole and fractional parts of the number. In cuneiform no spaces are left between sexagesimal places, because there are separate ciphers for ten signs and units, but like modern decimal notation it is a *place value* system. That is, relative size is marked by the order in which the figures are written, in descending order from left to right. For instance, 22 45 (= 1365) could never be confused with 45 22 (= 2722). On 2N-T 30, the number 1 45 has been written twice, because it is to be squared, but the resulting 3 03 45—better, 0;00 00 03 03 45 square rods— has not been recorded. Nevertheless, it is clear from the final answer that the scribe got the right result: he must have multiplied up by 60 (shekels in a square rod) and then by 180 (grains in a shekel). The exact sexagesimal answer is 9;11 15 grains, which the scribe had to approximate as 9⅓ grains (i.e., 9;12), because the absolute systems did not use sexagesimal fractions but only the simple unit parts $1/n$.

In sum, a good idea of the mathematical sense of 2N-T 30 can be gained from its internal characteristics alone: its contents could be summarised simply as $0;00 \ 01 \ 45^2 = 0;00 \ 00 \ 03 \ 03 \ 45$. But that would leave many questions unanswered: Who wrote the tablet, and under what circumstances? Who, if anyone, did he intend should read it? Why didn't he record the sexagesimal result but only the version converted into area measure? What else did he write? What else did he do with his life? Such questions, once one starts to ask them, are potentially endless, but the tablet alone cannot answer them: we have look beyond, to its textual, archaeological, and socio-historical context if we want to know more.

1.3 THE CONTEXTS: TEXTUALITY, MATERIALITY, AND SOCIAL HISTORY

In the early 1990s Jens Høyrup pioneered the study of syntax and technical terminology of Old Babylonian mathematical language, with two aims in view.[19] First, a close reading of the discourse of mathematics reveals, however approximately, the conceptual processes behind mathematical operations—which translation into modern symbolic algebra can never do. This line of enquiry is most profitable on more verbose examples than 2N-T 30, particularly those that include instructions on how to solve mathematical problems. Many examples are presented in later chapters. Such work can be carried out, and largely has been, on modern alphabetic transliterations of the ancient texts with little regard to the medium in which they were originally recorded. But the idea of close examination can also apply to non-linguistic features: the disposition of text and numerals on the tablet, the presence and presentation of diagrams, the use of blank space, and so forth. In the case of 2N-T 30 one might be interested in the spatial separation of word problem and solution from the numerical calculation, as well as the fact that there are no traces on the tablet of the scribe's actual working or sexagesimal answer.

Second, looking at clusters of linguistic and lexical features enables unprovenanced tablets to be grouped and separated along chronological and geographical lines. This approach can also be supported by tracing the distribution of problem types and numerical examples within the known corpus and comparing the techniques of calculation used to solve them. Doing that for 2N-T 30 yields six other Old Babylonian tablets very like it (but with different numerical parameters) and eight tablets containing just the squaring calculation without the setting and solution expressed metrologically (table 1.4 and figure 1.4).

The large majority of the tablets are from Nippur, like 2N-T 30, but it is difficult to tell whether that is historically significant or is more an outcome of the history of excavation and publication. However, it probably is noteworthy that all the Nippur tablets are square, while those from Ur are both round (technically speaking Type R and Type IV respectively: see

TABLE 1.4
Old Babylonian Calculations of Squares

Tablet Number	Provenance	Length of Square	Sexagesimal Calculation	Tablet Description	Publication
NCBT 1913	Unknown	<58 rods, 4 cubits>	$58\ 20^2 =$ $56\ 42\ 46\ 40$	Round. The problem is not set metrologically, but the answer is given in both SPVS and area units. Horizontal line at the end; erasures.	Neugebauer and Sachs 1945, 10
NBC 8082	Unknown	1 rod, 4 cubits	$1\ 20^2 = 1\ 46\ 40$	Square with rounded corners; lower right corner missing. Numbers auxiliary to the calculation are written down the left-hand side of the tablet, separated from the rest by a vertical line.	Neugebauer and Sachs 1945, 10 (no copy); Nemet-Nejat 2002, no. 13
2N-T 472 = UM 55-21-76	Nippur	1 cubit, 2 fingers	$[5\ 20^2 = 28\ 36\ 40]$	Square with rounded corners; upper half missing.	Neugebauer and Sachs 1984 (no copy); Robson 2000a, no. 7
CBS 11318	Nippur	1 cubit	$5^2 = 25$	Square with rounded corners; lower left corner missing.	Neugebauer and Sachs 1984
2N-T 116 = IM 57846	Nippur	⅔ cubit, 9 fingers	$[9\ 40^2 = 1\ 33\ 26\ 40]$	Tablet shape unknown.	Neugebauer and Sachs 1984 (no copy)

2N-T 30 = IM 57828	Nippur	⅓ cubit, ½ finger	$1\ 45^2 = \langle 3\ 03\ 45 \rangle$	Square with rounded corners.	F. R. Steele 1951, pl. 7; Neugebauer and Sachs 1984
UM 29-15-192	Nippur	2 fingers	$20^2 = 6\ 40$	Square with rounded corners.	Neugebauer and Sachs 1984
UM 29-16-401	Nippur	—	$44\ 26\ 40^2 = 32\ 55\ 18\ 31\ 06\ [40]$	Square with sharp corners? (Now all broken.) Horizontal line above and below answer.	(See table B.10)
3N-T 611 = A 30279	Nippur	—	$16\ 40^2 = 4\ 37\ 46\ 40$	Square with rounded corners.	(See table B.10)
UET 6/2 211	Ur	—	$16\ 40^2 = 4\ 37\ 46\ 40$	Round.	Robson 1999, 251
CBS 3551	Nippur	—	$7\ 36^2 = 57\ 47$ (*sic*, for 45) 36	Square with sharp corners. Erasures. Horizontal line below answer.	(See table B.10)
N 3971	Nippur	—	$7\ 05^2 = 50\ 10\ 25$	Square with rounded corners. Erasures.	Robson 1999, 275
CBS 7265	Nippur	—	$5\ 15^2 = 27\ 33\ 45$	Rectangular, portrait orientation.	(See table B.10)
HS 232	Nippur	—	$4\ 50^2 = 23'\ 21\ 40$	Square with sharp corners? Horizontal line below answer.	Friberg 1983, 82
UET 6/2 321	Ur	—	$1\ 33\ 45^2 = \langle 2\ 26\ 29\ 03\ 45 \rangle$	Round. The answer has been erased.	Robson 1999, 251

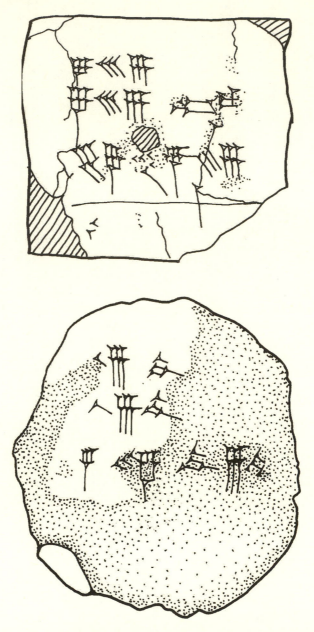

Figure 1.4 Two Old Babylonian exercises in finding the area of a square. (CBS 3551 and UET 6/2 211. Robson 2000a, no. 4; 1999, 521 fig. A.5.7.)

table 4.5). None of them has anything written on the back. Further, there is a striking similarity in textual layouts: all the Nippur tablets which set the problem in metrological terms write it in a box in the bottom right corner, just like 2N-T 30. Even more importantly, every single one of the sexagesimal calculations is laid out in the same way, with the length written twice in vertical alignment and the answer immediately underneath. Only the sharp-cornered square tablets mark it with a horizontal ruling, and 2N-T 30 is the only one of its kind with no sexagesimal solution. Some show traces of erased signs that suggest the remains of scratch calculations.

A further search would yield a list of mathematical problems about squares with the parameters now missing, and a compilation of simple geometrical problems about squares with diagrams but no answers.[20] A large tablet from Susa also catalogues (amongst other things) thirty statements and answers of the squaring problem, with lengths running systematically from 1 finger to 4 rods, just like the entries in a metrological list (see §4.2).[21] The hunt could go on, but enough evidence has already accumulated to show that elementary exercises in finding the area of a square were common in their own right, quite apart from squarings carried out in the course of solving other sorts of problems. That strongly suggests that the purpose of at least some cuneiform mathematical activity was pedagogical.

This discussion has already touched on the fact that the shape of tablets, as well as what is written on them, can be important. Once historians start to consider the material culture of mathematical cuneiform *tablets* as well as the features of ancient mathematical *texts,* then they become increasingly sensitive to their identities as archaeological artefacts with precise findspots and belonging to complex cultural assemblages. If mathematical tablets come from a recorded archaeological context, then they can be related to the tablets and other objects found with them as well as to the findspot environment itself.

Controlled stratigraphic excavation has been carried out to great effect in Iraq for over a century now. In the years before the First World War, the German teams led by Robert Koldewey and Walter Andrae at Assur, Babylon, and elsewhere set impeccable standards which have been followed, more or less, by most academic archaeological teams from all over the world ever since. However, scrupulous recording of the context of artefacts quite simply generated too much data for pre-computerised analyses to manage, however much painstaking and time-consuming work was put into their publication (it is not uncommon for final reports to appear decades after the last trench was scraped). Tablets in particular have often become separated from their findspot information. It is only in the last few years with the advent of mass-produced relational database programmes that large finds of epigraphic material have been satisfactorily analysed

contextually. The tablet 2N-T 30 constitutes a good example. It was excavated in 1948 from an archaeological site in the ancient city of Nippur in
southern Iraq, now in the middle of desert but once on a major artery of
the Euphrates. In the early days of exploration, expeditions were sent out
to Iraq with the express goal of recovering as many artefacts as possible for
their sponsoring institutions. The University of Pennsylvania organised
several large-scale trips to Nippur in the 1880s and '90s, bringing back
tens of thousands of tablets for its museum and depositing many more in
the Imperial Ottoman Museum of Istanbul (for at that time Iraq was still
a province of the Ottoman empire). Modern archaeology no longer oper-
ates like that; instead, it focusses on small areas which are carefully chosen
with the aim of answering specific research questions. After the Second
World War, the Universities of Philadelphia and Chicago jointly initiated a
new series of excavations at Nippur, which ran on and off until the Gulf
War brought all archaeology in Iraq to a halt. One of their early aims was
to understand better the area where their Victorian precursors had found
a spectacularly large trove of Sumerian literary and scholarly tablets—not
least because there had been a major controversy over whether the tablets
constituted a putative temple library or not, an issue which could not be
resolved at the time because the process of excavation had never been recorded.[22] The later excavators chose, in 1947, to open a trench in one of
the old diggings, on a mound the Victorians had dubbed 'Tablet Hill', in
the twin hopes of recovering more tablets of the same kind and of learning
more about their origins. Area TB, as they called it (T for Tablet Hill, B
because it was their second trench on the mound) measured just 30 by 40 m.
When in the second season they reached Old Babylonian levels, they found
not the monumental walls of temple architecture but well-built and spacious
houses, densely grouped together. The dwelling they labelled House B had
been rebuilt many times over the course of its useful life. In the second
layer from the surface they found fifty-three tablets in the central courtyard
and four of its six rooms (loci 10, 12, 17, 31, 45 in figure 1.5) as well as
the remains of domestic pottery. Tablet 2N-T 30 was in room 12.[23]

Unfortunately, although the excavation was published in the early 1960s,
and has been reanalysed since, there has been no systematic study of the
tablets found there. Partly this is a result of the division of finds between
Philadelphia, Chicago, and the Iraq Museum in Baghdad, but it is more a
symptom of the fact that tablets in general have not tended to be treated as
archaeological artefacts. Nevertheless, there are enough data in the original excavation notes and in an unpublished catalogue of the 2N-T tablets
to enable the assemblage of tablets found in House B to be reconstructed.
It turns out that 2N-T 30 is just one of eight mathematical tablets found
there (table 1.5), most of which are elementary calculations of squares or
regular reciprocal pairs (for which see §4.3) while the other two are extracts

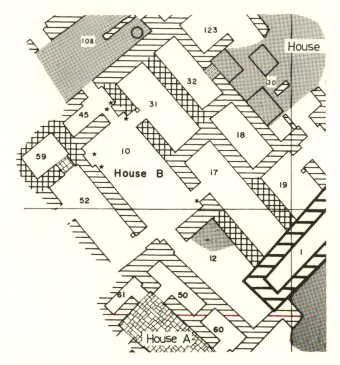

Figure 1.5 House B, Area TB, Level II.1, in Old Babylonian Nippur, excavated in 1948. Mathematical tablets were found in rooms 10, 12, and 45. (Stone 1987, pl. 30, courtesy of The Oriental Institute of The University of Chicago.)

from standard pedagogical lists (for which see §4.2). Two of the calculations are written on the same tablets as Sumerian proverbs from a standard scholastic compilation now known as Sumerian Proverbs Collection 2, which mostly set out appropriate behaviours for scribes (see further §4.4). Extracts from that proverb collection are found on nine further tablets, and elementary Sumerian literary compositions on nine more. All but five of the remaining tablets contain standard lists of cuneiform signs, Sumerian words, or other well-attested items in the curricular repertoire of Old Babylonian Nippur.[24] The curricular context of mathematics is examined in more detail in chapter 4, but a reasonable provisional conclusion would be that somebody wrote 2N-T 30 as part of a general scribal education that was quite standardised across the city, and that House B was probably either his own home or his teacher's.

Situating individual objects, and thereby sub-corpora of mathematics, within an archival and physical environment thus enables consideration of the *people* involved in creating, learning, and transmitting mathematical ideas, and to relate those ideas to the wider Mesopotamian world. The

TABLE 1.5
Mathematical Tablets Found in House B of Area TB, Old Babylonian Nippur

Excavation Number	Museum Number	Findspot	Description	Publication
2N-T 27	IM 57826	Room 12, level II-1	Calculation of regular reciprocal pair (7 55 33 20 ~ 7 34 55 18 45); erased exercise on reverse; rectangular tablet, c. 5.5 × 7 cm	Unpublished
2N-T 30	IM 57828	Room 12, level II-1	Problem and calculation of square; square tablet, c. 6.5 × 6.5 cm	(See table 1.4)
2N-T 35	IM unknown	Room 12, level II-1	Fragment of metrological list or table, c. 4 × 4 cm	Unpublished
2N-T 115	IM 57845	Room 45, level II-1	Calculation of regular reciprocal pair (9 28 53 [20] ~ 6 19 [41 15]); illegible exercise; multiplication table (times 1 40) on reverse; fragment c. 5.5 × 4.5 cm	Neugebauer and Sachs 1984
2N-T 116	IM 57846	Room 45, level II-1	Calculation of square; c. 6.5 × 7.5 cm	(See table 1.4)
2N-T 131	IM 57850	Room 10, level II-1	Fragment of table of squares, c. 5.5 × 3 cm	Unpublished
2N-T 496	IM 58966	Room 10, level II-2	Calculation of regular reciprocal pair (16 40 ~ 3 36) and Sumerian Proverb 2.42; rectangular tablet, c. 8.5 × 8 cm	Al-Fouadi 1979, no. 134; Alster 1997, 304; Robson 2000a, 22–3
2N-T 500	A 29985	Room 10, level II-1	Calculation of regular reciprocal pair (17 46 40 ~ 3 22 30) and Sumerian Proverb 2.52; rectangular tablet, c. 8 × 7.5 cm	Gordon 1959, pl. 70; Alster 1997, 55; Robson 2000a, 21–2

socio-historical context of ancient Iraqi mathematics can be split into three concentric spheres: the inner zone, closest to the mathematics itself, is the scribal *school* in which arithmetic and metrology, calculational techniques, and mathematical concepts were recorded as a by-product of the educational process. As House B illustrates, schools were not necessarily large institutions but could simply be homes in which somebody taught young family members how to read, write, and count. Who was mathematically educated, and to what degree? Who taught them, and how? The answers to these questions are only partially answerable at the moment, and differ depending on the time and place, but attempts will be made chapter by chapter for all the periods covered in this book.

Beyond school is the sphere of *work*: the domain of the professionally numerate scribal administrator who used quantitative methods of managing large institutions and, in later periods, of the scholar who used patterned and predictive approaches to the natural and supernatural worlds. Tracking the continuities and disjunctions between work and school mathematics can enable us to estimate the extent to which mathematical training equipped scribes for their working lives. This is a potentially enormous topic, so for manageability's sake the book focuses on the domains of land and labour management, though it could equally well have highlighted livestock, agriculture, manufacturing, or construction. Equally, other areas of professionally literate intellectual culture are analysed for traces of mathematical thinking in literature, divination, and of course astronomy.

The outer sphere extends beyond the literate to various aspects of Mesopotamian *material culture*: an ethno-mathematical approach can, for instance, identify the external constraints on mathematical thinking, or use the detailed insights of the mathematical texts to reveal the conceptualisation of number, space, shape, symmetry, and the like in other aspects of ancient life, thereby encompassing the lives and thoughts people with no professional mathematical training. The theory and methods of ethnomathematics were developed by scholars such as Ubiratan D'Ambrosio and Marcia Ascher for the study of cultures without writing;[25] but an ethnomathematical approach is just as applicable to the areas of literate societies that are beyond the reach of written mathematics. It proves a particularly useful tool for understanding ancient Iraq, in which only a tiny minority of professional urbanites could read and write.

But the ancient world was not just composed of social institutions: it was populated by individuals with families, friends, and colleagues. So this book also spotlights individual creators, transmitters, and users of mathematical thoughts and ideas. That does not mean 'great thinkers' along the lines of Euclid or Archimedes or Ptolemy, but rather school teachers and pupils, bureaucrats and accountants, courtiers and scholars. Some of this evidence takes the form of names on tablets, but even anonymous writings can be grouped by handwriting, stylistic idiosyncrasies, and subject matter. Even when a tablet's origins are unknown, its orthography (spelling conventions), palaeography (handwriting), layout, shape, size, and other design features can all give important indications of the time and place of composition as well as clues about its authorship and function.

Such prosopographic work is necessary primarily because large numbers of tablets have no known context, having been acquired on the antiquities market by dealers, museums, and collectors. Before the creation of the modern Iraqi state and the drafting of strong antiquities laws in the 1920s[26] even the big Western museums employed agents to undertake opportunistic or systematic searches for 'texts' which, they were optimistic enough to suppose, could 'speak for themselves' without recourse to archaeological context. More recently, since the Gulf War of 1991 and again since the Iraq War of 2003, many more cuneiform tablets have flooded the international antiquities market. Some can still be found on sale today, despite increasingly stringent international legislation. This glut was caused first by the looting of provincial museums after the Gulf War, later fuelled by the Iraqi economic crisis of the 1990s, and more recently provoked by the invasion and insurgency since 2003. International sanctions against Iraq followed by gross post-war mismanagement led both to the virtual collapse of the State Board of Antiquities and Heritage, which has responsibility for the protection of some ten thousand identified archaeological sites nationwide, and to sudden and desperate poverty across larges swathes of a previously prosperous population. Now, as in former times, the perpetrators have been both impoverished local inhabitants who exploit the tells for profit like any other natural resource of the area, and naive or unscrupulous purchasers whose paramount goal has been the recovery of spectacular objects for public display or private hoarding. Almost every cuneiform tablet on sale today, whether in a local antiques shop, through a major auction house, or on the web, is stolen property. In most countries it is technically a criminal offence to trade them. But legality apart, the example of 2N-T 30 has shown that the archaeological context of a cuneiform tablet is as crucial to the holistic decipherment of that tablet as the writing it carries. While historians have to do the best they can with the context-free tablets that came out of the ground in the bad old days, anyone at all committed to the sensitive understanding of the past should resist the temptation to buy. Cuneiform tablets and other archaeological artefacts must be allowed to remain undisturbed until they can be excavated and studied with the care and attention that they and the people who wrote them deserve.

CHAPTER TWO

Before the Mid-Third Millennium

From about 6000 BCE, long before writing, Neolithic villagers used simple geometric counters in clay and stone to record exchange transactions, funded by agricultural surpluses. As societies and economies grew in size and complexity, ever more strain was placed on trust and memory. By the late fourth millennium the intricacies of institutional management necessitated both an increasing numerical sophistication and the invention of written signs for the commodities, agents, and actions involved in controlling them (§2.2). Very gradually over the third millennium this numerate, literate urbanism enabled literary and mathematical writings to evolve from the immediate needs of temple bureaucracy (§2.3). Visual culture, as well as accounting practice, strongly influenced the language and content of mathematics (§2.4). For much of this period, however, the evidence is slight, patchy, and ambiguous (§2.1). That may not simply be the outcome of archaeological happenstance but perhaps also an artefact of the way professional numeracy was taught through direct practice rather than by formal education (§2.5).[1]

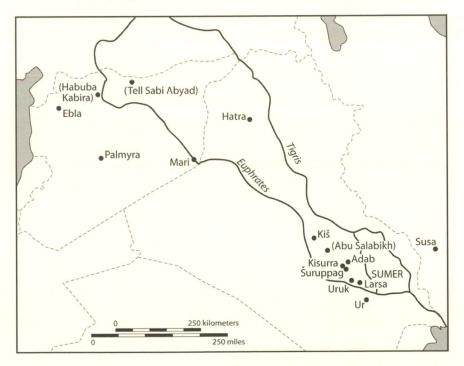

Figure 2.1 The locations mentioned in this chapter, including all known findspots of mathematical tablets from the late fourth and early third millennia.

2.1 BACKGROUND AND EVIDENCE

People have been living in the Middle East for at least 500,000 years, when Neanderthals occupied caves in the Zagros Mountains of northern Iraq. Anatomically modern humans have inhabited the region for about 100,000 years.[2] From some time between 15,000 and 10,000 bc[3] the slow process of sedentary settlement began. As Roger Matthews explains, communities constructed dwellings, shrines, storerooms, and cemeteries on which their economic, social, and spiritual lives were increasingly focussed.[4] Over the next few millennia, hunting and gathering gradually and patchily gave way to agriculture and then animal husbandry, but never entirely died out. By 7000 bc three domesticated cereals—barley, emmer wheat, and einkorn— were widespread. Hunting dogs were the first animals to be domesticated, probably even before the existence of agriculture. Sheep, goats, and then pigs, none of which are direct competitors for food eaten by humans, were exploited for meat and secondary products. Cattle were the last of the major domesticates to appear in the Middle East, some time after 6500 bc.[5]

As farming techniques became more efficient and cultivars higher yielding, the protection and equitable distribution of harvest surpluses also became increasingly necessary. It is in these circumstances that number was first recorded, in the form of tiny 'tokens' made of clay or stone and shaped into simple geometrical forms. Such tokens are found in villages and settlements all over the Middle East, from western Syria to central Iran. Throughout the fourth millennium increasing surplus wealth, managed through token technology, drove a massive population growth, especially in the region of southern Iraq, which was then a mix of fertile river plain and abundantly verdant marshes. Permanently dry land, fit for habitation, was hard to come by, so large settlements grew in single spots, gradually rising above the extraordinarily flat landscape on the accumulated rubble and rubbish of centuries. These increasingly large cities, with populations in the tens of thousands, were far too huge and impersonal to be run along the old lines of family, trust, and memory.[6] In southern Iraq the city of Uruk, after which the period is now named, grew to 250 hectares in size by 3000 BCE, with at least 280 further hectares of inhabited hinterland.[7] The temple administrators of Uruk adapted token accounting to their increasingly complex needs by developing the means to record not only quantities but the objects of account as well. Thus numeracy became literate for the first time in world history.

That is not to say that the genre of mathematics immediately came about as a consequence. For most of the third millennium it is often difficult—and perhaps even wrong-headed—to distinguish confidently between mathematics as an intellectual, supra-utilitarian end in itself, 'written for the purpose of communicating or recording a mathematical technique or aiding a mathematical procedure to be carried out'[8] in the course of scribal training, and numeracy as the routine application of mathematical skills by professional scribes. However, some rules of thumb can usefully be applied. Generally speaking, working documents are dated, record the work of real people known from other documents, and may be sealed. Pedagogical mathematics exercises in the main have none of these characteristics, but a working document may not have all of them either. Further, such exercises are often on crudely written or misshapen tablets, or on the same tablets as other schoolwork; there may be mistakes in the layout, spelling, or arithmetic; the text may be unfinished; or there may be complete or partial duplicates. Perhaps most diagnostically, however, the numerical parameters are often unrealistic in their size, roundness, or regularity.[9] Thus the lists of early mathematical tablets given here (tables B.1–B.4) are neither definitive nor exhaustive. It is highly likely that there are further undiscovered mathematical exercises amongst the tens of thousands of published quantitative documents from the long third millennium, and exactly what constitutes that genre of early mathematical exercises is often a matter of personal interpretation, open to vigorous debate and thorough reassessment.

The question of identification is most acute for the very oldest known tablets, from the late fourth millennium. As explained in §2.2, the very large majority of the five thousand tablets excavated from archaic Uruk contain quantitative data and little else; thus distinguishing mathematical exercises from accountants' records is hard, to say the least. However, a dozen of those tablets seem to bear mathematical exercises, including nine on the areas of irregular quadrilaterals (table B.1). A further two tablets, of unknown provenance, apparently carry metrological tables and another three may bear accounting exercises. Two are from the earliest attested script phase, so-called Uruk IV, the rest from the slightly later Uruk IIII period (table B.2).[10]

The area problems are exemplified by the badly damaged tablet W 20044,20 (figure 2.2). The surface of each tablet is divided into four cases with a length measurement written in each, accompanied by either a horizontal or a vertical line. On many of the tablets those lengths are given in round numbers. None records a resultant area (as here, the reverse of the tablet is usually blank) but when calculated using the so-called surveyors' method—multiplying the mean lengths and widths—the resultant areas are often conspicuously round too. Some of the tablets also bear two or three word signs (as here), but nothing that is clearly recognisable as a transaction or official title. They may just, however, be the names of the student scribes to whom the exercises were assigned, as in the later Sargonic period (§3.1).

On W 20044,20 the large D-shaped impressions each represent 60 length units (c. 360 m) and the small o-shaped impressions represent 10 length

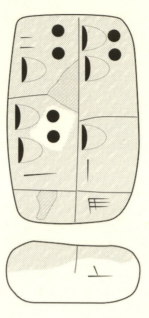

Figure 2.2 One of the world's earliest-known mathematical exercises, a problem in finding the area of an irregular quadrilateral field, from late fourth-millennium Uruk. (W 20044,20, courtesy of the CDLI project.)

units (c. 60 m); a single length unit of around 6 m (later called the rod) would be recorded with a small D. Thus the tablet reads something like:

60 + 20 horizontal   120 + 20 <vertical>
120 + 20 horizontal   60 vertical

The surveyors' method yields $(80 + 140)/2 \times (140 + 60)/2 = 110 \times 100 = 11,000$ square rods (c. 40 hectares). Expressed in the largest possible area units the hypothetical answer is 6 *bur* 2 *iku*, where 1 *bur* = 18 *iku* and 1 *iku* = 100 square rods.

Notoriously, none of the tablets from archaic Uruk was found in its original context; all had been re-used as building rubble in the central temple precinct shortly before 3000 BCE but had probably not been transported from very far away. The excavators divided the enormous area into a grid comprising 20×20 m squares, with seventeen such squares on each side. Tablets were found in over a hundred of them, apparently randomly distributed and with no correlation to the buildings constructed over them.[11] However, all nine of the area-problem tablets were found in just three adjacent excavation squares (along with many other tablets), underneath the so-called Great Court. It is possible, then, that they were dumped there from a single original source (although it is highly unlikely that such a hypothesis could ever be proved for certain).

For much of the third millennium the political infrastructure of the region remained fairly stable. Presumably all large administrations had to train their scribes in various numerate activities although very little writing of any sort survives from the early and mid-third millennium (so-called Early Dynastic I–II).[12] During the Early Dynastic III period just after 2500 BCE, individual city states such as Šuruppag (ED IIIa), and later Adab and Ebla (ED IIIb), tended to retain strong local identities while sharing a common culture, religion, and management structure. They regularly formed economic and military alliances with their neighbours for the sake of prosperity and security, while disputes over water and land resources were settled by warfare or arbitration by a third polity. At times even cities as far away as Ebla in western Syria participated in this regional exchange, diplomacy, and cuneiform literacy.[13]

Writing gradually accrued new powers as it was adopted for a variety of institutional purposes, from recording legal contracts to issuing administrative orders and documenting political events.[14] Paranomasia, or punning, enabled scribes to record the sounds of all words of the Sumerian language, not only those that could be represented pictorially. At the same time, visual changes were made: signs arranged inside rectangular cases were straightened out into lines of writing; the script became ever more abstract and 'cuneiform' or wedge-shaped. Impressed numerals that imitated the form of accounting tokens were gradually replaced by incised cuneiform notation. While many individuals and families doubtless earned their livings through personal enterprise based on trust and memory, never coming into contact with the written record, institutional economies continued to run on centrally managed quantitative models, planned and carried out by literate and numerate professionals. Literacy and numeracy thus remained overwhelmingly in the hands of bureaucrats managing large-scale quantities of land, labour, and livestock, yet also began to develop into independent genres that, by the last third of the millennium, can clearly be identified as literature and mathematics (see chapter 3).

The largest corpus of Early Dynastic mathematics is from the city of Šuruppag, also known by its modern name, Fara, in the scholarly literature. Around five hundred tablets were excavated from the 100-hectare site, almost all of them written within a year of a city-wide fire in about 2450 BCE.[15] Giuseppe Visicato's study of scribes of Early Dynastic Šuruppag has revealed a strict hierarchy of control, with the *ensi*$_2$ 'city governor' managing *saĝĝa* 'administrators', *saĝ-sug*$_5$ 'accountants/land surveyors', and, surprisingly, about a hundred *dub-sar* 'scribes', *aĝrig* 'stewards', and *um-mi-a* 'scholars'. Some of those individuals are attested only as the authors of single legal contracts, copyists of one or two literary or lexical works, or the recipients of rations—but that still leaves over forty named literate, numerate officials working for the bureaucracy of Šuruppag in a single year.[16] Visicato estimates that the city's institutions produced at least two thousand tablets annually, or just fifty per bureaucrat—who cannot, therefore, have spent their entire working lives poring over their documents.

The sixty-odd school tablets whose findspots are known come, in the main, from two widely separated buildings: an office in the south of the city from which donkeys appear to have been disbursed but which also housed a 'library'; and a house to the northwest that has been tentatively characterised as a scribal centre, as only school tablets have been found in and around it. None of the mathematical tablets are provenanced, however. Five can clearly be identified as mathematical: a metrological table and four problems stated and answered; a further half-dozen may also be mathematical exercises of some sort but their reading is uncertain (table B.3). Two problems give variant solutions to finding the number of workers that can be fed by a large quantity of grain; another calculates the grain needed to feed a set number of workers. The metrological table and the fourth of the problems both concern the relationship between the sides and areas of large squares. This is also the concern of the only mathematical tablet known from the nearby city of Adab, a few generations later (table B.4).

Contemporary with Adab are the central palace archives of the great Syrian city of Ebla. Italian excavations there in the mid-1970s yielded some two thousand tablets, mostly from a small archive room, L.2769, off a monumental courtyard in Palace G. The tablets were found as if they had slipped off shelves during or after the final destruction of Ebla, which must have happened under the reign of one king Ibbi-Sipiš, some time around or shortly after 2350 BCE.[17] Just four or five mathematical tablets were found there: two set out solutions to problems about dividing large quantities of grain, one equates weights of metal (copper and tin) with grain, one may be a set of word problems, and one lists large number signs (table B.4). This tablet and the one from Adab contain the earliest unequivocal names of particular individual scribes of mathematics: the former says that 'Nammah wrote the calculation', while the latter ends, 'Established (?) by the scribes of Kiš. Išmeya'. Nothing else is known at all about either of these scribes; equally, nothing else is known about mathematical activity in Early Dynastic Kiš, although some fifty contemporaneous administrative tablets come from that city. Kiš and Ebla were in direct diplomatic contact at the time.[18]

Some lexical lists (standard lists of words and cuneiform signs written out by trainee scribes as part of their formal training) also yield useful evidence for numeracy in this early period. So-called Word List D (Food/Grain) from late fourth-millennium Uruk and its successor, Early Dynastic Food (known from Šuruppag and Ebla, as well as the cities of Abu Salabikh and

Susa), begin with a list of number signs, probably associated with different sizes of vessels or loaves of bread.[19] A tablet from Ebla (where the inhabitants were not native Sumerian speakers) writes out the numbers 1–10 in syllables—our best systematic evidence for how numbers were actually pronounced in the third millennium. It demonstrates a quintal, or base 5, system:[20]

1—*diš* or *aš*
2—*min*
3—*eš*
4—*līmu*
5—*ya*
6—*yaš* (from *ya* + *aš*)
7—*umin* (from *ya* + *min*)
8—*ussa* (probably not from *ya* + *eš*)
9—*ilīmu* (*ya* + *līmu*)
10—*ḫaw*

For the most part, though, numbers were written as numerals, which means that we will probably never know how the larger numbers were pronounced. The impressed number signs of Uruk, which resemble the mark an accounting token might make if pressed into the clay, were gradually superseded in the Early Dynastic cities by a mixture of impressed (token-like) and incised (scratched) number signs, the exact relationship between which is not always clear. But whatever their appearance, number signs at this early period were never simply context-free numerals but were always metrologically identifiable. That is, while there was a single *word* for 'ten' there was no single *numeral* but different signs for 'ten-discrete-objects', 'ten-units-of-grain', and 'ten-units-of-land'. Equally, the value a sign took depended on its metrological context. So a small o-shaped impression could signify 'ten-discrete-objects', 'eighteen-large-land-units', or 'thirty-small-grain-units', as necessary. The origins of this system can be traced back to the Neolithic Middle East, millennia before the invention of writing, in close relationship with the development of institutional management and, later, the first city states.

2.2 QUANTITATIVE MANAGEMENT AND EMERGING STATEHOOD

The intertwined origins of numeracy, literacy, and the state in early Mesopotamia have been widely recognised for several decades now. Yet their exact relationship remains a matter of hot dispute. At one extreme, Denise Schmandt-Besserat argues that early literacy was no more than a bureaucratic accounting tool whose sole precursor was an increasingly complex and uniform system of clay counters or tokens—tiny spheres, cylinders, cones, discs, and tetrahedra—used all over the Middle East from the eighth millennium. At the opposite pole, Jean-Jacques Glassner sees no role for numeracy in the origins of cuneiform, evidencing private and familial uses as well as institutional ones.[21]

The very earliest attested tokens come from five early eighth-millennium sites in northern Syria and western Iran, predating both farming and the use of pottery in the region.[22] Of course, in many contexts small clay objects are very difficult to interpret: there is no way of knowing why prehistoric villagers might have made or kept them. Evidence for the function of tokens thus comes through particular types of archaeological context, as for instance at Tell Sabi Abyad, a late Neolithic village in north-central Syria which was destroyed by fire around 6000 BCE.[23] Five densely clustered rectangular houses, composed of many small rooms, and four round ones, known anachronistically by the Greek word *tholoi,* were uncovered in Dutch excavations at the so-called Burnt Village in 1991–2. Two houses and one *tholos* yielded clay tokens, carefully fashioned into simple geometric shapes (figure 2.3). Houses II and V were primarily used for storing and processing cereals, as witnessed by knee-high deposits of grain in some rooms and large numbers of grinding slabs, mortars, and pestles in others. In each case, the tokens were found in just one or two rooms, where they had been carefully stored along with hundreds of clay stamp-sealings from bags, baskets, jars, and doors as well as a variety of other small stone, ceramic, and bone artefacts. *Tholos* VI was used for both storage cloth manufacture. None of the three constructions seems to have been inhabited domestically.

These sealings, amongst the earliest attested anywhere in the Middle East, were designed to stop unauthorised entry into whatever they marked, thus showing that people were storing and controlling access to goods of high value. Although over sixty different designs of seal impressions are known at Sabi Abyad, suggesting that a large number of individuals were involved in the process, not a single actual stamp seal has been recovered. The excavators conclude that either the seals were made of perishable material such as wood or bone—or that the objects were sealed at some other place and came to Sabi Abyad through trade or exchange.[24] We should probably understand the tokens, so carefully stored with the sealings, as records of the transactions, but exactly which quantities and commodities they represented it is impossible to say. It is likely, however, that systematisation was local at best.[25] Particular shapes and sizes must have had particular meanings to the individuals involved in trade and storage at Sabi Abyad but, although the general signification of tokens was well understood across the region, the clusters of tokens could themselves be meaningfully deciphered only by someone who remembered the transaction they invoked.

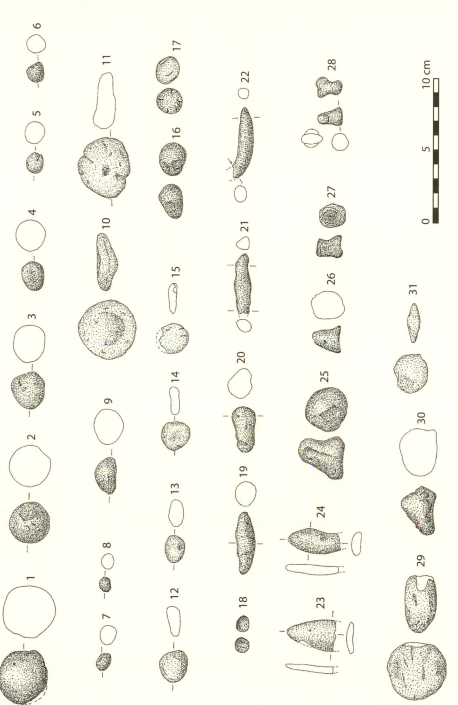

Figure 2.3 Small clay accounting tokens found in public buildings at Neolithic Tell Sabi Abyad in northern Syria. (Akkermans and Verhoeven 1995, fig. 14, courtesy of Peter Akkermans and the Archaeological Institute of America.)

During the course of the fourth millennium people from the area around Uruk came north up the Euphrates to dwell alongside the local population in economically and socially autonomous enclaves and to found new settlements, presumably set up to exploit local resources. Habuba Kabira South, for instance, was a walled town of about 10 hectares on a strategic bend of the Syrian Euphrates, excavated in 1971–6.[26] Communal living was organised on a completely different scale from that of the Neolithic village of Sabi Abyad. Inhabited for about 120 years some time between 3400 and 3000 BCE, the town was centrally planned and managed, with a grid system of paved and drained streets and very regular architecture. Rations were distributed, here as at Uruk and elsewhere, by means of mass-produced, standardised bowls. Uruk-style administrative technologies—cylinder seals and clay sealings, tokens, and clay envelopes in which to store them—were also ubiquitous. The two-hundred-odd baked clay tokens display an unprecedented variety of shapes; over half were incised or otherwise marked before firing.[27] Two broken clay envelopes containing two and six tokens respectively were found in one of the largest houses of the city, along with other administrative debris including flat pieces of clay impressed with both seals and token-like marks (figure 2.4).[28]

Habuba Kabira and similar northern Uruk 'colonies' thus witness the merging of counting and sealing technologies, with the invention of the sealed clay envelope in which clusters of tokens could be stored. Impressions of the tokens to be put inside were first made on the outside of the envelope, and the whole thing sealed—now with cylindrical seals that could cover the whole surface. But the redundancy in this system rapidly became apparent: why go to all the trouble of storing tokens when one has already made tamper-proof records of them? Thus recording quantities of objects took another step towards abstraction as empty envelopes were transformed into flat tablets of clay, whose surfaces bore impressions of tokens and seals.[29]

Even before writing, then, people from southern Iraq were recording numbers of stored commodities to protect them from theft or to document transactions. Yet we probably should not call this activity mathematics, or even arithmetic, as there is no evidence yet of consistent number bases.[30] Further, this procedure was just as dependent on memorising the correlations between individual transactions and individual tablets as the Neolithic token system of Sabi Abyad. It was still not possible to record the objects of accounting, or the agents, circumstances, or types of activity. All that had changed was the move from ephemeral clusters of tokens to permanent impressions of them in clay. Indisputably, though, this was the precursor of a system that became incredibly sophisticated and flexible when writing was thrown into the mix, somewhere in southern Iraq at about the time that Habuba Kabira was abandoned in the very late fourth millennium.

 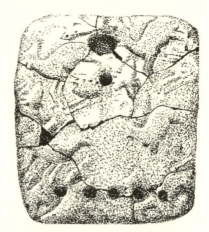

Figure 2.4 Tablets impressed with token-like marks found in the fourth-millennium town of Habuba Kabira on the Syrian Euphrates. (Strommenger 1980, fig. 56, courtesy of Eva Strommenger.)

German excavators, working regularly at Uruk since 1928, have shown that its central administration was based in the monumental temple complex of some 30 hectares at the heart of the city. It was grounded on a strong religious ideology, expressed through imagery on objects large and small, focussed on the chief priest and the anthropomorphic deity he worshipped.[31] Just as in the northern colonies, at Uruk excavators found tokens, sealings, clay envelopes, and ration bowls all around the central religious precinct—along with some five thousand proto-cuneiform tablets which had been thrown away after their useful life was over.[32]

Nevertheless, on internal grounds it is possible to isolate three stages of the tablets' development.[33] 'Numero-ideographic tablets' are almost identical to those found at Habuba that bear impressions of cylinder seals and tokens. But they are also marked with one or two signs, incised with a pointed tool, which represent discretely countable objects such sheep, jugs of beer, and textiles. These notations, so far attested only in the big cities of the south, are thus the first explicit record of the material objects of accounting. Many are clearly pictographic, schematically representing containers and the like; it is thus probable that the remainder are too, even if the objects signified or the conventions used to represent them cannot yet be identified.[34]

The following stage, known conventionally as 'Uruk IV', is currently attested only in Uruk itself. These two-thousand-odd tablets, mostly unbaked and unsealed, exhibit a range of some nine hundred incised word signs and five different numerical systems for counting different types of commodity.

They may record one or more transactions of a single type, whether income, expenditure, or internal transfer. The front side of a tablet may be divided into box-like cases, each of which records a separate item of account, typically with a total or other derived figure on the reverse. Analysis of the individual entries on the front, combined with summations on the back, has established that the same numerical signs could take different quantitative values, depending on what was being counted. Five major metrological systems have been identified, with eight variants or subsystems.[35]

But for the first time it was not only the objects and quantities of account that were recorded, but also the human and institutional agents and the type of transaction involved.[36] And unlike at Neolithic Sabi Abyad, where it seems that tokens were used to account for *traded* goods, here in Uruk the accountants dealt only in the *domestic* economy: fields and their crops, livestock and their products, and the temple personnel themselves. And once again accounting technology had undergone a concomitantly large shift through the invention of incised symbols, mostly pictograms, to represent the commodities, land, and labour under central administrative control. Numbers, however, continued to be represented by impressions of tokens—crucially, this is how we can infer so confidently the function of tokens in prehistoric periods.

The mature stage of literate accounting at Uruk is represented by some 2500 tablets from the city itself plus a further 800 from settlements across southern Iraq and southwest Iran. Conventionally known as 'Uruk III', this phase probably dates to a century or so around 3000 BCE. The forms of the incised ideograms show much greater standardisation than the preceding Uruk IV stage, and have lost much of their pictorial quality. Document formats can be much more complex too, with 'secondary' accounts summarising and consolidating the contents of many 'primary' records, often over several years.[37]

Take, for example, the brewing of beer, an essential staple for many societies with no reliable access to clean drinking water. In southern Iraq brewers used barley which had been processed in two different ways: malt, which involves soaking, germinating, and drying the barley grains; and groats, which are hulled and crushed barley grains. On the front of the tablet MSVO 3,6 (figure 2.5), a typical 'primary' tablet, four different types of beer, counted by the jug, are allocated to several different officials, with perhaps the number of days of brewing work entailed given on the back. While individual objects such as beer jugs were counted in groups of 10 and then 60, barley was accounted for with a special set of four capacity units related to each other by the ratios 1:6:60:180, with the smallest unit equivalent to about 24 litres and the largest to some 4320 litres.[38]

A typical 'secondary' beer-account, such as MSVO 3,11 (figure 2.6), calculates the malt and groats necessary to brew each of the four beer types

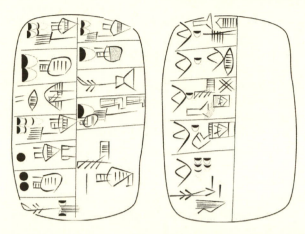

Figure 2.5 A 'primary' account listing four different types of beer distributed to officials, probably from the late fourth-millennium city of Uruk in southern Iraq. (MSVO 3,6, courtesy of the CDLI project.)

in the 'primary account', for one of the officials listed on that tablet as well as several more. Malt and groats were notated with the same number system as barley, but marked with tiny dots and diagonal dashes respectively. First the quantities of groats and malt needed for each of the four types of beer were calculated separately, then the total number of beer jars summed;

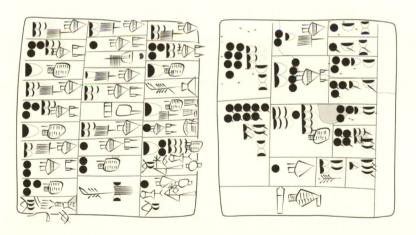

Figure 2.6 A 'secondary' account calculating the amount of groats and malt needed to brew four different types of beer for distribution to officials, probably from the late fourth-millennium city of Uruk in southern Iraq. (MSVO 3,11, courtesy of the CDLI project.)

finally, the overall amounts of groats and malt for all beer types were totalled.[39] It is not apparent whether these were theoretical calculations,
made in advance of production in order to requisition the necessary raw
ingredients, or an accounting after the fact. Either way, tablets like these
comprised a tiny part of the Uruk bureaucracy that traced production right
from sowing the field with barley, to harvest, to the final production and
consumption of the beer.

In short, by the end of the fourth millennium southern Mesopotamian
city states had implemented an extensible and powerful literate technology
for the quantitative control and management of their assets and labour
force. In doing so, they had created in parallel a new social class—in Uruk
called the *umbisaĝ* 'accountant/scribe'—who was neither economically
productive nor politically powerful, but whose role was to manage the
primary producers on the elite's behalf.[40]

2.3 ENUMERATION AND ABSTRACTION

As the production of accounts entailed complex multi-base calculations,
trainee scribes had to practice both writing and calculating, and they did
so increasingly systematically. Whereas fewer than 2 percent of Uruk IV
tablets have been identified as scribal exercises, some 20 percent of the
more mature Uruk III corpus, about seven hundred tablets, are the product
of bureaucratic training.[41] The exercises take two forms: on the one hand,
highly standardised lists of the objects of account; and on the other, apparently *ad hoc* exercises in accounting for commodities and calculating the
area of land (see §2.1).

The calculations themselves were not written down but, as in many other
ancient societies, presumably performed by the manipulation of counters—
the descendants of the age-old accounting tokens.[42] But to aid their learning of complex metrologies, trainee scribes tabulated the key relationships
between the major units of account. Only two such tables are known (both
unprovenanced), suggesting that scribes for the most part committed number facts to memory, as their Old Babylonian successors were also trained
to do (§4.2). MSVO 3,2 for instance (figure 2.7) begins by explicating the
relationships between the fractional grain capacity units and the larger ones
and then moves on to show the (equal) quantities of barley and malt needed
for brewing three types of beer and for making other grain products.

The first four lines can be thought of rather like a grain multiplication
table, with the multipliers written in the bisexagesimal system (used to count
discrete grain products amongst other things) and the multiplicands and
products in the dry capacity system, marked with annotations for groats:

10[1] lots of ½-units of groats	5 units of groats
10 lots of ⅓-units of groats	3⅓ units of goats

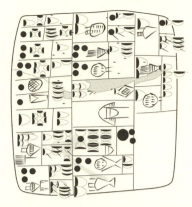

Figure 2.7 A pedagogical exercise detailing the relationships between different grain-accounting units, and between the metrological systems for grain and beer jugs, probably from the late fourth-millennium city of Uruk in southern Iraq. (MSVO 3,2, courtesy of the CDLI project.)

| 20 lots of ¼-units of groats | 5 units of groats |
| 30 lots of ⅕-units of groats | 6 units of groats |

The section on beer, which begins with the penultimate entry in the first column, states:

5 large jugs of beer A	7½ units of groats, 7½ units of malt
30 small jugs of beer A	18 units of groats, 18 units of malt
120 jugs of beer B	36 units of groats, 36 units of malt

and so on. The essential metrological relationships in grain and beer accounting are thus enumerated and abstracted, using convenient round numbers wherever possible. The exact identities of the two types of beer are not known, but they are the same as some of those accounted for in MSVO 3,6 and MSVO 3,11 (§2.2, figures 2.5 and 2.6).[43]

Scribal students learned how to write numbers in context by means of exercises such Word List C, which enumerates the basic commodities of the Uruk administration and is known from nearly sixty fragmentary manuscripts. Lines 18–25 exemplify the list's use of single-numeral round numbers:

10 (vessels of) milk
1 (vessel of) cream
10 cows
1 bull
10 ewes
1 ram

10 nanny goats
1 billy goat[44]

Trainee scribes also copied and memorised thematic lists of nouns related to the accounting practices of Uruk: commodities such as containers,
textiles, metal and wooden objects; fish, birds, and animals; professional
and administrative titles; names of cities and other geographical terms.
The list of professions now known as Lu A, for instance, is attested in over
160 exemplars while others, such as a list of bovines, are apparently unique
manuscripts that were rarely used for training. While the *thematic* range of
these word lists is co-extensive with the bureaucrats' remit—there are no
lists of gods, for instance, or wild animals—their *lexical* range is much
wider: they record many words that were never in practice used by the
administrators, so far as we know. The point was, though, that they could
have been: as Niek Veldhuis argues, the new writing system had to accommodate all the possible uses that the accountants of early cities might want
to put it to.[45]

In Early Dynastic cities such as Šuruppag, Adab, and Ebla some five
centuries later, scribal training continued to rely on much the same types
of exercise as before, though with the addition of new genres. Novice bureaucrats still copied and memorised highly standardised lists of words—
often the same exercises that had been created by the Uruk accountants in
the late fourth millennium, even when (as in the case of the professions list
Lu A) many of the words they contained were no longer in use. New lists
of nouns were also added, such as lists of gods, but these too were now
necessary knowledge for palace and temple administrators. Word List C
continued to be copied, and arithmetical exercises set whose primary goal
was still the correct manipulation of lengths and areas or the fair distribution of rations.

Now, though, arithmetical formats developed that did not simply imitate account documents—although doubtless many scribes were also
trained on the job with no pedagogically designed exercises to practice on.
It is only rarely that we can identify unequivocally mathematical training
exercises—through replication, and from suspiciously round, large, or
small numerals. For instance, from Šuruppag come two different answers
to a problem about rationing a warehouse full of grain between a large
group of workers at 7 *sila* (c. 7 litres) each. One student calculated a round-
number answer with a small amount of grain left over; the other attempted
an exact solution. In real administrative circumstances no worker ever received 7 *sila* of grain but always some multiple of 5 or 10 *sila*. Rather, 7
may have been chosen precisely for its mathematical difficulty, being the
smallest integer that does not neatly divide any of the number bases used
in grain metrology.

These two exercises, and the calculations from Ebla, were solved not by top-down division of the large quantities of grain given in the statements of the problems, but through repeated bottom-up multiplication and addition of smaller units, as Duncan Melville and Jöran Friberg have shown.[46] This is particularly clear in the Ebla calculations, which show the scribe's working. For instance, TM.75.G.2346 reads in its entirety:

4 *gubar* 4 *sila* of grain
    1 hundred *sila*
20½ *gubar* 8 *sila* of grain
    5 hundred *sila*
4¹¹½ *gubar* 4 *sila* of grain
    1 thousand *sila*
2 hundred, 10 minus 2 *gubar* 8 *sila* of grain
    5 thousand *sila*
4 hundred, 20 minus 2 *gubar* 4 *sila* of grain
    1 myriad *sila*

Now, as the Ebla accountants took 1 *gubar* = 20 *sila* (c. 20 litres) it is clear that this is not a list of metrological relationships, as in that case the first two lines would read 5 *gubar* = 1 hundred *sila*, or perhaps 4 *gubar* 4 *sila* = 1 24 *sila*.[47] Instead we have, expressed in *sila*, an increasingly accurate ratio of 5:6 going down the text:

$$84/100 = 0.84$$
$$418/500 = 0.836 \qquad (84 \times 5) - 2$$
$$834/1000 = 0.834 \qquad (418 \times 2) - 2$$
$$4168/5000 = 0.8336 \qquad (834 \times 5) - 2$$
$$8364/10{,}000 = 0.8364 \qquad (4168 \times 2) - 2$$

While the right-hand side simply doubles or quintuples from entry to entry, the left-hand side is generated by following the same procedure and then subtracting 2. In the last line the scribe has made an error, however (underlined)—perhaps by mis-transcribing the result of first, doubling stage of the generating algorithm as 8366 *sila* instead of 8336 before subtracting 2 and converting the result into *gubar*. What was the calculation for, though? The text itself gives no extra information. But as Alfonso Archi has recognised, it may relate to distributions of seed grain for sowing fields, which were sometimes administered at ⅚ of the expected rate for reasons that remain unclear.[48]

Three of the mathematical tablets from the cities of Šuruppag and nearby Adab concern lengths and the square areas that result from multiplying them together. Two systematically work through the length units, one tabulating them in descending order in headed columns, one setting out the lengths and areas in ascending order in alternating lines of text. The third is a problem with an erroneous answer (which should therefore be susceptible to an analysis of how it was calculated).

In short, the arithmetical lists, tables, and exercises together show an increasing pedagogical interest in the properties of numbers for their own sake, beyond the immediate needs of administrative accounting but always stemming from that context. Textual formats had not yet stabilised (as shown by the two different listings of metrological data) nor had methods of solution (as witnessed by the two grain division problems from Šuruppag). Indeed, it is arguable that the genre of mathematics itself was not yet fully formed: none of the exercises or lists invokes scenarios or requires arithmetical techniques that are not directly relevant to the needs of professional accounting. The boundaries between pedagogically motivated exercises and professionally numerate documents were not yet clearly drawn.

These genre difficulties are not unique to the mathematics-numeracy nexus: the boundaries between early Mesopotamian lists and literary works are also very difficult to define. As Gonzalo Rubio has noted, 'the same text can be regarded as a list of field names, canals and cult places by one scholar, whereas another would interpret it as a literary text'.[49] There are two basic problems at the heart of the ambiguity. First is the matter of comprehensibility: scribes did not yet write all the grammatical elements of the Sumerian language, making it difficult for modern readers to determine authorial intention. Second is the matter of literary style. As Miguel Civil has put it:

> A frequent rhetorical device in Sumerian literature is what could be called 'enumeration'. A text may consist mainly of a listing of the terms of a lexical set. Each term is encased in a fixed repeated formula and provided with a comment. The enumeration is inserted in a narrative or laudatory frame.[50]

In this enumerative light, Rubio argues, we should view Sumerian lexical lists and literary works not as antithetical but as two complementary and mutually productive artefacts of scribal culture.[51] Further, enumeration by its very definition entails counting as well as listing; it was at the very heart of cuneiform literacy and numeracy. We have already seen the process of enumeration and abstraction at work in the two metrological lists of the Uruk III period (§2.2) as well as in the Early Dynastic tables. But where the Uruk lists summarise key relationships across a variety of metrological systems, their Early Dynastic counterparts systematically work through the relationships between length and area. Thus out of recorded quantitative bureaucracy emerge, via the humble pedagogical list, the first explorations of the aesthetic qualities of numbers and words— namely the beginnings of mathematics and literature—in the middle of the third millennium BCE.

2.4 SYMMETRY, GEOMETRY, AND VISUAL CULTURE

While no mathematical diagrams are known from before the Sargonic period (§3.2), principles of geometry and symmetry played an important role in early Mesopotamian visual culture. Most famously, a gaming board from the Early Dynastic cemetery at Ur is decorated with abstract geometrical designs within square grounds (figure 2.8). Some five non-mathematical tablets from Šuruppag and two from Abu Salabikh also have very geometrical doodles in the blank spaces on the reverse.[52] None of these Early Dynastic images has any writing associated with it; so under what circumstances is it legitimate to treat them as mathematics? And, under such conditions, what useful information can be gleaned from them?

The discipline of ethno-mathematics—the anthropological examination of mathematising concepts in non-literate culture—has been a growing field of research over the last few decades. Hitherto, however, it has been applied almost exclusively to societies or social groups which have no tradition of literate numeracy. As a result there has been much debate about the degree of mathematical intentionality in, for instance, complex geometrical textile designs or gambling games compared to the anthropological observers' imposition of mathematical interpretations on such artefacts and activities.[53] For Mesopotamia, however, we are in the happy position of being able to compare the detailed textual evidence of literate mathematics with apparently mathematising tendencies found in other texts and artefacts.

Whether one's philosophical stance is that mathematics is created or discovered, it is nevertheless demonstrable that ways in which mathematics is conceptualised, described, and discussed are culturally bounded: this book is littered with such examples. Thus we should expect to find commonalities between mathematics and other modes of discourse within a single culture, and we can use those points of comparison in two ways. First, we can seek the subject matter of Mesopotamian mathematics in the natural and built environments of Mesopotamia, albeit in a more nuanced manner than simply describing it as a 'reflection of everyday life'.[54] Second, we can then analyse the detailed and formal discourse of ancient mathematics to help illuminate the conceptualisation of related documents and artefacts. This approach depends on the postulate that the ancient practitioners and authors of mathematics had a world view that was not radically different from that of other Mesopotamian scribes and artisans. This assumption is defensible on the grounds of the all-pervasiveness of numeracy in cuneiform culture: as we have already seen, most scribes were trained in methods of quantitative documentation and control, which they used for the large majority of their working lives.

Very roughly a third of the extant corpus of mathematical word problems concern two- or three-dimensional figures in some way, and some 150

Figure 2.8 A gaming board from the Early Dynastic Royal Cemetery of Ur, decorated with symmetrical figures. (BM ANE 120834. Courtesy of the University of Pennsylvania Museum of Archaeology and Anthropology and the Trustees of the British Museum.)

problems or rough calculations (from across the whole of the cuneiform time frame) are illustrated by geometrical diagrams. This interrelationship between the visual, the textual, and the numerical is particularly rich. Close examination of how geometrical figures were drawn, named, described, and calculated can help to elucidate the way that they, and the relationships between them, were conceptualised. This is important not only for the intellectual history of mathematics but also for understanding ancient concepts and categorisations of shape, space, and symmetry in Mesopotamian art, architecture, and design.[55]

As Donald Washburn and Dorothy Crowe have convincingly argued, every society shows preferences for particular patterns over others.[56] By classifying those patterns according to group theoretical principles one can compare different 'cultures of symmetry' with one another. That is not to say that the symmetries embedded in the patterning of a cultural artefact somehow reflect group theoretical intent or knowledge on the part of its maker; rather, symmetry groups serve as an ethnographer's tool for analysing often complex designs. Where Washburn and Crowe are interested in cross-cultural comparison of pattern choices across time and space, here we shall compare the symmetries of mathematical diagrams with decorative design on non-mathematical objects.

Washburn and Crowe treat point or finite symmetries (single motifs), strip or one-dimensional symmetries (such as on a decorated belt or ribbon),

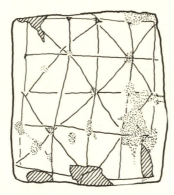

Figure 2.9 A geometrical doodle on the back of an administrative list from the Old Babylonian city of Mari on the Syrian Euphrates near the Iraqi border. (A 2541 reverse. Drawing by the author, after Ziegler 1999, no. 37.)

and field or two-dimensional symmetries (such as woven into a cloth or carpet), including both monochrome and multi-coloured artefacts. This little study, however, concerns only the simplest case: monochrome finite symmetries, which comprise rotation about a point and reflection about an axis. On the back of A 2541, an administrative roster of female musicians from the Old Babylonian city of Mari, is a roughly sketched square inscribed with other geometrical figures (right triangles and more squares) (figure 2.9).[57] Ignoring its rough and ready execution, it exhibits 90° rotational symmetry—that is, rotating the image through 90°, clockwise or anti-clockwise, would produce an image identical to the original one—and four-fold reflective symmetry in the vertical, horizontal, and both diagonal planes. It is as symmetrical as a monochrome square design can be. Another mathematical sketch, TSŠ 77 from Old Babylonian Kisurra (table B.10), comprises four identical circles inscribed in a square; it too is completely symmetrical.[58]

The Old Babylonian collection of geometrical problems BM 15285 (see also §4.3) originally contained around forty illustrated problems about calculating the areas of figures within squares (figure 2.10). No answers or solutions are given, just the questions themselves, including detailed descriptions of the illustrations. For instance, problems (7) and (8)—which accompany identical diagrams in the second column of the obverse—read:

(7) The square-side is 1 cable long. Inside it I drew a second square-side. The square-side that I drew touches the outer square-side. What is its area?

(8) The square-side is 1 cable long. Inside it <I drew> 4 wedges and 1 square-side. The square-side that I drew touches the second square-side. What is its area?

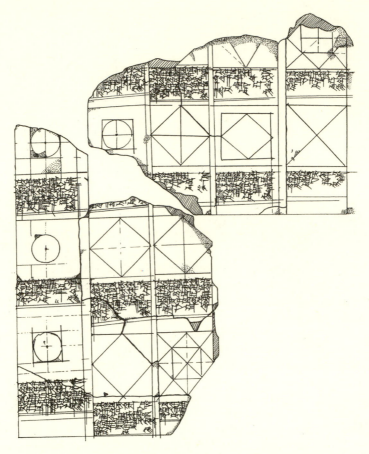

Figure 2.10 A compilation of geometrical problems from the Old Babylonian period. (BM 15285. Drawing by the author.)

Note that the lengths of the squares, as stated in the text, are c. 360 m long, while the diagrams themselves are just a few centimetres across.

The exercises are ordered didactically, moving from simple problems about circles, squares, and right triangles to more complex arrangements of figures based on the circle, many of which do not feature in modern European geometry and for which there are no modern technical terms. Much of that terminology derives from the visual culture of early Mesopotamia, as Anne Kilmer has shown.[59] None of the textual descriptions is complete in itself, in that the problems cannot be solved without reference to the images. Although shapes of each kind are counted, their sizes are never given, nor, mostly, are their relative positions. The only spatial preposition used is 'inside', Akkadian *ina libbi*; the adjective *kīdûm* 'outer' is used just once. 'Above', 'below', 'upper' and 'lower', 'left' and 'right' are

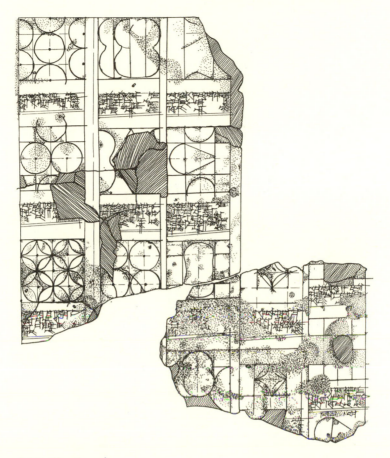

Figure 2.10 (*Continued*).

never used, even though they occur in other mathematical contexts. The sole verb of spatial relationship is *emēdum* 'to touch, lean, be in contact with', used when the corners of a square or four points of a circle touch the mid-points of an outer square.

Neither is there any textual description of symmetry, even though the images clearly show that symmetry is an important feature of these constructions. All but one of the fourteen extant illustrations on the obverse exhibit reflective symmetry in four planes, and thus four-fold rotational symmetry too. The other shows two-fold rotation and reflection. Even the fifteen more complex combinations on the reverse, as far as they survive, are all reflective in at least one plane. In short, two-thirds of the squares show full, four-fold reflectional and rotational symmetry, just like the geometrical sketches from Mari and Kisurra. Six of the remainder exhibit horizontal and vertical reflectional symmetry with 180° rotation, and just three have

TABLE 2.1
Three Symmetry Cultures Compared

Rotation	90°	180°		None		
Reflection[a]	HVD₁D₂	HV	None	H or V	None	Total
Old Babylonian mathematical tablet BM 15285	20 (69%)	6 (21%)	—	3 (10%)	—	29
Early Dynastic gaming board PG 513	4 (67%)	1 (16%)	—	1 (16%)	—	6
Modern Latin capital letters	—	4 (15%)	3 (12%)	12 (46%)	7 (27%)	26
Inkan tunic from 15th- or 16th-century Andes	—	10 (45%)	7 (32%)	3 (14%)	2 (9%)	22

[a] H = horizontal axis; V = vertical axis; D₁, D₂ = two diagonal axes.

only one axis of reflection. None of the diagrams on BM 15285 exhibits rotation but no reflection; and none is asymmetrical (table 2.1).

There are remarkable similarities with the decorations on the Early Dynastic gaming board (figure 2.8). Not only are its patterns based on the subdivision of squares into four or sixteen smaller ones, just like BM 15285, TSŠ 77, and A 2541, but four of the six arrangements, namely two-thirds of them, show four-fold rotational and reflectional symmetry too. Even the rosette is based on equal divisions of the square.

Lest these similarities in symmetry culture seem arbitrary or trivial, let us consider two comparanda. First the twenty-six capital letters of the modern Latin alphabet: A B C D E F G H I J K L M N O P Q R S T U V W X Y Z. None is fully symmetrical by early Mesopotamian standards, but four have two axes of reflection and 180° rotation. A further twelve letters reflect about the vertical axis or the horizontal axis. Three exhibit 180° rotational symmetry, while seven are entirely asymmetrical (table 2.1). The symmetry culture is utterly different. But this may not be a fair comparison: good typography surely entails the instant recognition of clearly identifiable symbols and, after all, the letters are not quite square. Let us take instead an Inkan tunic, woven in the Peruvian Andes some time around 1500 CE—and thus from a culture completely removed in time, space, and contact from early Mesopotamia. The detail shown here (figure 2.11) comprises forty squares in twenty-two patterns (ignoring colour variation).[60]

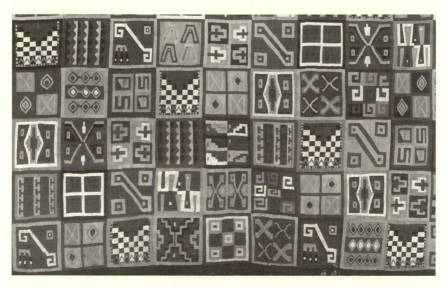

Figure 2.11 A detail of an Inka tunic from the Peruvian Andes, composed of woven squares in symmetrical designs. (Dumbarton Oaks B-518.PT, detail, © Dumbarton Oaks, Pre-Columbian Collection, Washington, DC.)

Treating those designs as monochrome, it turns out that none of them is fully symmetrical in the early Mesopotamian style, but rotational symmetry is much more prevalent. Seven patterns have two axes of reflection and 180° rotation; seven exhibit 180° rotation only; three show one axis of reflection; and two are completely asymmetrical (table 2.1). In short, elegant and satisfying as the design is, and however formally similar in its composition through figures inscribed in squares, it hails from a completely different symmetry culture than that of the early Mesopotamian mathematical and decorated objects we have just been examining.

2.5 CONCLUSIONS

Up to the period around 2400 BCE mathematics did not have a very strong self-identity. Its terminology, subject matter, methodology, and conceptualisation were adopted directly from the culture of numerate bureaucracy from which it developed, and which it directly served. Numeration (as distinct from number words) was heavily context-dependent, with commodity-specific metrologies that had been developed in the fourth millennium based, presumably, on various socially agreed norms for storage containers, measuring equipment, and the like, and the quantitative relationships between them. Weight was apparently the last of the metrological systems to be formalised, some time before the time of the Šuruppag tablets in the mid-third millennium.[61] Numbers, then, were attributes of sets of countable objects (4 beer jugs, 600 sheep, middle-size grain storage jars) or properties of measurable objects (a field boundary of 10 rods—which can itself be thought of as a set of standard-length measuring rules). In the late fourth millennium, systematic unit fractions existed only as the smallest units of grain metrology; the sexagesimalisation of numeration is explored further in §3.4.

The primary focus of pedagogical mathematical exercises at this period was thus the intersections between the most important metrological systems: grain capacities and discretely counted objects such as people or beer jugs on the one hand, and lengths and areas on the other. A nascent mathematical genre began to assert itself through the enumeration of key meeting points (the Uruk metrological lists, §2.2; the Early Dynastic length-area tables, §2.3) and the exploration of arithmetically elegant or difficult metrological intersections (the Uruk field-area calculations, §2.1; the Šuruppag grain-division exercises, §2.3). The relationship between lengths and areas became abstracted and mathematised through the Early Dynastic focus on squares of all sizes, from a few cubits to several kilometres. By contrast, real fields under centralised scribal management at this time, while they could be approximately square, tended to be long and narrow to maximise access to irrigation systems (the interplay between administrative and mathematical depictions of shape is discussed further in §3.3).[62] On the other hand, the square was an important component of the aesthetic repertoire of early Mesopotamian visual culture (§2.4), often used as the vehicle for patterns heavily based on reflective symmetry. Indeed, motifs based on interactions between circles and squares are first attested on Mesopotamian pottery of the sixth millennium BCE, remained prominent in first-millennium decorative style, and eventually became incorporated into architectural decorative schemes of the Roman-Sasanian and early Islamic Middle East.[63]

It might be tempting to see the lack of a clearly defined mathematics, with a separate self-identity from administrative accounting, as somehow a failure on the part of early Mesopotamian bureaucrats—especially when compared to the pedagogical standardisation attested by the lexical lists of words from the same period. However, Jean Lave and Etienne Wenger argue, from contemporary ethnographic studies of apprentices in a variety of work situations, that learning takes place most effectively when it is situated in the social and professional context to which it pertains, through interaction and collaboration with competent practitioners, rather than through abstract, decontextualised classroom learning.[64] Learners become part of a 'community of practice' that inculcates not only the necessary technical skills but also the beliefs, standards, and behaviours of the group. Through gains in competence, confidence, and social acceptance, the learner moves from the periphery towards the centre of the practice community, in due course becoming accepted as a fully fledged expert. It is perhaps in this light, then, that we should understand the process of becoming professionally numerate in early Mesopotamia—an idea to which we will return in later chapters. But if situated learning is so effective, the development of supra-utilitarian mathematics for the training of scribes in the late third and early second millennia BCE becomes a puzzle that cannot be solved on purely pragmatic grounds, as we shall see in the following two chapters.

CHAPTER THREE

The Later Third Millennium

The key mathematical development of the later third millennium is unquestionably the sexag-esimal place value system (SVPS). It was not an isolated phenomenon but grew out of other changes: the gradual abandonment of impressed numerals and the expansion, standardisa-tion, and integration of metrological systems (§3.4). At the same time, accountants developed ever more sophisticated quantitative methods of predicting and managing institutional labour requirements, which varied according to the workers' relationship to the institution (§3.3). Land too could be managed more efficiently with the advent of visual representations of two-dimensional space (§3.2). On the face of it, surprisingly little evidence of school mathe-matics survives from this period of great conceptual change, but this may to some degree simply reflect patterns of archaeological recovery as well as the practices of scribal apprenticeship (§3.1).[1]

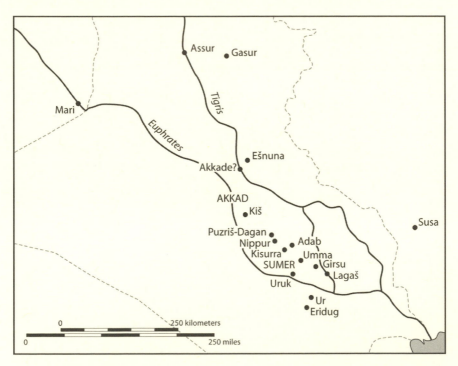

Figure 3.1 The locations mentioned in this chapter, including all known findspots of mathematical tablets from the late third millennium.

3.1 BACKGROUND AND EVIDENCE

When king Sargon gained power in northern Babylonia in around 2340 BCE there was little to suggest that his reign would mark a decisive turning point in the political or intellectual history of ancient Iraq. Little is known from contemporary sources about the life and deeds of this monarch, although his name and fame lived on for nearly two millennia in the cuneiform tradition. But the personality traits and exploits of individual kings are less important to the concerns of this book than the social, organisational, and conceptual innovations that their reigns brought about. On the one hand, there is little in the settlement archaeology to suggest that there were major changes or disruptions to traditional life-ways or habitation patterns. Most immediately striking is a change in formal language policy in favour of the Semitic language Akkadian (named after Sargon's capital city, Akkade) over Sumerian. To what extent this linguistic shift in official writings reflects an ethnic shift in power structures is difficult to determine; however, it certainly does not point to an Akkadian 'invasion' of Sumerian culture, as many older overviews have it. Many documents continued to be written in Sumerian, including all extant mathematical problems and exercises.[2]

As Sargon and his immediate descendants gradually unified the whole of southern Iraq (below the approximate latitude of modern-day Baghdad) their chief administrators instituted a raft of centralising and simplifying bureaucratic reforms to accommodate an ever-growing number of hitherto independent city states. In a state-wide system of year names, dates were recorded with reference to major royal, cultic, or military events of the previous year. Metrologies were unified to become increasingly interdependent (table A.2 and §3.4), while numerals were frequently written with incised cuneiform instead of impressed signs. New styles of book-keeping developed to accommodate the much wider remit of imperial bureaucracy, by utilising approximations and estimates (§3.3). These changes did not come into effect immediately on the accession of Sargon but were implemented largely under his grandson Naram-Suen. The large majority of the six thousand extant Sargonic administrative documents are mostly from the south and mostly date towards the end of the period, c. 2250 BCE.[3]

Twenty-two Sargonic mathematical tablets have been published to date. No arithmetical lists or tables are yet in the public domain. A small round tablet bearing a geometrical diagram was found in a sanctuary of the god Enlil's temple at Nippur (see further §3.2); one from Ĝirsu contains model accounts (table B.6). A very interesting exercise from Ešnuna lists weights of silver in descending order from 1 mina to 1 shekel, each assigned to imaginary individuals with names beginning with *diĝir* 'god'. The remaining tablets, mostly from Ĝirsu, set (and usually solve) problems in lengths, areas, and volumes, which can be grouped into five different types (table B.5). They mimic working administrative documents to varying degrees.

Problem type (a), to find the short side of a rectangle, given the long side and area (five tablets), is most distant from professional praxis, both in the scenario of the problem and in that the area of the rectangle is always a round or single area unit. Problem type (b), to find the area of a square given its side (five tablets), is most reminiscent of the Early Dynastic length-square area exercises (§2.3), although unlike those the Sargonic lengths are not whole numbers. Both types, especially (b), use the Sumerian verb *pa* or *pa₃* 'to find', the earliest attested instance of terminology that is exclusively mathematical rather than administrative. For example, PUL 31 (figure 3.2) reads simply:

The long side is 4(*ĝeš*) and 3 <rods>: <what is> the short side of a 1(*iku*) field? Its short side is to be found.

Problem types (c)–(e) are to find the area of a quadrilateral field (six tablets), sometimes with derived agricultural data (two tablets), or the volume of a rectangular prism (one tablet). They attempt to imitate the structure and terminology of administrative records by adding circumstantial information about people and places. The unrealistic size of the numerical parameters, the clumsiness of orthographic and/or arithmetical execution, and the lack of credible contextual data mark them out as pedagogical exercises.

Many of the problems, of all types, are associated with named individuals and/or professional designations. Until very recently it was assumed that these were the names of students to whom the exercises were assigned.[4] However, it looks increasingly likely that, in some cases at least, these were added to the exercises in imitation of work assigned to overseers or fields identified by their owners. For example, Ist L 2924 (figure 3.3) reads:[5]

16 East 20 minus 3
37 North 31½
Its area is (*blank*).
Baran the household administrator.
40 North 25½
30 East 25½
Its area is (*blank*).
Uruna the household administrator.
4 (*iku*) 1(*ubu*) field [. . .]
Dudu [[(*the official*)]]

In each case, the two parallel measurements are written either side of the cardinal direction to which they pertain, with no metrological units attached. The resultant areas have not been written on the tablet but were probably supposed to be calculated using the still ubiquitous surveyors'

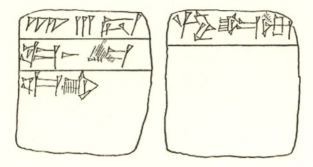

Figure 3.2 A Sargonic mathematical problem about finding the short side of a 1-*iku* field. (PUL 31. Limet 1973, no. 39, courtesy of La Société d'Édition 'Les Belles Lettres'.)

Figure 3.3 A Sargonic mathematical exercise in finding the area of irregular quadrilaterals, designed to look like a genuine surveyor's document. (Ist L 2924. Genouillac 1910, no. 2924.)

method (§2.1). The area given at the bottom is significantly smaller than either individual field and so cannot represent their total.

The majority of these word problems have been identified in the period since mid-2001, mostly amongst already published but unidentified material and some from unpublished private collections. No doubt more will continue to turn up. We should thus expect the known corpus of Sargonic mathematics to grow further in the coming years and our concomitant understanding of it to change.

With the retreat of central authority as the Sargonic empire fell in the twenty-second century, there is a striking dearth of written remains except at individual small polities such as Lagaš-Ǧirsu and Uruk. But a new empire eventually formed from the remnants of the old. The Third Dynasty of Ur (or Ur III for short) was founded at Ur in the far south of Mesopotamia in the late third millennium and collapsed around a century later. The period is conventionally dated to between 2112 and 2004 BCE but although we can be fairly sure of the time span, the absolute dates are not certain. Its second king, Šulgi, ruled for forty-seven years at the beginning of the twenty-first century. He initiated an aggressive twenty-year campaign of expansion in the second half of his reign until his empire stretched from southern Mesopotamia far north and west, encompassing Assur, Susa, and many other far-flung cities. The influence of the empire was felt even further afield: far up the Euphrates to Mari and deep into Iran. There was relative stability for around thirty years after Šulgi's death, until the system started to fall apart under his great-grandson Ibbi-Suen.

The Ur III state has been characterised as one of the most totalitarian in history; it was certainly one of the most bureaucratic. To cope with the upkeep of new territories and the vastly increased taxation revenues they brought in, large-scale administrative and economic reforms were executed over the same period, producing what Piotr Steinkeller has characterised as 'a highly centralised bureaucratic state, with virtually every aspect of its economic life subordinated to the overriding objective of the maximisation of gains'.[6] There is almost an overabundance of administrative records from the Ur III period, dating for the most part from the fifty or so years between the sweeping reforms of the second half of Šulgi's reign and the rapid decline of the empire after the accession of the last king, Ibbi-Suen. By the end of 2006, some 57,000 administrative tablets from this period had been catalogued on the CDLI database and the numbers are growing all the time.[7] The vast majority of tablets—almost exclusively in Sumerian with a smattering of Akkadian loan words—are from just five cities: Ur itself (over 4,200 tablets), Ǧirsu (some 19,000), Umma (over 14,000), Puzriš-Dagan/Drehem (over 11,000), and Nippur (nearly 3,400).[8] In most cases, the sites were not properly excavated so that the administrative archives have to be pieced together again from museum collections around the world, with little or no archaeological information to assist in their reconstruction.

On the face of it, the Ur III mathematical assemblage is extraordinarily poor compared with that of the immediately preceding Sargonic period, given the vast numbers of contemporaneous administrative documents: five reciprocal tables—a new genre (§3.4)—and three accounting exercises (table B.7).[9] Jens Høyrup has argued extensively that this is no mere accident of historical discovery.[10] Aligning himself with Bob Englund's view of the Ur III state as managing a 'Kapo economy' of enforced labour control, he concludes that the Ur III school promulgated 'a mathematics teaching *not* based on problem solution, a mathematics teaching deliberately eliminating even the slightest appeal to independent thought on the part of the students'.[11] His arguments are based not only on this paucity of evidence from the Ur III period itself, but also on the algebraic terminology of the succeeding Old Babylonian period. This, he suggests, shows no trace of a Sumerian, and therefore Ur III, legacy. However, these arguments are not watertight.

First, Englund's model of a centralised slave economy, on which Høyrup's arguments are predicated, is not universally accepted; equally convincing, for instance, is the model of an entrepreneurial culture in which surviving accounts depict exactly those activities which were contracted out of the system to external risk-bearing overseers.[12] We shall return to this question in §3.3.

Second, as already mentioned, the vast quantities of extant tablets are almost all administrative in nature and almost all from four or five illicitly excavated hoards. There is little direct archaeological or textual evidence at all for any sort of scribal training in the Ur III period, yet it would be fatuous to argue that scribal training did not take place. Niek Veldhuis notes that the few securely identified Ur III lexical lists 'reproduce the third millennium tradition', with lists of birds, fish, and professions.[13] Richard Zettler reveals that excavations at the Ur III–period House J in Nippur yielded half a dozen practice tablets, bearing elementary writing exercises that had mistakenly been published as Old Babylonian ones, but only some of which have parallels in the later tradition.[14] Gonzalo Rubio characterises the Ur III literary corpus, currently comprising about a dozen tablets from Nippur, as similarly heterogeneous. Some compositions are also known from the Old Babylonian period, in more or less similar recensions; some have parallels in Early Dynastic literature; others are apparently unique.[15] Thus it is highly likely that unprovenanced Ur III pedagogical exercises, including mathematics, have been similarly misattributed to other periods or languish unpublished as peripheral oddities.

Even the much-quoted hymns of self-praise to Šulgi, the second king of the dynasty, which enumerate his educational and mathematical achievements, cannot be taken as unproblematic evidence of Ur III scribal education. All but one extant manuscript can be securely provenanced to scribal schools of the late eighteenth century BCE (see §4.4).[16] How those hymns might have been modernised or otherwise adapted in the meantime is, for the moment, impossible to determine.[17] In any case, the mathematical operations described there—Sumerian $ĝa_2$-$ĝa_2$ 'addition', zi-zi 'subtraction', $šid$ 'counting', and $niĝ_2$-$šid$ [better, $niĝ_2$-kas_7] 'accounting'—cannot unproblematically be 'presumed to have been the [only] standard terms . . . in the Ur III school', as Høyrup maintains.[18]

Finally, Høyrup's arguments presuppose a simple linguistic relationship between Sumerian and Akkadian, whereby technical terms coined in the Ur III period yet surviving into Old Babylonian must necessarily have been Sumerian. Conversely any terminology in Akkadian must post-date the Ur III period. Thus, on Høyrup's view, 'the absence of adequate [Sumerian] equivalents for the whole metalanguage (logical operators, interrogative phrases, announcement of results, algebraic parenthesis and indication of equality) shows that the whole *discourse of problems* was absent from the legacy left by the Ur III school.'[19] Yet take, for instance, the 'logical operators', Akkadian *šumma* 'if' and *aššum* 'since, because'. Sumerian generally does not utilise one-word equivalents which could serve as convenient logograms within an Akkadian text. Rather, conditionality and causality are embedded within the syntactic structure of the sentence.[20] So, while Høyrup appears to have a case that the discourse of *algebra* was absent from the Ur III school, that was more a consequence of inadequate exploitation of the newly developed sexagesimal place value system (SPVS, §3.4) than the putative 'atrocious testimony of one of the most oppressive social systems of history'.[21]

3.2 MAPS, PLANS, AND ITINERARIES: VISUAL AND TEXTUAL REPRESENTATIONS OF SPATIAL RELATIONSHIPS

While abstract designs and depictions of individual objects and figures are common in very early Mesopotamian art (§2.4), visual representations of the natural and built environment first appear only in the Sargonic period. On the one hand, naturalistic images of landscape start to be depicted in the art historical record,[22] while on the other hand large-scale formalised maps of field systems and plans of buildings are first attested.[23] One of the earliest known examples is from Sargonic Gasur near the modern city of Kirkuk in northern Iraq (figure 3.4).[24] Its visual design is purely qualitative, showing the positional relationship between stylistically depicted hills, watercourses, and settlements. The quantitative information is all in the textual labels, from the cardinal directions on the edges of the tablet to the

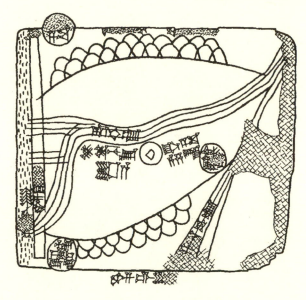

Figure 3.4 A map of the area around Gasur, near Kirkuk in northern Iraq, drawn up some time in the Sargonic period. The cardinal directions are marked on the edges of the map, with east at the top. The central area, below the Rahium river, is described as '20(*bur*) – 1(*eše*) of irrigated gardens, belonging to Azala'. (SMN 4172. Meek 1935, no. 1, courtesy of Harvard University Press.)

calculated areas of the central topographical features. The text gives qualitative information too, in the names of the rivers and the circular towns.

Ur III field plans, over thirty of which are known,[25] privilege the textual over the visual to an even greater extent. The terrain is sliced into easily calculable triangles and quadrilaterals, whose exact lengths and derived areas are represented not in the lines of the diagram itself but in the annotations covering it (figure 3.5). In other words, like the map from Gasur, these field plans are topological rather than metrical. Furthermore, perimeter lengths are used as the basis of area calculations rather than, say, measurements across the middle of fields; this concern with perimeters is visible in pedagogical geometrical exercises too. The areas of the fields are calculated with the same method—the so-called surveyors' formula—used since late fourth-millennium Uruk (see §2.1). In two particularly complex examples the central quadrilateral areas were computed twice in different ways, demonstrating that some scribes, at least, knew that the method was not exact. The two series of values were recorded upside down with respect to each other and the mean of the two totals calculated afterwards.[26]

A similarly linear, descriptive approach to landscape can also be seen in Ist Ni 2464, a royal decree of new provincial boundaries, in which the borders

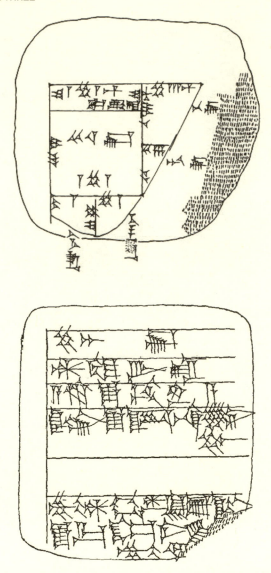

Figure 3.5 A field plan drawn up in 2045 BCE, with a modern scale reconstruction. The reverse gives the total area, minus the 'claimed land': 4(*bur*) 2(*iku*), c. 27 ha. It states that the land, belonging to the temple of goddess Ninurra in Umma, was measured 'at the command of the king', Amar-Suen. (YBC 3900. Obverse, Stephens 1953, 1, courtesy of the American Oriental Society; reverse, Clay 1915, no. 22.)

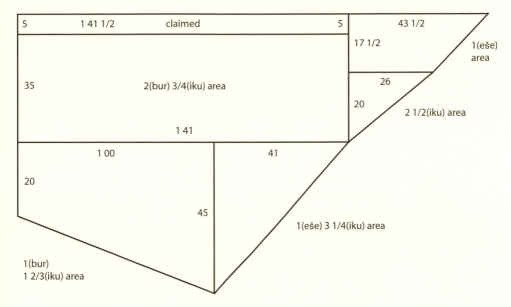

5	1 41 1/2	claimed		5	43 1/2	

17 1/2

1(eše) area

35   2(bur) 3/4(iku) area

26

20

2 1/2(iku) area

1 41

1 00   41

20

45

1(eše) 3 1/4(iku) area

1(bur)
1 2/3(iku) area

Figure 3.5 (*Continued*)

are demarcated by (apparently straight) lines from one landmark to another.[27] Quantitative descriptions are given only in the rare cases where there are no manmade features on the landscape to serve, as this extract shows:[28]

> From the Numušda Tower to the Numušda Shrine, from the Numušda Shrine to the Hill Tower, from the Hill Tower to the Šer-ussa canal, from the Šer-ussa canal to Ibilum village, from Ibilum village to the Abgal canal, cross the Abgal canal and then go 9(ĝeš) 20 rods (c. 3.2 km)—it is the . . . of the boundary: its northern side.
>
> From the . . . of the boundary to Me-Belum-ili: its eastern side.
>
> From Me-belum-ili to the bank of the Abgal canal at the mouth of the Ilum-bani canal, cross the Abgal canal, then from the mouth of the Ida'um canal to the Imnia canal: its southern side.
>
> From the Imnia Canal to Nagarbi, from Nagarbi to Marsh-town, from Marsh-town to the Hill, along the Hill back to the Numušda Tower: <its western side>.
>
> King Ur-Namma has decreed the boundary of the god Meslamta-ea of Apiak.

The repercussions for early Mesopotamian mathematics are complex. As for other numerate genres, it is usually fairly easy to distinguish between 'real' house and field plans drawn up by working scribes and school mathematics problems. 'Real' plans are complex and annotated with accurate measurements, and information about the date, administrator, and circumstances of recording. Scribal exercises, on the other hand, are often visually and numerically simple, with measurements fixed to be in nice arithmetic ratios or to give whole-number areas. They also tend to be anonymous and undated.

The earliest known mathematical diagram containing quantitative textual data is on IM 58045 (table B.6), a small round tablet found in secondary archaeological context in a shrine of the god Enlil's temple in Nippur (figure 3.6). To judge from the stratigraphy and palaeography it almost certainly dates to the Sargonic period.

Like the field plans, IM 58045 presents a topological view of the shape under consideration, with all quantitative information contained in the surrounding text. Differences in length are minimised in the diagram, which shows a symmetrical trapezoidal area with a dividing line. Given that the round form of the tablet almost certainly signals a student's exercise, then we have to decide what the original problem might have been. If it were simply to find the area of the whole figure using the surveyors' method, then the central dividing line might have marked the mean length of the two vertical sides:

$$12 \text{ cubits} \times (17 + 7)/2 \text{ cubits} = 144 \text{ square cubits} = 1 \text{ } sar.$$

As in the area problems from late fourth-millennium Uruk (§2.1), the problem has been set to yield a simple unit answer.[29]

The hundred or so extant diagrams in Old Babylonian area geometry are constructed on the same principles. In this case we can compare visual and textual evidence to gain a deeper understanding of the conceptualisation of space that underlies this aesthetic. Old Babylonian area geometry is based on defining components: external lines (which may be straight or curved) from which the area of that figure is defined and calculated.[30] In many cases, the names for the defining component and the figure itself are identical. Both the circle and the circumference have the Akkadian name *kippatum* from the verb *kapāpum* 'to curve', while both the square and its side are called *mitḫartum*, from the reflexive stem of *maḫārum* 'to be equal and opposite'. The word for a rectangle and its diagonal is *ṣiliptum*, from *ṣalāpum* 'to strike through' (but of course, in this last case, a diagonal does not uniquely define its surrounding rectangle).

These naming conventions are not mere happenstance but the result of a fundamentally boundary-oriented conceptualisation of two-dimensional space. This understanding of area is most strikingly apparent in both diagrams and instructions on calculating the area of a circle. A typical passage, in an Old Babylonian word problem on finding the volume of a

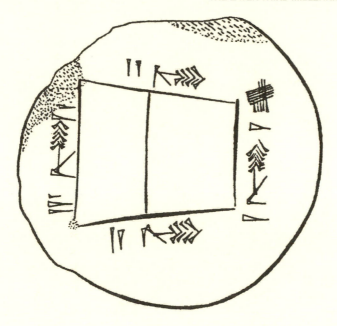

Figure 3.6 A geometrical diagram from the Sargonic period, marked with the lengths of the lines. The top and bottom are 2 reeds each, the left side 2½ reeds 2 cubits, and the right side 1 reed 1 cubit. (IM 58045. Drawing by Aage Westenholz in Friberg 1987–90, 541, courtesy of Aage Westenholz and De Gruyter Verlag.)

cylindrical log with longer cross-section at the bottom than at the top, reads:[31]

> Triple 1;40, the top of the log, and 5, the circumference of the log, will come up. Square 5 and 25 will come up. Multiply 25 by 0;05, the constant, and 2;05, the area, will come up.

In other words, even when the diameter of a circle is known—in this case '1;40, the top of the log'—the scribe chooses not to calculate the area directly by the procedure we know as $A = \pi r^2$. First he finds the circumference by 'tripling' the diameter and then working out the area in relation to that circumference.[32] Similarly, all known labelled diagrams of circles from the Old Babylonian period show the length of the circumference, its square, and the area, but never the diameter or radius of the circle—which are not even marked (figure 3.7). This absence encapsulates a fundamental distinction between the modern circle and the ancient. Whereas the modern Western tradition conceptualises the circle as a figure generated by a radius rotating around the centre (as in a pair of compasses, and our formula

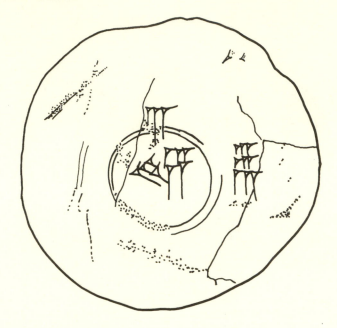

Figure 3.7 An Old Babylonian exercise in finding the area of a circle; neither radius nor centre are marked. Inside the circle is the number 45; at the top is a 3 and to the right a 9. (YBC 7302. Drawing by the author.)

$A = \pi r^2$), in the Old Babylonian period it was seen as the figure surrounded by a circumference.

Although the *tallum* 'diameter' features regularly in Old Babylonian problems about circles, the *pirkum* 'radius' is never mentioned. That does not mean, though, that the radius played no part in Old Babylonian geometry. It is found, for instance, in problems about semicircles. In those contexts, however, it functions as the short transversal of the figure, perpendicular to the *tallum*. In fact, this is the function of the *pirkum* in the context of all Old Babylonian geometrical shapes; it is never conceptualised as a rotatable line. That is not to say that the radius had not been discovered: BM 15285 (figure 2.10) shows several circles inscribed in squares, with clearly marked central compass points. It is rather that the radius was not central to the early Mesopotamian concept of a circle: the scribes *chose* not to use it.

In short, early Mesopotamian depictions of 'realia', whether two- or three-dimensional,[33] were not intended to be realistic drawings; neither were they scale replicas. Rather, they present an alternative view of numerical data which is visually logical and consistent with the textual description and which may also augment the text by providing information about spatial relationships. When interpreting visual artefacts such as surveyors' plans or learners' exercises, not only do words and numbers matter as much as lines, but the spatial relationship between text and image is important too. The outside-in concept of shape pervades both mathematical and non-mathematical descriptions of objects, visual and textual. The itinerary form, in which the reader is escorted around the boundaries of figures, through time and space, privileges perimeters over diameters and diagonals. The particular case of the circle entails that, without the concept of a rotating radius, the idea of measured angle cannot have existed in early Mesopotamia.[34] We should not expect, and do not find, accurate scale drawings. Rather, both mathematical diagrams and surveyors' maps and plans are topological; that is, they show qualitative spatial relationships but depend on accompanying text to bear all quantitative information.

3.3 ACCOUNTING FOR TIME AND LABOUR: APPROXIMATION, STANDARDISATION, PREDICTION

Already in late fourth-millennium Uruk accountants relied on standardised quantities and fixed numerical relationships, as in the case of groats and malt for beer production (§2.2). Throughout the course of the third millennium abstractions, approximations, and standardisations were applied to an ever-widening range of resources and assets, including time spent on production. The commodification of labour was used both as an economic management tool and, some have argued, a means of social control as well.

Some of the earliest pertinent evidence for quantitative labour management comes from Early Dynastic IIIb Lagaš (c. 2400 BCE). A group of forty-odd records documents three stages of monitoring repairs to irrigation canals and their dykes under the control of the temple of the goddess Bau. The first stage was to measure out what needed to be done (see table B.3 for metrological notation):[35]

> From the wall of the queen's estate to the goddess Nanše's temple is 2(ĝeš) 20 rods. This is the dyke where work need not be done. The tamarisk garden is next to that side.
> From the tamarisk garden to Nanše's temple is 1(ĝeš) ½ rods. This is the dyke where the work must be done.

Then, in separate documents, the work was assigned to named labour gangs, perhaps associated with neighbouring villages (the designations are unclear):

> 5 reeds in length, next to the Ĝišhur dyke: men of Birkakešdu were assigned.
> 3(ĝeš) 40 <rods>, 1 rope minus 1 reed in length: men of Ĝišhur were assigned.

2 ropes, 7 reeds in length: where the work was left unfinished.
The work is to be completed.

Finally, the completed work was surveyed, together with a record of the number of labourers assigned to it at a quota of 1 cubit each:

11 men. For each man 1 cubit of work was assigned. Its work is 1
    reed, 5 cubits. Ur-Šenirda (was responsible).
11 men. Its work is 1 reed, 5 cubits. Kaka (was responsible).
8 men. Its work is 1 reed, 2 cubits. Uwu (was responsible).
. . . (other entries) . . .
Total: 1 rope, 4 cubits of work performed. Dyke construction at the
    Ugig field of Nintud's temple. En-iggal the supervisor built it.

But by later standards the data are incomplete: no record is made of the *volume* of work done, the time taken to complete each section, or how much (if anything) the workers were paid. In these documents, then, the labourers are not a commodity to be accounted for; they are simply a resource to draw on. The calculations serve only to ensure that enough manpower will be provided in order to complete the task.

Fuller accounting methods are first attested in the early Sargonic period. Upon the accession of Sargon's successor Rimuš (c. 2278–2270) many of the southern cities rebelled. According to his inscriptions over 32,000 rebels were killed, some 23,500 captured, and at least 14,000 deported.[36] Thousands of the citizens of Umma (and other cities) were apparently put to forced labour; it happens that part of an administrative archive of a camp at nearby Sabum has survived. Three documents amongst the hundred or so record the labour expended on breaking something called *su-ru*, perhaps flint (later Akkadian *ṣurru(m)*).[37]

2 [⅙ *su-ru*: length 1 reed]; width 4 cubits, 10 fingers: height 1 cubit, 1
    half-cubit.
        Overseer: Ur-Ninsar, supervisor: Akalle.
1 <*su-ru*>: length 4 cubits; width 3 cubits; height 1 cubit, 1 half-cubit.
        Overseer: Aga, supervisor: Az.
[1] ⅔ <*su-ru*>: length 5 cubits; width 4 cubits; height 1 cubit, 1 half-
    cubit.
        Overseer: Saggani, supervisor: Tirku.
Total: 5 minus a 6th *su-ru* broken.
Work done: 2 months, 30 minus 3 days.
Year 4, month 10 minus 1, day [. . .].

Neither the volume of a *su-ru* nor the work-rate is explicitly stated, but both can be inferred from some simple calculations. The second entry shows that a *su-ru* is equal to $4 \times 3 \times 1\frac{1}{2} = 18$ cubic cubits (c. 2.25 m³). The total work time expended is 2 administrative months and 27 days, or

87 days; the standard work-rate is thus 1 cubic cubit, or $\frac{1}{18}$ *su-ru*, a day. However, the duration of the project in real time was almost certainly not 87 days. If each overseer managed a gang of 10, for instance, then the real time taken would be $^{87}/_{30}$, or roughly 3 days.

Ben Foster identifies the introduction of standard work-rates in the Sargonic period as part of a much larger phenomenon, which he calls 'Akkadian accountability', tracing its origins through Ebla to the Early Dynastic city of Kiš.[38] 'Secondary' administrative records (§2.2) in the Sumerian language, from both the Early Dynastic and Sargonic periods, tended to aggregate all the data from their constituent primary records onto a single large tablet without summary or approximation. By contrast Akkadian-language accounts relied on estimates, round numbers, and schematic standard rates of labour. They display a completely new type of numeracy, which privileges overview and estimation over detail and precision. The two administrative styles could co-exist in the same archive, however, serving rather different ends: approximation tended to be used for calculating the assets and rights of the wealthy and the institutionally powerful; precision for the dues of their dependents.[39]

By the Ur III period, these two types of accounting had fused into a very sophisticated system which combined accurate detail with standardised assumptions about land and labouring productivity. For each work group hundreds of daily records of activity were drawn up—probably after the event, rather than by a scribe on the spot—and combined into annual or bi-annual balanced accounts of materials used, work completed, and work owed to the state. Merchants, pottery workshops, milling teams, dairies, and fisheries were all held accountable in this way, as well as agricultural labourers.[40]

For instance AO 5676, an annual balanced account from Umma, was compiled by the central administration to account for the activities of a certain Ur-Ninsu and his team of twenty agricultural labourers over the preceding twelve months of Šu-Suen's second regnal year (c. 2036 BCE).[41] The tablet is huge, with six columns on the obverse and seven on the reverse (figure 3.8). It begins with a debits section, listing the workdays that Ur-Ninsu's team owed to the city for the year:

[27(*ĝeš*) 45] $\frac{1}{3}$ labourers for 1 day, deficit of the year that Šu-Suen became king.

10 labourers, the ploughmen and their sons;
(*5 named men, minus 2 working elsewhere*)—menial workers;
(*8 named men, minus 1 working elsewhere*)—ordinary workers:
for 12 months their wages are 2(*šar*) days.

From Month I to Month XII: 36 *gur* 1(*barig*) 2(*ban*) of grain. Wages at 6 *sila* each: its wages are 3(*ĝešu*) 13$\frac{1}{3}$ days—wages of hired men for the fields.

Figure 3.8 An annual balanced account recording the work performed by Ur-Ninsu's agricultural labourers, drawn up in the city of Umma for the year 2036 BCE. (AO 5676. Genouillac 1921, no. 5676.)

> From the mouth of the granary: 9½ labourers for 1 day, from Isarrum.
>
> Total: 2(šar) 3(ǧešu) 1(ǧeš) 8 and 10 shekels—labourers for 1 day. Head of the account.

The debits are divided into three sections: first, 1665⅓ workdays owed from last year, which has been carried over from a previous account; second, 7200 workdays owed by the established team of 20 men, calculated on the assumption that they all work for the full 360 days of the administrative year. Third, there is the casual, hired labour, whose 202⅚ workdays are back-calculated from the grain paid out to them. In total Ur-Ninsu owes 9068⅙ labour days to the city of Umma because he has been paid for them in the form of wages for his ordinary workers and hire for his casual labourers, either in the current year or in preceding ones.

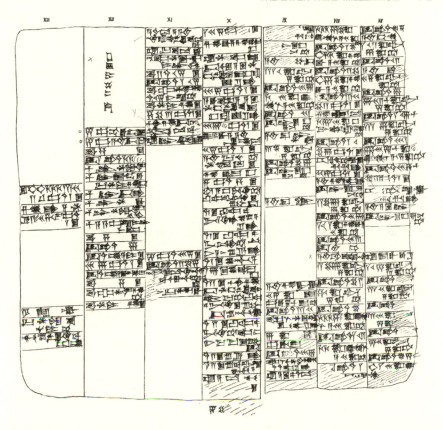

Figure 3.8 (*Continued*)

The following credits sections are organised according to the type of work done, field by field, roughly following the cycle of the agricultural year. The individual entries may summarise several different daily records. First is the spring harvest, in which the unit of accounting is the working day but there are no required work quotas. For instance:

> 1(*ĝeš*) 40 labourers for 1 day reaping grain; 1(*ĝeš*) 52 labourers for 1 day making sheaves and spreading; 1(*ĝeš*) 40 labourers for 1 day on duty at the threshing-floor—at the Gu-edina field.

The second credits section concerns preparation of the same fields for sowing seed:

> 2(*bur*) deep ploughing at 1½(*ubu*); harrowing, twice, at 4(*iku*) 1(*ubu*): its *eren*-workers' wages are 3(*ĝeš*) 12 days;
> 1(*bur*) deep ploughing at 1½(*ubu*); harrowing, twice, at 5(*iku*): its *eren*-workers' wages are 1(*ĝeš*) 33½ days, 6 shekels;
> . . . —at the Gu-edina field.

In this section the estimated rate of work is always recorded, and work-days calculated on the basis of a three-man team. The quotas can vary—here, harrowing at 4½ or 5 *iku* (c. 1.6–1.8 ha) per day on the same field—perhaps through negotiation between overseer and recording scribe. If the team worked faster than the quotas given they would make a profit, if slower a loss: the account was drawn up solely according to these theoretical work norms rather than by the time it actually took to complete the allotted tasks.

There then follows a short section on sowing the fields at a daily rate of 2 *iku* (c. 0.72 ha) per person. The land is categorised as domain land, which produced grain for the city itself, or plough land, which was cultivated on behalf of various state dependents. The fourth set of credits concerns maintaining the fields and keeping the weeds down as the new crops grow, assuming that a standard area was cleared each day:

> 2(*ĝeš*) 48 *sar* hoeing at 6 *sar* each: its wages are 36 days; 1(*ĝešu*) 1(*ĝeš*) *sar* cutting reeds at 20 *sar*: its wages are 33 days—wages and cattle fodder at the Gu-edina field.

In this section, the account displays even more variability in the work-rates: hoeing at 3–6 *sar* (c. 108–216 m²) a day; cutting down or pulling up various weeds at 15–30 *sar* (c. 540–1080 m²). Curiously, large areas are given here in sexagesimal multiples of the *sar*, not in *eše* and *iku*, a phenomenon we shall return to in §3.4. Tasks for which hired labourers were used, such as clearing the furrows and bundling up the cut reeds, were not quantified.

The next section lists waterways maintenance with a mixture of quantifiable and non-quantifiable daily rates, and the transport of grain from Umma to a nearby town and back again by boat. Finally, two last credit sections concern days away from work for holidays, half-time working, and *bala* duties (a sort of centralised livestock and labour tax):

> 5 labourers for 35 days: its wages are 2(*ĝeš*) 55 days—going to *bala*-work, on *bala*-work duty, [returning from] *bala*-work. [Via] Ur-Nungal.
>
> 36 labourers for 1 day—holidays of menial workers—for 4 (*sic*) months from Month X to Month XII.

The summary is in the final column. The work done by the hired, casual labourers has been added in a note at the top.

> Credit: 2(*ĝeš*) 45 hired men.

> Total: 2(*šar*) 33(*ĝeš*) 22⅓ workdays subtracted. Credit: 2(*ĝeš*) 24⅙ workdays.
>
> Account of Ur-Ninsu, domain overseer, for the year that the god Enki's boat was caulked.

Unusually, the 9202⅓ days of total labour accomplished is slightly more than the 9068⅙ workdays expected of Ur-Ninsu's team, including the shortfall from the previous year. In other words, the labourers worked faster—ploughed more land, sowed more seed, cleared more weeds—in the year than the accountant expected of them. Ur-Ninsu thus made a nominal profit equivalent to 865 litres of grain (at the standard hiring rate) in 2036 BCE.

Bob Englund sees this type of labour account, in which the workday is treated as a unit of account no different from a shekel of silver or a *sila* of barley, as the dehumanisation of workers into no more than expendable assets. He also guesses that 'the expected labor performance was in all likelihood simply beyond the capabilities of the normal worker. Moreover an incentive for the workers to produce more was non-existent: their re-muneration consisted of no more than the minimum amount of grain and clothing to keep them able to produce.' He even raises the spectre of a dead overseer's offspring being carted away by the city authorities in lieu of an unpaid deficit.[42] On this view, the over-zealous application of numeracy is a symptom of callous repression.

But there is another way of interpreting this account (and others like it), bearing in mind that it was drawn up *for the city authorities*, who had no reason or need to document the workers' welfare. What they did have an interest in was guaranteeing steady access to agricultural labour through-out the year, to ensure the productivity of the land they owned. Labour was always in short supply, while the agricultural cycle was much more labour-intensive at some times of the year (especially harvest) than others. On this view, annual accounts such as these are documentation of an an-nual contract: the overseer is an entrepreneur who agrees to take on the year's work for a fixed fee, as set out in the debits section of the account. He takes responsibility for employing and managing the labourers, and he bears the risk of profit or loss. In this way the state authorities minimise financial risk and have no direct responsibilities for the workforce. Re-cently published documents show that individuals with entrepreneurial contractual relationships with the Ur III authorities also made money in other ways: merchants did private business as well as fulfilling state con-tracts and were free to invest the balance of their state accounts elsewhere; chief herdsmen could lend substantial sums to the military elite and prob-ably did not dirty their hands with actual sheep.[43] Seen in this light Ur-Ninsu's fate is much less sinister than Englund fears; though how he in turn treated his workers is another matter entirely, and one to which we may never have access. In either case, Ur-Ninsu was not as closely moni-tored as Englund suggests: the work-rates in the debits sections do not, after all, record the exact time taken to complete the labour but rely on estimates that vary substantially, even within this single account.

For the most part it is impossible to tell how the work-rates, or delivery rates for productive units, were decided upon. A rare exception consists in dairy products. The tablet AO 5499 uniquely plots the theoretical breeding patterns in a small cattle herd over ten years, 2056–2047 BCE.[44] Because it is dated and attributed to one Idua, the chief administrator of a temple, it is presumably a serious, professional attempt to model reality rather than a school exercise. Equally, because simple arithmetical rules govern the model of growth, and because the herd is not assigned to a shepherd or otherwise located in the real world, it is probably not a straightforward record of actuality.

The initial herd consists of just four adult cows, which are presumably serviced by a visiting bull. Their reproduction algorithm can be inferred from the following years: when there is an even number of cows, half of them give birth each year; an odd number of cows reproduce as if there were one fewer in the herd. Male and female calves are produced in equal numbers; when there are odd numbers their sex ratio alternates from year to year. The calves reach maturity in their fourth year, when the females are added to the reproductive stock. The adult bulls remain with the herd too; but the animals never die. Adult females also produce milk products at an annual rate of 5 *sila* of ghee (clarified butter) and 7½ *sila* of cheese per animal, independent of the number of unweaned calves in the herd. Their value was calculated at a rate of 10 *sila* of ghee and 150 *sila* of cheese per shekel of silver; ghee was thus fifteen times as valuable as cheese in this model. Real herding contracts and accounts from 2023 to 2021 BCE, some thirty years later, assume identical standard rates of production, at 5 *sila* of ghee and 7½ *sila* of cheese—but it was increasingly difficult for cowherds to meet even a quarter of those quotas.[45] The accountants, however, made no attempt to adjust their expectations downwards in line with falling productivity.

While few of the Sargonic pedagogical exercises deal with agricultural labour (except problem type (d); see table B.5), it is central to the Ur III model accounts, which use much the same terminology and methodology as balanced accounts like AO 5676. Walker 47a (table B.7), for instance, concerns the work involved in constructing an irrigation dyke.[46] It was first published as an Ur III administrative document but must in fact be a school exercise, as it contains no date, totals, toponyms, or personal names, and appears to be unfinished. Further, some of the phrasing is rather odd, and a mistake has been made in one of the calculations.

1(*ĝeš*) 5 rods length, at ⅔ <*sar*> volume for 1 rod—piling up an embankment. Its earth is 43⅓ *sar*; its workers' wages are 4(*ĝeš*) 20 for 1 day.

1(*ĝeš*) 5 rods length, reed bundles on an embankment. A worker did 1 rod length in 1 day. Its workers' wages are 1(*ĝeš*) 5 for 1 day.

5(*iku*): a worker hoed 2½ *sar* in 1 day. Its workers' wages are 3(*ğeš*) 20 for 1 day in the first year.

5(*iku*): a worker hoed 3 *sar* in 1 day. Its workers' wages are 2(*ğeš*) 53 for 1 day in the second year.

Ploughing for 1 day . . . (*unfinished*).

No work-rates are stated for the first two tasks, but these can be reconstructed from the figures in the text. The first job is to pile up earth for an embankment at a daily rate of ⅙ *sar* (c. 3 m³). Reed bundles are then secured onto the same length of embankment at a rate of 1 rod a day. If, as in later school exercises, the volume of reeds is a quarter of that of the earth, or a fifth of the embankment as a whole, this work would also be carried out at a rate of ⅙ *sar* a day—identical to the rates given in Ur III administrative documents for reinforcing dykes with reeds.[47] In the final sections a 5-*iku* (c. 1.8 ha) area is hoed twice at different rates, just as in AO 5676. The last number of workers should be 2(*ğeš*) 46 ⅔, not 2(*ğeš*) 53, but it is not clear how the error arose. Standardised work-rates like these were later ubiquitous in Old Babylonian mathematics, not only for earthworks and waterworks, but also for manufacturing and carrying bricks of very standardised sizes. Indeed, nearly half of the extant Old Babylonian mathematical word problems can be solved only through knowledge of Ur III–style work-rates and standards, which were collected and organised into lists of constants (see §4.1).[48]

3.4 THE DEVELOPMENT OF THE SEXAGESIMAL PLACE VALUE SYSTEM (SPVS)

The very earliest literate accounts—from late fourth-millennium Uruk—used commodity-specific metrologies with a variety of different numerical relationships between the units (§2.2, table A.1). Those original metrologies continued in use throughout the third millennium and beyond, whether essentially unchanged (for instance areas), or undergoing periodical reform (for instance capacities). They continued to be written with compound signs, which bundled both quantity and unit into a single grapheme, just as the preliterate accounting tokens must have done. Gradually, over the course of the later third millennium, scribes began to write those compound metrological numerals with a cuneiform stylus, rather than impressing a round stylus into the clay in imitation of accounting tokens. Throughout the Sargonic period, and even into the early Ur III period, impressed and incised number notations appear side by side on the same tablets—a phenomenon that has not yet been systematically documented or explained.

The discrete counting system, however, which grouped individually countable objects such as people or sheep into tens and sixties, six-hundreds and

TABLE 3.1
The Evolution of the Discrete Counting System

System	1 (diš)	10 (u)	60 (ğeš)	600 (ğešu)	3600 (šar)	36,000 (šaru)
Impressed signs (from late fourth millennium)	◿	•	◿	◖•	●	◉
Cuneiform signs (from mid-third millennium)	𝈿	𝈿	𝈿	𝈿	◇	◈
Sexagesimal place value system (from late third millennium, for calculations only)	𝈿	𝈿	𝈿	𝈿	𝈿	𝈿

three-thousand six-hundreds, was perhaps the first to be cuneiformised. The D-shaped units and sixties were replaced by vertical wedges (the sixties larger and deeper than the ones), while the O-shaped tens became corner wedges (table 3.1). This system was also co-opted for certain non-discrete uses from very early on. The archaic Uruk accounts already used the discrete notation system to record length measures for larger units, and when weight metrology became necessary with the development of silver accounting in the Early Dynastic IIIa period, 60 was chosen as the ratio between the two units mina (c. 500 g) and shekel (c. 8 g) (table 3.2).[49] Archaeological specimens of weights themselves are attested from Early Dynastic IIIb onwards, with widely different standards for commodities such as silver and wool.[50] At that point area (and volume) metrology was also extended downwards by subdividing the *sar* into 60 shekels.[51] Occasionally the units 'small mina' (60 grains, thus ⅓ shekel or ¹⁄₁₈₀ mina) and 'small shekel' (¹⁄₆₀ shekel, or ¹⁄₃₆₀₀ mina) were also used.[52] In the Sargonic period the shekel was itself subdivided into 180 grains and the talent set at 60 minas. At that time area metrology also began to be used to count bricks, where 1 *sar* = 7200 bricks, regardless of their size.[53] But not every new metrological unit was sexagesimally structured. The smaller length measures, first attested in the Early Dynastic IIIb period, divide the rod into 2 reeds or 12 cubits, and the cubit into 30 fingers (table A.2).[54] None of these newly invented units of measure was recorded with compound metrological numerals, but always written as numbers recorded according to the discrete notation system followed by a separate sign for the metrological unit.[55] This has implications for our understanding of the material culture of early Mesopotamian calculation, as well as for the shifting conceptualisation of number.[56]

Parallel to the generalised sexagesimal fraction was the unit fraction—already attested in archaic Uruk grain metrology, which included notations for a half, third, quarter, fifth, sixth, and tenth of the smallest whole

TABLE 3.2
Metrological Reforms of the Mid-third Millennium

Period	Metrological Innovations	New Fractional Notations
ED IIIa (c. 2500 BCE)	Subtractive notations; shekels and minas for weights; *gur* of 240 *sila* capacity measure	Special signs for ⅓, ½, and ⅔
ED IIIb (c. 2400 BCE)	Small units for lengths; areas divided into shekels; small mina and small shekel for areas and weights; weights archaeologically attested; *gur* of 144 *sila* capacity measure in Lagaš	General unit fraction *igi*-n-*ğal*$_2$ '*n*th' for *n* = 3, . . . , 6
Sargonic (c. 2300 BCE)	Grains and talents for weights; *sar* measure for bricks; *gur* of 300 *sila* capacity measure	Special sign for ⅚
Ur III (c. 2050 BCE)	Small mina and small shekel abandoned; shekels and grains for fractions of days and other metrologies	Sexagesimal place value system for calculations

unit (table A.1). Special notations for ⅓, ½, and ⅔ had developed by Early Dynastic IIIa, and ⅚ by the Sargonic period.[57] In Early Dynastic IIIb Ğirsu the more general notation *igi*-n-*ğal*$_2$ '*n*th part' was first used for unit fractions of the smaller units such as shekel, *sila*, and *sar*, where *n* could take the values 2, 3, 4, 5, or 6. From Early Dynastic IIIa both additive and subtractive metrological expressions were common,[58] so that in records from Early Dynastic IIIb Ğirsu one finds such complex statements as 'total: ⅓ mina 1 shekel and a 4th part of washed silver' (that is, 21¼ shekels, c. 18 g), or even 'its area is 10 minus 1 *sar*, minus a 4th part' (that is, 8¾ *sar*, c. 315 m²).[59] All such notations were ubiquitous by the Ur III period at the end of the millennium.

These innovations conveniently extended the range and refined the granularity of the various systems of weights and measures, but they did not particularly help in converting between them—the focal point of almost all pedagogical mathematical exercises from the late fourth millennium onwards. At some point early in the Ur III period, the generalised sexagesimal fraction and the generalised unit fraction were productively combined to create a new cognitive tool: the sexagesimal place value system (SPVS, §1.2). This calculating device took quantities expressed in traditional metrologies and reconfigured them as sexagesimal multiples or fractions of a base unit, often at a convenient meeting point between metrological systems. For length measures, for instance, the rod was chosen because 1 rod square = 1 *sar*.[60] For height, however, the base unit was the cubit, because the basic unit of volume was not 1 rod cubed but 1 area *sar* multiplied by 1 cubit height.[61] The SPVS temporarily changed the status of numbers from properties of real-world objects to independent entities that could be manipulated without regard to absolute value or metrological system. Calculations could thus transform numbers from lengths into areas, or from capacity units of grain into discretely counted recipients of rations, without concern for the objects to which they pertained. Once the calculation was done, the result was expressed in the most appropriate metrological units and thus re-entered the natural world as a concrete quantity.

Early evidence for the fully-fledged SPVS is extremely rare, for two reasons. First, almost all calculations were performed mentally or with accounting tokens, as discussed further below. Second, when calculations (or their intermediate results) were written down they were supposed to be erased on completion. Good scribes never showed their working. Perhaps it is no coincidence that erasable, waxed wooden writing boards are first referred to on clay tablets in exactly the same year—Amar-Suen's eighth regnal year, 2039 BCE—as the earliest dateable sexagesimal scratch calculation, first identified by Marvin Powell in 1976 (figure 3.9).[62]

The first four lines of the unprovenanced tablet YBC 1793 are written in sexagesimal place value notation:

14 54
29 56 50
17 43 50
30 53 20

The next lines read: 'Total 1½ minas, 3½ shekels minus 7 grains of silver; various deliveries'. And that is indeed the total of the numbers given above. Similarly, the text continues, '7 minas, 19 shekels of silver; deliveries from the royal lustration priest'. That sum is the total of the sexagesimal numbers written in the second column:

2 54
45
28
17
2 28
27

At the bottom of the first column the two totals are added: 'Grand total 8⅚ minas, 2½ shekels minus 7 grains of silver, from the white silver; deliveries in Majestic Festival month, Year of the High Priest of Eridug.' At the top of the second column is a similar weight: '8⅚ minas, 4½ shekels [. . .] grains', followed by a line that is too damaged to read. The reverse is blank.

Figure 3.9 Scratch calculations in the sexagesimal place value system, on a draft of a silver account written in 2039 BCE. (YBC 1793. Photo by the author.)

The totals, written in weight metrology with special fraction signs and subtractive expressions, were doubtless intended to be entries in a silver account. Another Ur III scratch calculation, undated but from Ĝirsu, records tracts of land that have been re-measured. It explicitly states that one entry is taken from a writing board.[63] Presumably such documents were constructed in the compilation of 'secondary' records such as Ur-Ninsu's agricultural labour account (§3.3, figure 3.8).[64]

Indeed, Ur-Ninsu's account occasionally gives clues that sexagesimal calculations were involved in its construction. Normally land measures were recorded in non-sexagesimal metrology, as in the second credits section, where 2(*bur*) and 1(*bur*) of land are deep-ploughed at 1½(*ubu*) a day and harrowed twice at 5(*iku*) or 4(*iku*) 1(*ubu*) a day. But in the fourth credits section similar areas are written in sexagesimal multiples of the *sar*, using the discrete counting system: 1(*ĝeš*) 34 *sar* and 1(*ĝešu*) 1(*ĝeš*) *sar* instead of 1(*iku*) 1(*ubu*) 18 *sar* and 1(*eše*) 1(*ubu*) 10 *sar*. These areas are to be hoed at 6 *sar* a day and cleared of reeds at 20 *sar* a day respectively. The simplest explanation for this non-standard, sexagesimalising notation is that it enabled easy calculation of the days to be credited, by multiplying the areas worked by the reciprocals of the work rates (all values of which are found in the standard Ur III reciprocal table, below). The calculations of ploughing and harrowing times, by contrast, are much more involved, involving three team members, and several work rates a hundred times larger than for field clearance. It was probably felt in those cases that little efficiency could be gained from pre-converting the areas worked into sexagesimal multiples of the *sar*. Labour accounts first used this sexagesimalised notation for areas of hoed and weeded fields during the final years of king Šulgi's reign, c. 2050 BCE, around fifteen years before Ur-Ninsu's account was drawn up, and did so increasingly frequently thereafter.[65]

So much for developments in measuring, recording, and calculating within administrative praxis; how did those changes affect scribal training in numeracy? Most immediately, it is notable that new genres of mathematical text reflect new metrological and notational concerns. While the metrological table from Early Dynastic IIIa Šuruppag deals with lengths no smaller than the rod (and thus areas no smaller than the *sar*), its counterpart from Early Dynastic IIIb Adab works upwards from the newly created cubit and expresses the resultant square areas in terms of the new shekel fractions of the *sar*. Sargonic exercises on lengths and areas also explore the furthest reaches of those two systems. A tablet recently documented on the private market but whose present whereabouts is unknown provides a unique insight into how such calculations were performed.[66] It reads:

> The long side is 2(*ǧeš*) rods.
> <What is> the short side? <The area is> 2(*iku*) minus ½(*ubu*).
> In it you put a double-hand, a 6th part of it.
> You put a 4th part of it. It was found.
> Its short side is 5 seed-cubits, 1 double-hand, [5] fingers.

The scribe of this exercise was not yet very metrologically competent. The length units in the first line are written with horizontal instead of vertical wedges, while the area units in the second line are written with verticals instead of horizontals. Further, the answer makes sense only if the scribe wrongly set 1 *iku* at 60, not 100 *sar*. That is, he took the area to be 105 square rods, rather than 175, or—written sexagesimally—1 45. He had to divide that by the length of 2(*ǧeš*), which he could probably do by inspection, to get 0;52 30 (which he more likely thought of as 52½ shekels; we do not know). But whether he was using the fully fledged SPVS or, more probably, all-purpose shekels, he knew that the shekel was not the appropriate unit in which to express fractions of a rod. Seed-cubits are conveniently equal to 10 all-purpose shekels, so it is easy to see that the width is 5 seed-cubits, with 2½ shekels or ¼ seed-cubit left over. This was all done without writing, it seems. But for the smaller units, the scribe had to explicitly recall that 1 double-hand was 'a 6th part of it' (the seed-cubit). Thus he could express 'a 4th part of it' as 1½ double-hands—and, as 1 double-hand equalled 10 fingers, he then easily converted the half to 5 fingers.

So his problems arose exactly at the non-sexagesimal junctures of the procedure: he failed to remember that the *iku* is a non-sexagesimal multiple of the *sar*, and had to remind himself of the relationship between the seed-cubit and double-hand—which, at ¹⁄₃₆ rod, does not have a simple numerical relationship with the all-purpose shekel fraction. The rest he could do unproblematically by inspection.

Given the increasingly sexagesimal nature of most metrologies, and the problems that could arise when that regularity broke down, it is not surprising that by c. 1040 BCE administrative scribes were sexagesimalising everything behind the scenes. The Ur III scribal teachers threw out the old-style metrological tables and instituted a new type of list, the so-called reciprocal table, that explored the properties of the ubiquitous and all-powerful sixty. The reciprocal table systematically enumerates pairs of numbers whose product is sixty, either ignoring or explicitly stating which integers (those with factors other than 2, 3, and 5) do not have reciprocal pairs. For instance, BM 106425 from Ur III Umma (figure 3.10) reads:[67]

Sixty: Its 2nd part is 30
Its 3rd part is 20
Its 4th part is 15
Its 5th part is 12
Its 6th part is 10
Its 7;12th part is 8;20
Its 7;30th part is 8
Its 8th part is 7;30
«Its 7;30th part is 8»
Its 9th part is 6;40
Its 10th part is 6
Its 12th part is 5
Its 15th part is 4
Its 16th part is 3;45
Its 18th part is 3;20
Its 20th part is 3
Its 24th part is 2;30
Its 27th part is 2;13 20
Its 25th part is 2;24
Its 30th part is 2
Its 32nd part is 1;52 30
Its 36th part is 1;40
Its 40th part is 1;30
Its 45th part is 1;20
Its 50th part is 1;15 (*sic*)
Its 54th part is 1;06 40
Its 1 00th part is 1

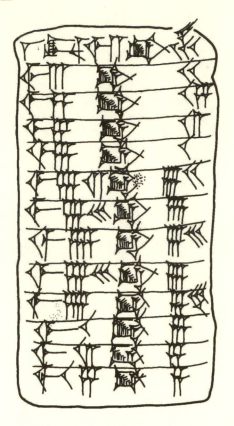

Figure 3.10 A table of reciprocals from Ur III Umma. The product of each pair of numbers is 60. (BM 106425. Robson 2003–4, fig. 1.)

Its 1 04th part is 0;56 15
Its 1 21st part is 0;44 26 40

Lists such as these crucially simplified top-down division. In Early Dynastic IIIa Šuruppag and Early Dynastic IIIb Ebla the trainee scribes had been taught to divide (say, a large quantity of grain) by multiplying up the divisor (say, a standard ration size by a number of recipients) until the original dividend was reached, if necessary adding the results of different multiplications through trial and error (§2.3). The scribe of the Sargonic word problem we have just examined probably used the all-purpose shekel to effect top-down division—though he only had to divide by two. But now, with the analysis of 60 into nearly thirty regular factors, the scribes could divide by almost any sexagesimally regular number, simply by multiplying the dividend by the divisor's reciprocal pair. Division was necessary not only for allocating grain rations (as already practiced in ED IIIa Šuruppag) but also, most crucially for the Ur III accountants, in order to calculate

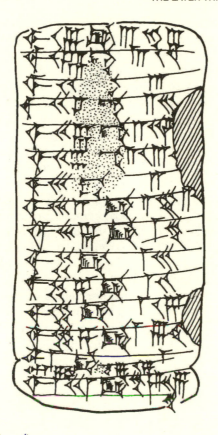

Figure 3.10 (*Continued*)

theoretical labouring days from areas or volumes worked and standardised work-rates or brick sizes, in the annual balanced accounts. And this is exactly what the trainee scribes actually practised (§§3.1, 3.3).

3.5 CONCLUSIONS

While the Ur III period has recently been presented as a time of atypical mathematical stagnation and repression (see §3.1), the evidence presented in this chapter combines to reveal it as a period of cognitive innovation hitherto paralleled only by the first commitment of numbers to writing in late fourth-millennium Uruk. And neither did the sexagesimal place value system appear out of nowhere: rather, it was the very state bureaucracy in which the scribes were embedded that both drove the need for it—by imposing increasingly high calculational standards on its functionaries through the demand for complex annual balanced accounts (§3.3)—and provided the tools with which it could be shaped (§3.4). Ur III numeracy combined Early Dynastic standards of precision with Sargonic ideas of approximation, standardisation, and prediction. Metrological sexagesimalisation was not in itself new but had been an increasingly dominant force in various centrally imposed reforms of weights and measures throughout the third millennium. Most crucial in this regard was the adoption of the shekel as a general-purpose ⅟₆₀ fraction alongside unit fractions, some of which had already been in restricted use since the late fourth millennium.

Indeed, it was the scribe's very closeness to, and even self-identification with, the bureaucratic system that pushed the boundaries of the system's quantitative capabilities. What little is known of late third-millennium scribal education suggests that a Lavian model of apprenticeship through 'situated learning' and 'legitimate peripheral participation' still largely pertained (§2.5).[68] With the exception of the standardised 'lexical lists' of bureaucratically useful words (many of which were by now centuries out of date), pedagogical exercises still consisted for the most part of *ad hoc* practice in directly relevant skills, with only brief experimentation with second-order instructions to 'find' solutions. Further, as the maps and itineraries demonstrate (§3.2), the scribes mentally situated themselves within the very landscapes that they managed and shaped at the same time as adopting a physically impossible bird's-eye viewpoint of panoptical control. Charlotte Linde and William Labov's experiments in modern New York, as related by Michel de Certeau, show that only a few people when asked to describe a familiar built environment will describe the spatial relationships between the key features in static, depopulated terms ('There are four rooms on the ground floor, connected by a central hallway'). The large majority prefer to narrate an embodied journey through that space ('You walk through the front door and turn left into the kitchen').[69] A similar cognitive strategy is behind the early Mesopotamian conceptualisation of areas as bounded by external, often walkable, and always measurable lines.

It retrospect, the Ur III period was also critical for the later development of Old Babylonian mathematics and professional numeracy. The SPVS and reciprocal technology enabled a sophisticated concrete algebra that entailed the manipulation of lines and areas of both known and unknown magnitude (§4.1, §4.3), while the subject matter and constants of quantity surveying and labour accounting remained central to the mathematical curriculum long after it ceased to be an important tool of institutional management. Further, the separation of quantitative and qualitative data through tabulation (§6.2), first witnessed sporadically in the Ur III period, developed into a key technique of numerate bureaucracy through the course of the second millennium. But, as chapter 4 explores, these developments in the concept of number are not in themselves sufficient to explain the veritable explosion of evidence for supra-utilitarian mathematics in the early second millennium BCE.

CHAPTER FOUR

The Early Second Millennium

The early second millennium, or Old Babylonian period, has yielded by far the largest number of mathematical tablets of the whole three millennia of cuneiform culture (§4.1), mostly from unidentified sites in southern Iraq. However, in the few cases where tablets have been excavated in context, they reveal vital information about training in literate numeracy, located within scribal schools run in private homes—which also accounts for why so many survive (§4.2). Even when tablets are unprovenanced, close attention to their linguistic, material, and social context can yield insights into their function that a solely mathematical analysis cannot (§4.3). Examination of the ways in which numeracy is portrayed in Sumerian literature, that other outstanding product of Old Babylonian scribal culture, helps to explain why early Mesopotamian mathematics took the particular form it did and no other (§4.4).[1]

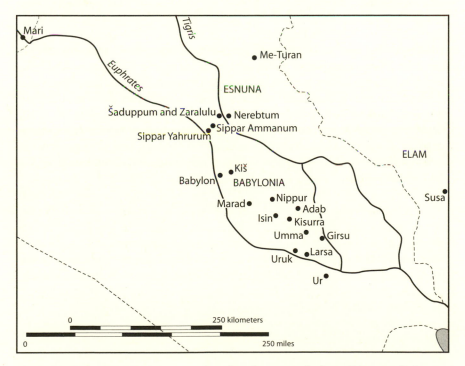

Figure 4.1 The locations mentioned in this chapter, including all known findspots of Old Babylonian mathematical tablets.

4.1 BACKGROUND AND EVIDENCE

The Ur III kingdom imploded over a ten-year period at the end of the third millennium, as one by one local governors asserted their independence. The most conspicuously successful was a former general of the Ur III state, one Išbi-Erra (r. 2017–1985), who founded a new dynasty at the city of Isin. For the following two centuries its main rival in southern Iraq was based in nearby Larsa. The two polities fought a near-constant war for access to water and to the religious centre of Nippur. Eventually, in 1793, the Larsan king Rīm-Sîn (r. 1822–1763) definitively conquered Isin territory, an event so momentous that each of his thirty remaining regnal years was named in its honour. But it was Isin, which portrayed itself as the natural successor state to the Ur III kingdom, that won the intellectual battle. At least, it was the royal hymns of that dynasty that dominated scribal education and scribal ideology until at least the mid-seventeenth century (§4.4). Meanwhile, further north, the small town of Babylon was growing in importance and influence. The year after Larsa's conquest of Isin young king Hammurabi (r. 1792–1750) ascended to the Babylonian throne. Over the next thirty years through acute political acumen, military prowess, and complex diplomatic manoeuvring he turned his rivals in neighbouring states against each other one by one— and then moved in to conquer once they were sufficiently weakened. Babylonia (southern Iraq) was a unified political entity by 1760, but remained so for only a few generations before disintegration set in once more.

There are many major archives from the period, both institutional and familial. Palaces and temples continued to monitor income, assets, and outgoings (though not as extensively as during the late third millennium); wealthy urban families kept legal records to show their ownership or rights to property or income, whether through lease, purchase, or inheritance. All such documentation required the involvement of numerate professionals.

In the early Old Babylonian (OB) period, elementary scribal training underwent a revolution.[2] The old lists of nouns were radically revised and new lists introduced that trained more abstract properties of the cuneiform writing system and the sexagesimal place value system. Thousands of arithmetical and metrological tables and several hundred mathematical word problems attest to this new regime, in which emphasis was more on the ability to manipulate imaginary lines and areas in almost algebraic ways than on the ability to count livestock or calculate work rates.

Old Babylonian mathematical texts are usually categorised as either tables or problems.[3] 'Tables', which might be more accurately described as lists of number facts written in Sumerian, are by far the most abundantly attested.[4] Reciprocal tables (§3.4) pair sexagesimally regular numbers up to 60, or sometimes beyond, with their inverses. But whereas in the Ur III period their product was always 60, now that was generalised to 1 or any power of 60. Reciprocal tables are often found at the head of the standard series of multiplications, which comprises around forty lists, each of up to twenty-five lines long—some thousand lines in total. The multipliers range from 50 down to 1;15, while the multiplicands are always 1–20, 30, 40, and 50. Sometimes the table ends with the square of the multiplier, as in the following example (figure 4.2).

25 steps of 1	25	[steps of 15]	6 [15]
steps of 2	50	[steps of 16	6 40]
steps of 3	1 15	steps of 17	7 05
steps of 4	1 40	steps of 18	7 30
steps of 5	2 05	steps of 19	7 55
steps of 6	2 30	steps of 20	8 20
steps of 7	2 55	steps of 30	12 30
steps of 8	3 20	steps of 40	16 40
steps of 9	3 45	steps of 50	20 50
steps of 10	4 10		
steps of 11	4 35	25 steps of 25	10 25
steps of 12	5	10 25 25 square-side	
[steps of 13]	5 25	Long tablet of Bēlānum	
[steps of 14	5 50]	Month XII, 7th day.	

A similarly standardised sequence of metrological units, from the capacity, weight, area, length, and height systems respectively, runs from tiny units to enormous ones, comprising around six hundred entries at its maximum extent (see table A.3 for OB metrological systems). Less frequently attested, non-standard arithmetical lists include reciprocal pairs halved and doubled from an initial pair; squares and inverse squares of integers and half-integers to 60; and cubes and inverse cubes from 1 to 30. Around four-hundred Old Babylonian mathematical 'tables' have been published (mostly standard arithmetical lists, table B.9, and some sixty metrological tables, table B.8), representing perhaps 5 or 10 percent of such material in accessible museums.

More recently four further genres have been identified in smaller numbers.[5] A dozen or so 'coefficient lists' give geometrical, metrological, and practical calculation constants for use in solving word problems. A similar number of 'catalogues', such as Plimpton 322 (see §4.3), list numerical variants on particular sets of word problems. Around fifteen known 'model documents' imitate professionally numerate genres such as accounts, field plans, and land surveys, while nearly 140 examples of rough calculations and/or diagrams are now known (table B.10).

However, by the far the most valuable genre to twentieth-century historians of Babylonian mathematics has been the word problem. About 160 tablets containing a total of several thousand problems have been published to date (table B.11). As the name suggests, they set out a mathematical

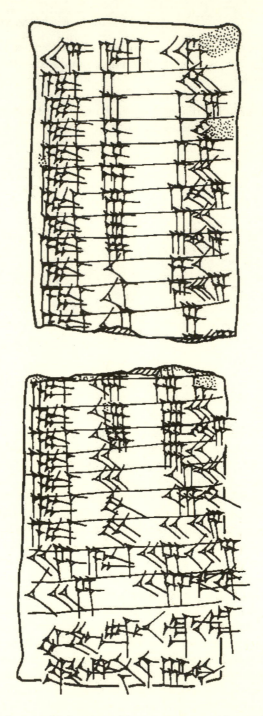

Figure 4.2 A verbose multiplication table on a Type III tablet. (Ashmolean 1922.178. Robson 2004a, no. 13.)

problem to be solved. They may also provide an answer, method of solution, and illustration for a problem, but do not as a rule show how the individual arithmetical operations were performed. A single tablet may contain any number of problems from one to a few hundred, which may or may not be thematically related to each other. Topics range from apparently abstract 'naïve-geometrical algebra', via plane geometry, to practical *pretexts* for setting a problem—whether agricultural labour, land inheritance, or metrological conversions. Even the most abstract problems may be dressed up with 'practical' scenarios. For instance, the eighth word problem in a collection on the unprovenanced tablet YBC 4663 reads:[6]

> 9 <shekels of> silver for a trench. The length exceeds the width by 3;30 (rods). Its depth is ½ rod, the work rate 10 shekels. Its wages are 6 grains. What are the length and width?
>
> You, when you proceed: solve the reciprocal of the wages, multiply by 0;09, the silver, so that it gives you 4 30. Multiply 4 30 by the work rate, so that it gives you 45. Solve the reciprocal of ½ rod, multiply by 45, so that it gives you 7;30.
>
> Break off ½ of that by which the length exceeds the width, so that it gives you 1;45. Combine 1;45, so that it gives you 3;03 45. Add 7;30 to 3;03 45, so that it gives <you> 10;33 45. Take its square-side, so that it gives you 3;15. Put down 3;15 twice. Add 1;45 to 1 (copy of 3;15), take away 1;45 from 1 (copy of 3;15), so that it gives you length and width.
>
> The length is 5 rods, the width 1½ rods. That is the procedure.

The procedure goes as follows:

1. Divide the total amount of silver (0;09 mina) by the wages paid to each worker (0;00 02 mina/day) to find the total number of work-days paid for: 4 30 days.
2. Multiply the number of workdays by the work rate for individual labourer (0;10 *sar*/day) to find the volume of the trench: 45 *sar.*
3. Divide the volume by the depth of the trench (6 cubits) to find its area: 7;30 *sar.*

So far, the solution has consisted in reducing the problem to one in two variables: given the difference between the length and width (3;30 rods) and the area of the trench (7;30 *sar*), find the length and width themselves. At this point the procedure enters the domain and vocabulary of cut-and-paste 'algebra', which Jens Høyrup has masterfully elucidated.[7]

4. Draw out a rectangle to represent the area of the trench, and mark off the difference between length and width (1;45 rods), along the length of the rectangle. Cut off half of it (figure 4.3 *top*).

5. Move that cut-off piece so that it forms a symmetrical L-shape with the part left behind; the total area is still 7;30 *sar*. The empty square framed by the L must have area 1;45 × 1;45 = 3;03 45 *sar*. So the total area of the whole figure, a square, is 7;30 + 3;03 45 = 10;33 45 *sar* (figure 4.3 *centre*).

6. The sides of this large square must be 3;15 rods. Subtract from it one side of the small square, 1;45 rods, to find the original width: 1;30 rods. Add the other side of the small square to the other side of the big square to find the original length: 5 rods (figure 4.3 *bottom*).

As this example illustrates, individual problems are posed in the form of a question in the first-person singular ('I'). All the information needed to solve it is stated in this question, apart from the values of any constants needed (as for the labouring work rate here). A diagram may be drawn above or to the left of the statement of the problem but is never referred to explicitly in the text. (The diagrams here are modern reconstructions.) The method of solution is given as a series of instructions in the imperative or in the second-person singular, introduced by the phrase 'you, in your working', or simply 'you'. The solution proceeds linearly like an algorithm or a recipe, in which the result of one step is used in the calculation of the next, or put aside until needed again. The instructions conclude with the phrase 'this is the procedure' or 'the procedure'. They are written in grammatically correct, though condensed, natural Akkadian with full sentences. Technical vocabulary, such as it is, varies across the corpus. (Here, 'combine' is a technical term for geometrical squaring.) While the SPVS is used for calculations, in the statements and numerical answers numbers are attributes of groups or objects that have quantity, measure, and dimension as in earlier periods (§3.4). Nothing equivalent to axioms or theorems is ever stated explicitly. Rather, the methodology is entirely inductive: solutions to specific problems serve as generic examples from which generalisations are inferred.

Direct, textual evidence for dating Old Babylonian mathematics comes almost exclusively from the colophons written at the ends of lists and tables. Just a dozen carry the names of identifiable years, all from the period 1852–1722 (table 4.1). Other tablets are approximately dateable (and of course exactly localisable) by their archaeological context (table 4.2), which extends the chronological range to c. 1860–1650. However, as tablets other than legal records were habitually recycled, and scribal establishments often had soaking bins in which to dump old tablets as well as fresh clay, these dates should not be taken as definitive start- and end-dates for the Old Babylonian mathematical tradition in any one place but rather the incidental outcomes of the construction and destruction of buildings and settlements.[8] Further, almost all the dated or dateable mathematics comprises lists, tables, and calculations. Just one compilation of word problems has detailed, published context: Haddad 104, from the 'scholar's house' in

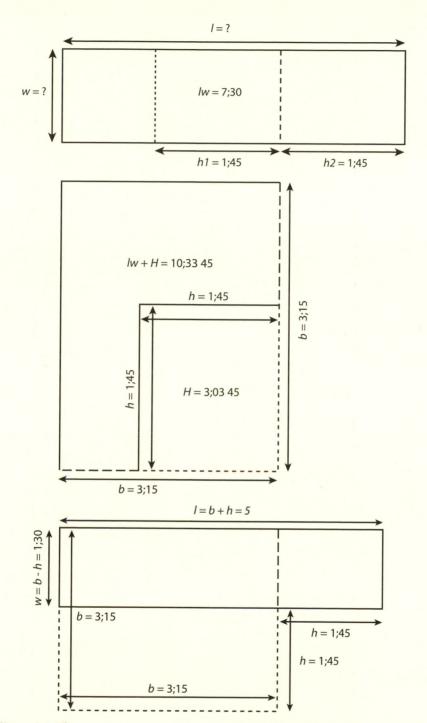

Figure 4.3 The cut-and-paste procedure described in the worked solution to YBC 4663 (8).

TABLE 4.1
Dated Mathematical Tablets from the Old Babylonian Period

Date	Year Name	Tablet	Description[a]	Publication
1852	Year of the wall of Sippar (Sumu-lael 29)	VAT 7892	Type III tablet: verbose multiplication table (× 5)	Neugebauer 1935–7, I 41 no. 52
1815	Year of Enki's temple in Ur (Rīm-Sîn 8)	YBC 11924	Type III tablet: verbose multiplication table (× 4)	(See table B.9)
1779	15th year after Isin was defeated (Rīm-Sîn 44)	VAT 15375	Type III tablet: verbose multiplication table (× 1 30)	Neugebauer 1935–7, I 42 no. 63
		YBC 4717	Type III tablet: verbose multiplication table (× 2 30)	Neugebauer 1935–7, II 36 no. 59a
1749–1712	Year that Samsu-iluna [. . .]	YBC 11138	Type III tablet: terse multiplication table (× 16)	Neugebauer and Sachs 1945, 21 no. 92,1
1749	Year that Samsu-iluna become king (Samsu-iluna 1)	AO 8865	Six-sided prism: metrological tables (length, height); arithmetical tables (inverse squares, inverse cubes)	(See table B.8)
1736	Year that Samsu-iluna the king <beat> the enemy king (Samsu-iluna 14)	BM 96949	Type I tablet: metrological list (capacities)	(See table B.8)
1727	Year of mighty Enlil (Samsu-iluna 23)	MLC 646	Type III tablet: terse multiplication tables (× 44 26 40, × 24)	Clay 1923, no. 36; Neugebauer 1935–7, I 54 no. 121
1723	[Year of] a white offering (Samsu-iluna 27)	MLC 1531	Type III tablet: metrological table (capacities)	(See table B.8)
1722	Year that Samsu-iluna the king by the command of Enlil <. . .> (Samsu-iluna 28)	MLC 117	Type III tablet: verbose multiplication table (× 10)	Clay 1923, no. 38; Neugebauer 1935–7, I 39 no. 33

[a] See table 4.5 for an explanation of the tablet typology.

TABLE 4.2
Archaeological Assemblages of Old Babylonian Tablets that Include Mathematics

Date	Findspot	Description	Publication
c. 1860	Uruk: a small room in Sîn-kāšid's palace	25 elementary school tablets, including 7 mathematical ones	Cavigneaux 1982
c. 1790	Ur: 'No. 1 Broad Street'	380-odd identified tablets of a reported 2000 school tablets and household records, including about 60 mathematical tablets	Charpin 1986, 451–2; Robson 1999, 245–72; Friberg 2000
c. 1780	Uruk, sherd-pit near the temple complex	180 school tablets, including 9 mathematical ones; 120 letters and business documents	Cavigneaux 1996; Veldhuis 1997–8
c. 1760	Me-Turan: 'scholar's house'	59 school tablets, of which 7 are mathematical; 22 magico-liturgical tablets; 90 household business documents and letters	Cavigneaux 1999
c. 1760	Nippur: Area TB House B	48 school tablets, of which 6 are mathematical (§1.3)	McCown and Haines 1967, 64–5; Stone 1987, 84–5; Veldhuis 2000b
c. 1740	Nippur: Area TA House F	Some 1400 school tablets, of which about 100 are mathematical	McCown and Haines 1967, 56–7; Stone 1987, 56–9; Robson 2001c; 2002b; 2007a
c. 1740	Ur: 'No. 7 Quiet Street'	46 school tablets, of which 4 are mathematical	Charpin 1986; Friberg 2000
c. 1650	Sippar Amnānum: gala-mahs' house	68 elementary school tablets, of which 8 are mathematical	Tanret 2002; 2003

Note: For more details, see Robson 2002b, 329. There are also more sparsely documented, or less well published, finds from OB Babylon (Pedersén 2005, 19–37 A1), Larsa (Arnaud 1994), Isin (Wilcke 1987), and Susa (see §6.1), as well as towns in the Diyala Valley (Isma'el forthcoming).

Me-Turan (table 4.2), far to the northeast of Babylonia in the kingdom of Ešnuna, which was destroyed on Hammurabi's conquest in 1762.[9] Iraqi excavations at other Ešnunan towns in the 1940s and '50s also yielded mathematical tablets of the same date, including a dozen small tablets of word problems found together in a private house in Šaduppûm, but their archaeology has never been fully published.[10]

Almost all the tablets studied and edited in the pioneering 1930s had been excavated illicitly and sold to museums by antiquities dealers. There was thus very little contextual or archaeological information to work with, apart from an occasional hearsay attribution to one ancient city or another. Fortunately tablets can also be grouped, provenanced, and dated on various textual and material grounds. Jens Høyrup, building on Albrecht Goetze's study of spelling conventions in Old Babylonian word problems, has analysed the technical terms in that corpus.[11] Many tablets can be assigned to approximate geographical areas as a result: Høyrup identifies or confirms distinct local identities for the mathematics of Old Babylonian Ešnuna, Larsa (?), Nippur, Sippar, Susa (see §6.1), Ur, and Uruk, as well as three further lexically and orthographically distinct but unlocalisable groups. However, he warns that the excavated tablets from Šaduppûm reveal that 'even the same locality might produce texts of widely diverging character during the same decades'.[12] On the other hand, some groups of tablets share so many textual, material, and museological features, quite apart from their mathematical content, that they were almost certainly written by a single individual. For instance, some arithmetical and metrological lists have dated and/or signed colophons that prove they were written in sequence within a short span of time (table 4.3).[13]

But just one autographed tablet containing word problems has been published to date. One Iškur-mansum, son of Sîn-iqīšam, signed a compilation of worked solutions on problems about brick walls and right-angled triangles, describing it as: 'Total: 25 processes'. In non-mathematical contexts the Akkadian word *kibsum* means a track, footprint, or animal spoor; but here it clearly refers to a series of mathematical steps rather than physical ones. The same word crops up in the colophons of several other compilations of word problems with worked solutions, which—as Goetze and Høyrup both noted—are also closely linked on palaeographic, orthographic, and terminological grounds (table 4.4).[14] Pieces of the tablets are now in three different museum collections, but all were acquired from dealers between 1899 and 1906.[15] It is very likely that they come from the city of Sippar, and their spelling conventions suggest a date in the late seventeenth century.[16] Indeed a teacher (Sumerian *dumu-e₂-dub-ba-a*) called Iškur-mansum was active in Sippar during the reign of Ammi-şaduqa (r. 1646–1626).[17]

TABLE 4.3
Clusters of Old Babylonian Lists and Tables Written by Named Individuals

Scribe and Date	Tablet Number	Description	Colophon	Publication
Lamassi-Dagan, 1852	VAT 7858	Type III tablet: verbose multiplication table (× 10)	Month VII, 9th day.	Neugebauer 1935–7, I 39 no. 32
	VAT 7895	Type III tablet: verbose multiplication table (× 9)	Month VII, 17th day. Long tablet of Lamassi-Dagan.	Neugebauer 1935–7, I 39 no. 34
	VAT 7892	Type III tablet: verbose multiplication table (× 5)	Month VII, 21st day, year of the wall of Sippar (1852).	(See table 4.1)
Suen-apil-Urim, 1816–1815	Ash 1923.447	Type III tablet: verbose multiplication table (× 24)	<Long> tablet of Suen-apil-Urim. Month XII, 9th day. Praise Nisaba and Ea!	(See table B.9)
	Ash 1923.451	Type III tablet: verbose multiplication table (× 24)	Month XII, 13th day; finished.[a] Long tablet of Suen-apil-Urim. Praise Nisaba!	(See table B.9)
	YBC 11924	Type III tablet: verbose multiplication table (× 4)	Long tablet of Suen-apil-Urim. Month VI, 11th day; finished.[a] Year of Enki's temple in Ur (1815).	(See table B.9)
Ubarrum	YBC 6705	Type III tablet: verbose multiplication table (× 9)	Long tablet of Ubarrum. Month VI, 25th day.	Neugebauer and Sachs 1945, 22 no. 99"
	YBC 6739	Type III tablet: verbose multiplication table (× 8)	Long tablet of Ubarrum. Month VI, 20th day; finished.[a]	Neugebauer and Sachs 1945, 22 no. 99,3a
Warad-Suen	YBC 4701	Type III tablet: metrological table (lengths)	Long tablet of Warad-Suen. Month IV, 16th day.	Figure 4.6 (previously unpublished)
	YBC 4700	Type III tablet: metrological table (heights)	Long tablet of Warad-Suen. Month IV, 23rd day.	Figure 4.6 (previously unpublished)

[a] It is not clear to me whether this refers to the end of the day, as usually construed, or the end of the exercise.

Table 4.4
Compilations of Word Problems Probably Written by Iškur-mansum, Son of Sîn-iqīšam, in Late Old Babylonian Sippar

Tablet Number	Description	Colophon	Publication
BM 85194 = 1899-4-15, 1	Worked solutions to problems about quantity surveying	Total: 35 processes of calculation.	Neugebauer 1935–7, I 142–93, II pls. 5–6
BM 85196 = 1899-4-15, 3	Worked solutions to problems about quantity surveying	Total: 18 processes.	Thureau-Dangin 1935; Neugebauer 1935–7, II 43–59
BM 85200 (= 1899-4-15, 7) + VAT 6599	Worked solutions to problems about trenches	30 processes.	Neugebauer 1935–7, I 193–219, II pls. 7–8, 39–40; Høyrup 2002b: 137–62
BM 85210 = 1899-4-15, 17	Worked solutions to problems about quantity surveying	20 processes. (*traces*) Nabû and Nisaba!	Neugebauer 1935–7, I 219–33, II pl. 9
BM 96954 (= 1902-10-11, 8) + BM 102366 (= 1906-5-12, 287) + SÉ 93	Worked solutions to problems about grain piles	[Missing. At least 30 problems originally.]	Jursa and Radner 1995–6, text B2; Robson 1999, 218–30
BM 96957 (= 1902-10-11, 11) + VAT 6598	Worked solutions to problems about bricks, and about 'gates'	Total: 25 processes. Iškur-mansum, son of Sîn-iqīšam.	Neugebauer 1935–7, I 274–87, II pls. 16–7, 44; Robson 1999, 231–44
VAT 6469	Fragmentary worked solutions	[Missing. Parts of 3 problems survive.]	Neugebauer 1935–7, I 269, II pls. 14, 43
VAT 6505	Worked solutions to problems about reciprocals	Total: 12 processes [. . .].	Neugebauer 1935–7, I 270–3, II pls. 14, 43; Sachs 1947a, 226–7
VAT 6546	Fragmentary worked solutions	[Missing. Parts of 2 problems survive.]	Neugebauer 1935–7, I 269, II pls. 14, 43
VAT 6597	Worked solutions to problems about inheritance	[Missing. Parts of 6 problems survive.]	Neugebauer 1935–7, I 274–7, II pls. 15, 43, 45

In short, there is reason to suspect that all ten tablets were written by Iškur-mansum. At first sight then, it looks as though this writer of worked solutions for at least 180 word problems should be celebrated as a productive and creative mathematician. However, an analysis of his frequent errors of repetition and omission shows that he was often copying from some other source—which he did not always understand. For instance, in BM 85196 (18), a simple problem on finding the base of a wall with triangular cross-section, he confuses volume and height because they have the same numerical value. A mis-transcribed number later in the procedure has no effect on the result, which suggests that he was not calculating as he wrote:

> A wall. The volume is 36, the height 36. In 1 cubit (height) the (total) slope is 0;00 50. <What are> the base and the top? You: solve the reciprocal of 36, the volume (*sic*, for 'the height'). You will see 0;01 40. Multiply 0;01 40 by 36. You will see 1. Put down 2 (*sic*, for '1'). Multiply 0;00 25, the slope (of one side), by 36. You will see 0;15. Tear out 0;15 from 1. You will see 0;45, the top. Add 0;15 to 1. You will see 1;15, the base. That is the procedure.

So, if Iškur-mansum, bungling copyist of Old Babylonian word problems, and Iškur-mansum, teacher in 1630s Sippar, were one and the same individual, let us hope for his students' sakes that mathematics was not his specialist subject.

4.2 METROLOGY, MULTIPLICATION, MEMORISATION: ELEMENTARY MATHEMATICS EDUCATION

Textual and material analyses thus reveal that Old Babylonian mathematics is not the homogeneous mass of undifferentiable, anonymous writings that it is often presented as. Further, archaeologically contextualised finds from Ur, Nippur, and Sippar have recently enabled close descriptions of the role of mathematics in the curricula of particular scribal schools.[18] House F in Nippur (table 4.2) has produced by far the most detailed evidence, if only because of the vast number of tablets excavated there (figure 4.4). This tiny dwelling, about 100 metres southeast of Enlil's temple complex Ekur, was used as a school early in the reign of Samsu-iluna (r. 1749–1712), after which some 1400 fragments of tablets were used as building material to repair the walls and floor of the building. Three tablet-recycling boxes (Sumerian *pu₂-im-ma* 'clay well') containing a mixture of fresh clay and mashed-up old tablets show that the tablets had not been brought in from

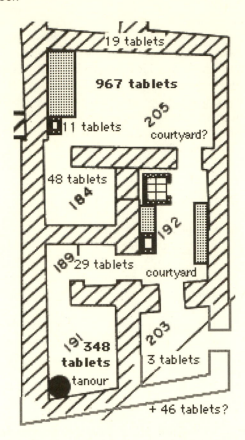

Figure 4.4 Archaeological plan of House F in Nippur, showing findspots of tablets and recycling bins. (Robson 2001c, fig. 3.)

elsewhere. Half carry elementary school exercises, and half bear extracts from Sumerian literary works. Both sets of tablets yield important information on mathematical pedagogy.

The elementary exercises were chosen from a standardised repertoire and—as Miguel Civil has shown—were written on tablets that were also very standardised in their shape and size.[19] Each different tablet type can be associated with a particular pedagogical function,[20] although their formats and functions varied somewhat from school to school, city to city, and indeed from exercise to exercise (table 4.5). Type III tablets were particularly favoured for multiplication tables in House F, for instance. Type IV tablets are barely attested there at all, although they account for twenty-one of the fifty-five tablets—nearly 40 percent—found in nearby House B

TABLE 4.5
Formats and Functions of Elementary Exercise Tablets in Nippur House F

Type	Description	Function (and Order of Use)	Number of Math/Total Tablets[a]
I	Large, multi-column tablet containing an entire exercise or a substantial part of one, generally carefully written	Final presentation of a completed exercise (4)	10/52
II	Large tablet: obverse contains carefully written model of 10–30 lines in left column; student's copy/ies on the right (and sometimes middle), often with multiple erasures; reverse contains an entire exercise or a substantial part of one, carelessly written over several columns	Obverse: first learning of a new passage (1) Reverse: rapid recall of a previously learned exercise (3)	38/246
III	Small, single-column tablet with a 20–30 line extract of an exercise, usually quite carefully written; sometimes ending with a catch-line giving the incipit of the next section	Reproduction of recently learned passage from memory (2)	31/52
IV	Small round tablet with a 2–3 line extract of an exercise, carefully written; may have model on obverse and copy on reverse (rare in House F)	Reproduction of recently learned passage (2)	0/4
P	Four or more–sided prism containing an entire exercise or a substantial part of one, generally carefully written	Final presentation of a completed exercise (4)	0/19

Note: Typology based on Civil 1979, 5–7.
[a] Includes only those House F tablets whose formats can be identified with certainty.

in Area TB (see §1.3). Type II tablets are not known at all from the schools in Ur. But for House F (and all other assemblages in which they occur) the Type II tablets are of crucial historical importance. For, as Niek Veldhuis has demonstrated, correlating the newly encountered extracts on the obverse sides with the already mastered compositions on the reverses enables the elementary curricula of individual houses to be reconstructed.[21] In House F the curriculum comprised four phases.

Mathematical learning began for the handful of students in House F in phase two, once they had mastered the writing of basic cuneiform signs in phase one. The six-tablet series of thematically grouped nouns—now known as Old Babylonian *Ur₅-ra*—contains sequences listing wooden boats and measuring vats of different capacities, as well as the names of the different parts of weighing scales. For instance:

1(*ĝeš*)-*gur* boat
50-*gur* boat
40-*gur* boat
30-*gur* boat
20-*gur* boat
15-*gur* boat
10-*gur* boat
5-*gur* boat
Small boat.[22]

Later in the same series come the stone weights themselves and measuring-reeds of different lengths. Thus students were first introduced to weights and measures and their notation in the context of metrological equipment, not as an abstract system (albeit always in descending order of size).

In House F systematic learning of metrological facts took place in phase three, along with the rote memorisation of other exercises on the more complex features of cuneiform writing (figure 4.5, table 4.6). This time the metrological units were written out in ascending order: first the capacities from ⅓ *sila* to 3600 *gur*; then weights from ½ grain to 60 talents; then areas and volumes from ½ *sar* to 7200 *bur*; and finally lengths from 1 finger to 60 leagues. The entire series, fully written, contained several hundred entries, although certain sections could be omitted or abbreviated. It could be formatted as a list, with each entry containing the standard notation for the measures only, or as a table, where the standard writings were equated with values in the sexagesimal system. Further practice in writing metrological units, particularly areas, capacities, and weights, came in the fourth curricular phase, when students learned how to write legal contracts for sales, loans, and inheritances.

Arithmetic itself was concentrated in the third phase of the House F curriculum, alongside the metrology. Again the students memorised a long sequence of facts, this time through copying and writing standard tables of reciprocals and multiplications. When the students first learned and copied each table they tended to write the entries in whole sentences: 25 *a-ra₂* 1 25,

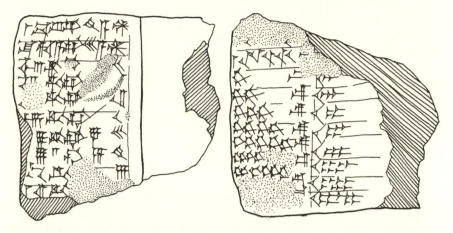

Figure 4.5 A Type II tablet from House F, with a verbose reciprocal table on the obverse (student's copy erased), a list of capacities on the reverse. (3N-T 594 = IM 58573. Drawing by the author.)

TABLE 4.6
3N-T 594 = IM 58573, a Type II Tablet from House F

Obverse i	Obverse ii	Reverse iii	Reverse ii	Reverse i
⅔ of sixty is 40	(erased)	(top missing)	(top missing)	(missing)
Its half is 30		5(ğešu) gur	12 [gur]	
Its 3rd part is [20]		1(šar) gur	13 [gur]	
[Its 4th part] is [15]		2(šar) gur	14 [gur]	
[Its 5th part] is 12		3(šar) gur	15 [gur]	
Its 6th part is 10		4(šar) gur	16 [gur]	
Its 8th [part] is 7;30		5(šar) gur	17 [gur]	
Its 9th [part] is 6;40		6(šar) gur	18 [gur]	
Its 10th part is 6		7(šar) gur	19 [gur]	
Its 12th part is [5]		8(šar) gur		
(rest missing)		9(šar) [gur]		

a-ra$_2$ 2 50 ('25 steps of 1 is 25, <25> steps of 2 is 50'), but when recalling longer sequences of tables in descending order they abbreviated the entries to just the essential numbers: 1 25, 2 50. Enough tablets of both kinds survive to demonstrate that the House F teacher presented students with the entire series of multiplication tables to learn (on Type II obverse and Type III tablets: Neugebauer's 'single multiplication tables'). Afterwards the students tended to rehearse only the first quarter of it in their longer writing exercises (on Type I and Type II reverse tablets: Neugebauer's 'combined multiplication tables'). It is also clear that the House F students did not learn the 48 times table, or the one for 44;26 40. They may also have omitted others that are otherwise well attested in Nippur: the teacher was apparently free to pick and choose. At no point in this elementary phase did the trainees do anything but memorise and reproduce by rote.[23]

It is impossible to tell how long all this took, but in any case that presumably varied greatly. Like many other pre-modern professional apprenticeships, it is likely that students started as soon as they were mature enough, perhaps as young as five or six and finished when they were competent to earn a living or when economic necessity drove them, typically in their early teens. But this is all guesswork. The clusters of signed Type III multiplication and metrological tables (table 4.3) suggest that students spent several days mastering each table before moving on to the next: Lamassi-Dagan left eight days between consecutive tables in the series and Ubarrum only five. Warad-Suen wrote his metrological tables seven days apart, probably learning another (of the longer lengths) in between (figure 4.6, table 4.7). Suen-apil-Urim's two 24 times tables show that one single table could be written out more than once, on different days, which perhaps helps to explain the very short time between Lamassi-Dagan's 9 and 5 times tables. If Suen-apil-Urim learned his tables at a fixed rate, then it would have taken him a whole year to cover the entire sequence (but presumably that was not his sole occupation for the duration!). These tablets are all unprovenanced but the Yale and Ashmolean tablets are almost certainly from the city of Larsa, south of Nippur. As neither illicit digs nor official excavations at Larsa have yielded any Type II tablets, which played such a prominent role in teaching and learning in Nippur scribal schools, these fragments of evidence for the duration of scribal training are not straightforwardly applicable to mathematics education in House F.

Nothing is known about the identities of the teacher or students in House F, as their writings are all anonymous. But it must have been a small class, as there is physically so little space in the courtyard where they worked. Further north, in Sippar, whole families of professional scribes can be traced; here it is likely that fathers trained their offspring themselves. For instance one Abba-ṭābum, active during the reign of Sumu-abum (r. 1894–1881), had two daughters, Inana-amagu and Niĝ-Nanna. Both women became scribes as well, writing legal tablets for the priestesses of Šamaš's temple and their associates.[24] Although we know nothing about

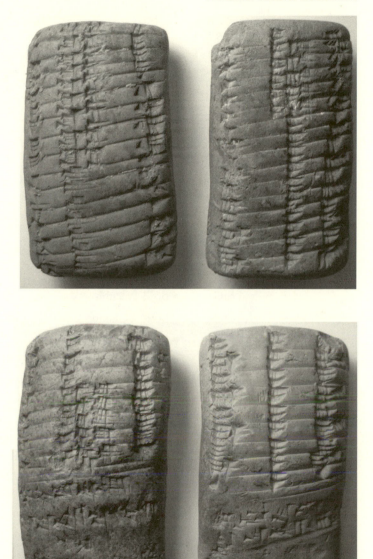

Figure 4.6 Two Type III tablets written by Warad-Suen. *Left*, metrological table of heights; *right*, metrological table of lengths. (YBC 4700, YBC 4701. Photos by the author, courtesy of the Yale Babylonian Collection.)

TABLE 4.7
Warad-Suen's Metrological Tables

YBC 4700		Possible Missing Table		YBC 4701	
[½] rod	0;30 (rods)	10 rods	10	1 finger	0;02 (cubits)
½ rod 1 cubit	0;35	20 rods	20	2 fingers	0;04
½ rod 2 cubits	0;40	30 rods	30	3 fingers	0;06
½ rod 3 cubits	0;45	40 rods	40	4 fingers	0;08
½ rod 4 cubits	0;50	50 rods	50	5 fingers	0;10
½ rod 5 cubits	0;55	1 cable	1 00	6 fingers	0;12
1 rod	1	2 cables	2 00	7 fingers	0;14
1½ rods	1;30	3 cables	3 00	8 fingers	0;16
2 rods	2	4 cables	4 00	9 fingers	0;18
2½ rods	2;30	5 cables	5 00	⅓ cubit	0;20
3 rods	3	6 cables	6 00	½ cubit	0;30
3½ rods	3;30	7 cables	7 00	⅔ cubit	0;40
4 rods	4	8 cables	8 00	1 cubit	1
4½ rods	4;30	9 cables	9 00	1⅓ cubits	1;20
5 rods	5	10 cables	10 00	1½ cubits	1;30
5½ rods	5;30	11 cables	11 00	1⅔ cubits	1;40
6 rods	[6]	12 cables	12 00	2 cubits	2
6½ rods	6;30	13 cables	13 00	3 cubits	3
7 rods	7	14 cables	14 00	4 cubits	4 «4»
7½ rods	7;30	½ league	15 00	5 cubits	5
8 rods	8	⅔ league	20 00	½ rod «1 cubit»	6[a]
8½ rods	8;30	⅚ league	25 00	½ rod 2 cubits	7
9 rods	9	1 league	30 00	½ rod 3 cubits	8
9½ rods	9;30			½ rod 4 cubits	9
10 rods	10			½ rod 5 cubits	10

Long tablet of Warad-Suen	Long tablet of Warad-Suen
Month IV, 16th day	Month IV, 23rd day
(*traces of erased signs*)	(*traces of erased signs*)

[a] This conflation of two lines leads to systematic numerical errors in the rest of the table.

their education, it is difficult to imagine that they were not home-schooled. Two centuries later in the same city, a *gala-maḫ* priest called Inana-mansum had a scribe named Šumum-līṣi call in to educate his son Bēlānum, some time during the 1640s. Šumum-līṣi was often at the house anyway, documenting Inana-mansum's various sales and loans, but he also wrote tablets for other clients, sometimes on the same day; and Inana-mansum also employed other scribes, sometimes on the same day as he used Šumum-līṣi.[25] As Bēlānum was destined for the priesthood, not a scribal career—he was ordained in 1641, taking the professional name Ur-Utu—his education was a limited one. He concentrated mostly on simple handwriting and spelling exercises, as well as metrological lists and basic calculations but no multiplication tables.[26] When, in later life, he remodelled the courtyard of the house he left the tablet recycling bin and its contents in situ, simply getting the builders to pave over it. Several centuries later, in the thirteenth-century city of Emar on the Syrian Euphrates, a diviner and scribal teacher named Ba'al-mālik bought four toddlers from their financially desperate parents for the sum of 60 shekels. The two boys of the family, Ba'al-bēlu and Išma'-Dagan, went on to become junior diviners themselves, writing school exercises (but no mathematics, so far as is known), probably under Ba'al-mālik's instruction.[27]

Less reliable evidence for the circumstances of Old Babylonian scribal schooling comes from Sumerian literary works. They are unreliable because they were curricular literature, memorised and written out by the scribal students themselves as part of their education, not as historical accounts of how schools really were. After all, the conditions of school life was one subject that trainee scribes did not need to be taught! Many aspects of these proverbs, dialogues, and stories are obviously humorous. Undoubtedly there is much more humour that is no longer accessible, for instance in the disparity between schooling as presented in the literature and as it was really experienced by the young scribes. For example, there are three student manuscripts from Nippur of a fictional letter from one Nabi-Enlil to Ilum-puzura, which begins:

> Say to Ilum-puzura: thus speaks Nabi-Enlil the scribe, son of Saĝ-Enlil. What is this that you have done? The boys should smell the scent of Nippur!
>
> Three years ago I returned to the man. There where they lived, in the master's house—in the first place, in my opinion it was not pleasant and, further, it was cramped; I told Pī-Ninurta. Because it was my master's house, I did not open my mouth. Now listen—there where they are living, it is not a proper scribal school. He cannot teach the education of a scribe there. He cannot recite even twenty or thirty

incantations, he cannot perform even ten or twenty praise songs. But in his presence, in my master's house, I cannot open my mouth.

Don't you know that the scribal school in Nippur is unique? I told you that . . . is ignorant; but afterwards you neglected my words. If they learn the scribal art at my hands, then Nippur will be built in Isin.[28]

Ilum-puzura seems to have put his sons' or pupils' names down for the wrong school. Nabi-Enlil reports that he visited that school a few years earlier, somewhere outside Nippur, and was shocked: the house was too small and the teacher—himself apparently Nabi-Enlil's former teacher—woefully ignorant. It came nowhere near to the standards of schooling expected at Nippur. Nabi-Enlil berates Ilum-puzura for his decision, and goes on to offer to educate the children himself—not in Nippur but in nearby Isin. The colourful language indicates the composition's humorous intent. But where exactly does that humour lie? Did the young Nippur copyists think it was wildly funny that anyone should be educated away from home? Or did they recognise all too well the description of cramped and uncomfortable living conditions? It is impossible to know. Maybe the punch-line was Nabi-Enlil's offer of a Nippur education in rival Isin (Oxford to Nippur's Cambridge, or Harvard to Nippur's Yale). The school stories remain an important window into the scribes' world; but rather than depict the reality of scribal education they reveal how the scribes liked to view themselves and the educational process. Pride in the profession, disdain for the incompetent or self-important, and a sharp sense of humour were all celebrated and passed on to the next generation. Such literature contained bigger messages too, as discussed in §4.4.

4.3 WORDS AND PICTURES, RECIPROCALS AND SQUARES

Elementary mathematics education in House F consisted solely of the memorisation and reduplication of metrological and arithmetical number facts, and writing them in context. The students did not practice calculations until they were also learning Sumerian literature, often on the same tablets. At Ur and in House B at Nippur (§1.3), however, students made calculations on the same tablets as Sumerian proverbs, at the end of their elementary training.[29] In seventeenth-century Sippar the young Bēlānum encountered calculations as part of his elementary education.[30] Calculations are typically written on Type III or Type R tablets, in very standardised formats (table 4.8). Sides of squares are written one above the other with the resultant area underneath (§1.3); in long sequences of multiplications, multiplicands are written in vertical alignment to the left of a vertical ruling with their product to the right of it.

TABLE 4.8
Formats and Functions of Advanced School Tablets in House F and Elsewhere

Type	Description	Typical Contents
L	Small, single-column tablet in landscape orientation, i.e., with the writing parallel to the longer side	Table of powers; Plimpton 322; or extract from Sumerian literary work (rare)
Mn	Large, multi-columned tablet (where n = the number of columns per side); same as elementary Type I	Compilation of word problems, with or without worked solutions; catalogue of word problems; or Sumerian literary composition(s)
R	Small, roughly square tablet	Mathematical rough work: calculation; diagram; or hastily written extract of worked solution
S	Small, single-column tablet in portrait orientation, i.e., with the writing parallel to the shorter side; same as elementary Type III	Single word problem with worked solution; extract from Sumerian literary work; and/or calculation

Note: Typology based on Tinney 1999, 160.

One particularly popular exercise was to find the reciprocal of a sexag-esimally regular number that was not included in the standard table of reciprocals (§3.4, §4.2). Wherever they are found, the exercises are laid out in a consistent way, and the parameters are consistently chosen from a series of pairs generated from doubling and halving the pair 2;05 ~ 28;48 (where ~ signals reciprocity). House F has yielded two such examples. On 3N-T 605 (table B.10), a Type S tablet, the student has simply written:

4;26 40
its reciprocal is 2;13 20.

A double ruling underneath signals the end of the exercise; the rest of the tablet is blank. The answer given, however, is half of the first number, not the expected 13;30. A second attempt, on 3N-T 362+, is more successful (table B.10). This is another Type S tablet, on which the calculation is writ-ten after the first twenty-two lines of a Sumerian literary composition now called *A Supervisor's Advice to a Younger Scribe* (see §4.4). The numbers read:

17 46 40   9
2 40   «2» 22 30
3 22 30   [2]

```
6 45      [1 20]
9          6 40
8 53 20
17 46 40
```

This is clearly an exercise in finding that 3;22 30 is the reciprocal of 17;46 40 and then working back to the original number to check the correctness of the solution. To understand the function of the intermediate numbers we can draw on a fragmentary set of word problems about the same topic, VAT 6505, which was probably copied by Iškur-mansum in seventeenth-century Sippar (§4.1, table 4.4). (Fortunately, they are correct.) Each of the six surviving solutions is damaged, but as they all describe the same method they can be restored with some confidence. Iškur-mansum's examples are the first six pairs of the 2;05 ~ 28;48 sequence; 3N-T 362+ is the tenth (and 3N-T 605 the eighth). Needless to say, the method of solution is the same whatever values are used, so we can imagine Iškur-mansum or the House F teacher writing the following:

> What is the reciprocal of 17;46 40? You, in your proceeding: solve the reciprocal of 0;06 40. You will see 9 00. Multiply 9 00 by 17;40. You will see 2 39 00. Append 1 00. You will see 2 40 00. Solve the reciprocal of 2 40. You will see 0;00 22 30. Multiply 0;00 22 30 by 9 00. You will see 3;22 30. Your reciprocal is 3;22 30. That is the procedure.

'Raising' and 'appending' are the Old Babylonian terms for geometrical multiplication and addition, which signal that, contrary to the purely numerical appearance of the calculation, the method in fact utilises cut-and-paste geometrical 'algebra' (figure 4.7). The product of any reciprocal pair is by definition 60 (or any power of 60). We can therefore imagine 17;46 40 as the side of rectangle of area 1 00; the task is to find the length of the other side (figure 4.7a). We can measure off a little bit of the known side, so that it has a length that is listed in the standard reciprocal table: in this case 0;06 40. We can thus draw a second rectangle on that 0;06 40 with area 1 00 and length 9 00 to create an L-shape (figure 4.7b). The procedure then tells us to multiply ('raise') the length of the new rectangle by the remainder of the known reciprocal to create a new rectangle of area 2 39 00 (figure 4.7c). Appending it to the second reciprocal rectangle of area 1 00 gives a total area of 2 40 00 (figure 4.7d). This compound rectangle is 2 40 times bigger than the original reciprocal rectangle of area 1 00. Thus 9 00 is 2 40 times bigger than the length we need to find—which must, therefore, be 3;22 30. The problem is repeated backwards, with an iteration to find the reciprocal of the intermediate value 6 45 because it is not in the standard list.

Even though no word problems themselves were found in House F, the calculations thus demonstrate that some mathematics (and not just

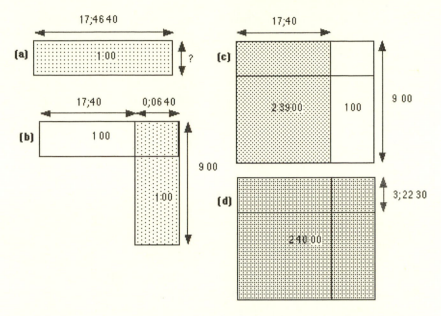

Figure 4.7 The cut-and-paste procedure for finding the reciprocal of a sexagesimally regular number.

arithmetic and metrology) was taught there. Word problems with worked solutions, such as those written by Iškur-mansum or the one discussed in §4.1, served as model answers—perhaps written out by the teacher as an aide-mémoire, or perhaps by a student to show that he or she had mastered a procedure. Collections of word problems without numerical answers or worked solutions, such as those on BM 15285 (§2.4), are more obviously text books, from which students could be assigned a series of problems. BM 15285 is explicitly pedagogical in structure, moving from simple problems about squares, circles, and right triangles on the obverse to more complex (and culturally embedded) figures on the reverse (see also §7.1). Some sequences of problems on that tablet show the same figure but describe it in different ways. For instance problems (7) and (8) both describe a square inscribed with a second square whose corners bisect the sides of the outer one (§2.4, figure 2.10). In order to solve either of those problems, one has to find the length of the inner square, which is diagonal to the outer one. Many Old Babylonian mathematical practitioners (and even Iškur-mansum) knew that the square on the diagonal of a right triangle had the same area as the sum of the squares on the length and width: that relationship is used in the worked solutions to word problems on cut-and-paste 'algebra' on seven different tablets, from Ešnuna, Sippar, Susa, and an unknown location in southern Babylonia.[31] Finding the length of that diagonal square-side is a rather different matter. There is no evidence that Old Babylonian scribes had any concept of irrationality; but there were iterative cut-and-paste procedures whereby square-sides could be calculated or approximated.[32] However, that was not always necessary, as key mathematical relationships were recorded in lists of constants that were grouped thematically according to geometrical figures or different types of labouring work rates.[33]

For instance the student who wrote YBC 7289—a Type IV tablet with a diagram of a square of side 30 that is very reminiscent of those on BM 15285—did exactly that (figure 4.8).[34] He found the length of the square's diagonal by multiplying the the side by the constant 1;24 51 10, exactly as stated in the coefficient list YBC 7243:[35]

1;24 51 10, the diagonal of a square-side.

In the diagram, the constant is written on the diagonal and its length, 42;25 35, underneath. If the student were competent, he or she could have found it through halving the constant by inspection. In any case, it is just a student exercise in calculating lines and areas, not an anachronistic anticipation of the supposed classical Greek obsession with irrationality and incommensurability as is sòmetimes implied (see §§9.2–4).[36]

Scribal teachers drew on a fund of commonly circulating word problems with appropriate parameters—such as the problems on reciprocal pairs in the 2;05 ~ 28;48 sequence—and also created their own. Indeed, very few word-for-word duplicates of worked solutions are known, even when the same subjects and parameters are found in different cities.[37] And although the parameters in model solutions often remained unchanged as they travelled, teachers drew up long lists of alternative viable parameters for single exercise types to enable them to set the same one to more than one student at once, or to give a single student several opportunities to practice the same method of solution. About a dozen such lists survive.[38] Further, a large group of Type IV tablets from Ur includes two sets of about half-a-dozen structurally very similar calculations, probably solving problems about labouring work rates and the capacities of storage vessels.[39]

Plimpton 322, undoubtedly the most famous of Old Babylonian mathematical artefacts, is also a teacher's catalogue of parameters (figure 4.9).[40] In its current state it comprises a four-column, fifteen-row numerical table on a tablet measuring about $13 \times 9 \times 2$ cm. Its notoriety rests on the fact that the central two columns list the widths and diagonals of fifteen right-angled triangles, apparently sorted in descending order of acuteness. The first surviving column (there was at least one to the left of it) seemingly lists the ratios of the square on the diagonal to the square on the width of each triangle (table 4.9).

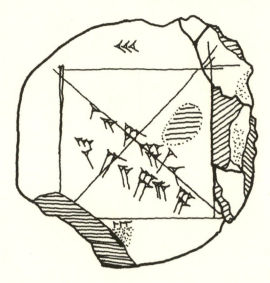

Figure 4.8 A Type IV tablet showing a student exercise in finding the diagonal of a square. (YBC 7289. Drawing by the author.)

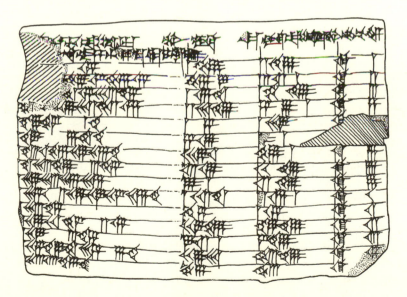

Figure 4.9 A teacher's tabular catalogue of parameters for word problems about diagonals and rectangles. (Plimpton 322. Robson 2001a, fig. 1.)

TABLE 4.9
The Parameter Catalogue Plimpton 322, Probably from Old Babylonian Larsa

The Tākiltum-*square of the Diagonal [From Which 1 is] Torn Out, so that the Short Side Comes Up*	*Square-side of the Width*	*Square-side of the Diagonal*	*Its Name*
[1 59] 00 15	1 59	2 49	1st
[1 56 56] 58 14 56 (*sic*, for 50 06) 15	56 07	3 12 01 (*sic*, for 1 20 25)	2nd
[1 55 07] 41 15 33 45	1 16 41	1 50 49	3rd
[1] 53 10 29 32 52 16	3 31 49	5 09 01	4th
[1] 48 54 01 40	1 05	1 37	5th
[1] 47 06 41 40	5 19	8 01	6th
[1] 43 11 56 28 26 40	38 11	59 01	7th
[1] 41 33 45 14 3 45	13 19	20 49	8th
[1] 38 33 36 36	9 (*sic*, for 8) 01	12 49	9th
[1] 35 10 02 28 27 24 26 40	1 22 41	2 16 01	10th
[1] 33 45	45	1 15	11th
[1] 29 21 54 2 15	27 59	48 49	12th
[1] 27 00 03 45	7 12 01 (*sic*, for 2 41)	4 49	13th
[1] 25 48 51 35 6 40	29 31	53 49	14th
[1] 23 13 46 40	28	53 (*sic*, for 1 46)	15th

Although it is the only known headed table in all of Old Babylonian mathematics, Plimpton 322 bears a striking resemblance to administrative tables written in around 1800 BCE in the kingdom of Larsa (§6.2)—the provenance given by Edgar Banks, the dealer who sold it to George Plimpton in the 1920s.[41] The headings at the top of the table, often overlooked by modern interpreters, state explicitly what each column contains. The last column, a line-count, is headed like the other tables from Larsa with the signs MU.BI.IM 'its name'. The two columns immediately preceding that are headed IB$_2$.SI$_8$ SAG 'square-side of the width' and IB$_2$.SI$_8$ *si-li-ip-tim*, 'square-side of the diagonal'. But the real key—though by far the most difficult to understand for modern readers—is the heading of the first surviving column, which states:

[*ta*]-*ki-il-ti ṣi-li-ip-tim* [*ša* 1 *in*]-*na-as-sà-ḫu-ú-ma* SAG *i-il-lu-ú*

The *tākiltum*-square of the diagonal [from which 1 (00) is] torn out so that the width comes up.

Two technical terms in this description signal definitively that the subject-matter is cut-and-paste geometry: *nasāhum*, 'to tear out', is the verb of geometrical subtraction and *tākiltum*, a noun derived from the verb of geometrical squaring *šutākulum* 'to combine', indicates a square constructed in the course of a cut-and-paste procedure. For instance, the word problem YBC 6967 is, like Plimpton 322, almost certainly from late nineteenth- or early eighteenth-century Larsa.[42] It contains instructions for finding a regular reciprocal pair given their difference, using exactly the same procedure as the second half of YBC 4663 (8) (figure 4.3):

[A reciprocal] exceeds its reciprocal by 7. What are [the reciprocal] and its reciprocal?

As before, the product of any pair of reciprocals is by definition 60. We can thus conceptualise them as the unknown lengths of a rectangle with area 60.

You: break in half the 7 by which the reciprocal exceeds its reciprocal: 3;30 (will come up). Combine 3;30 and 3;30: 12;15 (will come up).

Following the instructions, we can move the broken piece of the rectangle to form an L-shaped figure, still of area 60, around an imaginary square of area 12¼.

Append [1 00, the area,] to the 12;15 which came up for you: 1 12;15 (will come up). What is [the square-side of 1] 12;15? 8;30.

Together, therefore, they comprise a large square of area 72¼ and side 8½.

Put down [8;30 and] 8;30, its equivalent, and tear out 3;30, the *tākiltum*-square, from one (of them); append (3;30) to one (of them). One is 12, the other is 5. The reciprocal is 12, its reciprocal 5.

We remove the vertical side of the imaginary small square from that of the large composite square, reverting to the smaller side of the original rectangle, a side whose length is 5. We find the longer side of the rectangle by adding the horizontal side of the imaginary square onto that of the large composite square and arrive at the answer 12.

If instead we choose a reciprocal pair whose product is not 60 but 1, their product can be imagined as a much longer, narrower rectangle than in figure 4.3. But the half-difference of the reciprocals can still be found and the rectangle rearranged to form an L-shaped gnomon, still of area 1. Its outer edges will still be the lengths of a large square, and its inner edges the lengths of a small square. That is, we will have a composite large square that is the sum of 1 (itself a square) and an imaginary small square.

TABLE 4.10
From Reciprocals to Rectangles in Plimpton 322 (With Errors Corrected)

Reciprocal Pair	Half-difference	Half-sum	Square of Half-sum	Square-side of Width	Square-side of Diagonal	Square-side of Length	Its Name
2;24 0;25	0;59 30	1;24 30	1;59 00 15	1 59	2 49	2 00	1st
2;22 13 20 0;25 18 45	0;58 27 17 30	1;23 46 02 30	1;56 56 58 14 50 06 15	56 07	1 20 25	57 36	2nd
2;20 37 30 0;25 36	0;57 30 45	1;23 06 45	1;55 07 41 15 33 45	1 16 41	1 50 49	1 20 00	3rd
2;18 53 20 0;25 55 12	0;56 29 04	1;22 24 16	1;53 10 29 32 52 16	3 31 49	5 09 01	3 45 00	4th
2;15 0;26 40	0;54 10	1;20 50	1;48 54 01 40	1 05	1 37	1 12	5th
2;13 20 0;27	0;53 10	1;20 10	1;47 06 41 40	5 19	8 01	6 00	6th
2;09 36 0;27 46 40	0;50 54 40	1;18 41 20	1;43 11 56 28 26 40	38 11	59 01	45 00	7th
2;08 0;28 07 30	0;49 56 15	1;18 03 45	1;41 33 45 14 03 45	13 19	20 49	16 00	8th
2;05 0;28 48	0;48 06	1;16 54	1;38 33 36 36	8 01	12 49	10 00	9th
2;01 30 0;29 37 46 40	0;45 56 6 40	1;15 33 53 20	1;35 10 02 28 27 24 26 40	1 22 41	2 16 01	1 48 00	10th
2 0;30	0;45	1;15	1;33 45	45	1 15	1 00	11th
1;55 12 0;31 15	0;41 58 30	1;13 13 30	1;29 21 54 02 15	27 59	48 49	40 00	12th
1;52 30 0;32	0;40 15	1;12 15	1;27 00 03 45	2 41	4 49	4 00	13th
1;51 06 40 0;32 24	0;39 21 20	1;11 45 20	1;25 48 51 35 06 40	29 31	53 49	45 00	14th
1;48 0;33 20	0;37 20	1;10 40	1;23 13 46 40	28	53	45	15th

This set of three squares, all generated by a pair of reciprocals, obeys the 'Pythagorean' rule, just like the entries in Plimpton 322.

In this light the heading of column I describes the area of the large square, composed of 1 plus the small square. There is apparently one minor terminological discrepancy: in Plimpton 322 *tākiltum* refers to the area of the large composite square, while in YBC 6967 it means the side of the small imaginary square. But as Old Babylonian mathematics names squares and their sides identically (§3.2), that is not problematic. The word itself does not suggest that its meaning should be restricted to either the little square or the big square but its pattern of attestation is restricted to exactly these cut-and-paste geometrical scenarios. The entries in columns II and III are the sides of the small squares (widths) and the sides of the large squares (diagonals), all scaled up or down by common factors 2, 3, and 5 until co-prime values are reached.

Indeed it was Evert Bruins who first proposed that the entries in Plimpton 322 were derived from reciprocal pairs, running in descending numerical order from 2;24 ~ 25 to 1;48 ~ 33 20 (although, writing before Høyrup's work, he did not conceptualise this geometrically).[43] These pairs could well have been the contents of the missing column(s) to the left of the tablet (table 4.10). Although only five of them occur in the standard reciprocal list, the other ten are widely in evidence elsewhere in Old Babylonian mathematics (as constants, for instance) or could have been calculated using known methods, such as doubling and halving. None of them is more than four sexagesimal places long, and they are listed in decreasing numerical order, thereby fulfilling the current conventions of administrative tabular formatting. Similarly, none of the widths or diagonals in columns II and III is more than three sexagesimal places long—in practice the maximum level of arithmetical complexity for setting students problems about rectangles (or right triangles) and their diagonals. The resultant lengths are often only one place long. Column I and the missing reciprocal columns were probably for the teacher's use only.[44] Although the resultant parameters are simple, the elegance of their construction and the deft handling of often complex sexagesimal arithmetic show that not every scribal teacher was as incompetent as poor old Iškur-mansum.

4.4 MEASUREMENT, JUSTICE, AND THE IDEOLOGY OF KINGSHIP

One of the House F reciprocal calculations was on the same Type S tablet as an extract from the Sumerian literary work now known as *The Supervisor's Advice to a Younger Scribe* (see §4.3). It ironically constructs an ideal image of a humble, conscientious scribe through the self-description of a boastful, bossy supervisor. He begins:

One-time member of the school, come here to me:
Let me explain to you what my teacher revealed.
Like you, I was once a youth and had a mentor.
The teacher assigned a task to me—it was man's work.
Like a springing reed, I leapt up and put myself to work.
I did not depart from my teacher's instructions;
I did not start doing things on my own initiative.
My mentor was delighted with my work on the assignment.
He rejoiced that I was humble before him and he spoke in my favour.[45]

He continues in this vein for some time, then rebukes a junior colleague for failing to live up to this ideal. The young man resentfully replies:

Whatever you revealed of the office of scribe has been given back to you.
You placed me in charge of your household and I have never served you
    with idleness.
I have assigned work to the maidservants, servants, and attendants in
    your household.
I have kept them happy with rations, clothing, and oil rations.
I have assigned the order of their work to them,
So that you do not need to follow the servants around in your master's
    house.
I am doing things from the break of day; I follow them round like sheep.[46]

The scribe finally gains his superior's approbation and is rewarded with the right to teach others, under the guidance of Nisaba, patron goddess of scribes.

Some eighty different literary compositions are known from House F, in around seven hundred manuscripts.[47] Almost all of them are also attested elsewhere, but House F is one of the most important sources of classical Sumerian literature, both for the sheer size of the assemblage and for its archaeological context. Yet it is also very unusual in that the House F compositions are (not surprisingly) obsessed with scribal professionalism, literacy, and numeracy. In modern terminology, they comprise much the same genres—narrative myths and epics; hymns to deities, kings, and temples; city laments; fictional letters (see §4.2); compositions about scribal life—as the rest of the Sumerian literary corpus. But the eighty House F works use words for literate and numerate professions, actions, and equipment ten times as frequently as the other three-hundred-odd known compositions do.[48] What is more, the fictional actors in the House F compositions exhibit a striking gender distribution: goddesses and anonymous (male) humans are by far the most likely to be associated with literacy and numeracy, followed by historically attested kings, gods, and named humans. Epic heroes such as the Sumerian Gilgameš, by contrast, have nothing at all do with the scribal arts (cf. §6.4).

The transition from elementary scribal education to more advanced work on Sumerian literature was made with a group of four hymns, identified by Steve Tinney and now called the Tetrad.[49] Three are praise poems to long-dead kings of the Isin dynasty—Iddin-Dagan (r. 1974–1954), Lipit-Eštar (r. 1929–1924), Enlil-bāni (r. 1860–1837)—each of which contains at least one description of the goddess Nisaba fostering scribal skills, either in the king himself or in the students of the e_2-dub-ba-a 'tablet house'. The fourth, a hymn to Nisaba herself, celebrates her prowess as scribe. *Lipit-Eštar hymn B* is addressed directly to the king:

Nisaba, woman sparkling with joy,
Righteous woman, scribe, lady who knows everything:
She leads your fingers on the clay,
She makes them put beautiful wedges on the tablets,
She makes them (the wedges) sparkle with a golden stylus.
A 1-rod reed and a measuring rope of lapis lazuli,
A yardstick, and a writing board which gives wisdom:
Nisaba generously bestowed them on you.[50]

The opening description of Nisaba puts equal stress on her femininity and on her wisdom. She is twice associated with the verb 'to sparkle' and the mensuration equipment she gives Lipit-Eštar is also made of high-value, sparkling materials. This is not just feminine glamour, however: Irene Winter has cogently demonstrated the divine qualities of lustre and radiance in early Mesopotamia.[51] Indeed, the opening line of *Nisaba hymn A* addresses the goddess as 'Lady sparkling like the stars of heaven, holding a lapis lazuli tablet!'[52] And later in the same hymn the image recurs:

(Enki) has opened up Nisaba's House of Wisdom.
He has placed the lapis lazuli tablet on her knees,
For her to consult the holy tablet of the heavenly stars.[53]

But what does king Lipit-Eštar do with the wisdom that Nisaba bestows on him through literacy and numeracy? He immediately establishes and dispenses justice, as the next lines of his hymn show:

Lipit-Eštar, you are Enlil's son.
You have made righteousness and truth appear.
Lord, your goodness covers everything as far as the horizon.[54]

And so on, for the next twelve lines. The hymns to Enlil-bāni and Iddin-Dagan follow very similar schemas. The message of the Tetrad is remarkably consistent and straightforward, as befits its elementary pedagogical character: Enki, the great god of wisdom, has granted heavenly wisdom to Nisaba, who brings literacy and numeracy to kings and scribes in order for them to ensure just rule.

Where the Tetrad focuses on the royal acquisition of literacy and numeracy through Nisaba's heavenly guidance for the administration of justice, another curricular grouping (now called the House F Fourteen)[55] also shows both goddesses and anonymous humans in professional action. *The Supervisor's Advice to a Scribe* belongs to this group. Thus the message moves on from the divine origins and ultimate purpose of literacy and numeracy to correct deportment in its professional deployment. Scribal identity and good conduct count for more than individualism and boastfulness.

Some 40 percent of the Sumerian literary compositions found in House F have not been assigned to curricular clusters.[56] In *Enki and the World Order* Enki determines Nisaba's destiny:

My illustrious sister, holy Nisaba,
Is to receive the 1-rod reed.
The lapis lazuli rope is to hang from her arm.
She is to proclaim all the great divine powers.
She is to fix boundaries and mark borders. She is to be the scribe of the
    Land.
The gods' eating and drinking are to be in her hands.[57]

Similarly, in *Enlil and Sud* the great god Enlil bestows literacy and numeracy on his bride Ninlil as a wedding present:

The office of scribe, the tablets sparkling with stars, the stylus, the tablet
    board,
Reckoning and accounts, adding and subtracting, the lapis lazuli measur-
    ing rope, the . . . ,
The head of the peg, the 1-rod reed, the marking of the boundaries, and
    the . . .
You have been perfected by them.[58]

In both passages, literacy is subservient to mensuration: Nisaba and Ninlil are given the means to measure land justly and accurately, resulting (in Nisaba's case) in the equitable distribution of the harvest. *Inana's Descent* sheds a different light on the importance of mensuration to the great goddesses' self-identity. When the powerfully sexual Inana enters the Underworld to visit her sister Ereškigal, at each of its seven gates she is stripped of one aspect of her divinity in order for her to enter as one of the powerless, desexualised dead. It is only at the penultimate gate that she can bear to give up her reed and rope to the doorkeeper:

The 1-rod reed and lapis lazuli measuring rope were snatched from her
    hand.

'What is this?'
'Be silent, Inana, a divine power of the underworld has been fulfilled.
'Inana, you must not open your mouth against the rites of the
    Underworld.'[59]

Male deities, by contrast, have very little to do with writing, counting,
or measuring in curricular Sumerian literature except for Utu (Akkadian
Šamaš), the sun and god of justice, who is occasionally associated with
weighing. So the scribal students of House F in eighteenth-century Nippur
were taught to connect writing and land measurement with goddesses,
above all Nisaba, the just kings of centuries before, and the self-effacing
professional scribe who ensures the smooth, fair running of households
and institutions. But what did this mean in practice? Did these literary tropes
have any bearing on the lives, ideals, and practices of working scribes?

The literary ideal of metrological justice certainly had a counterpart in
contemporary royal rhetoric. Sets of laws were promulgated by various
kings of the Ur III and Old Babylonian periods, including Lipit-Eštar, an
anonymous king of Ešnuna, and most famously Hammurabi himself.[60]
The prologue to an early Ur III law code, written on behalf of either Ur-
Namma (r. 2112–2095) or his son Šulgi (r. 2094–2047), explicitly states
the king's direct involvement with the establishment of fair metrology:

> I made the copper *bariga* and standardised it at 1(*ĝeš*) *sila*. I made the
> copper *ban* and standardised it at 10 *sila*. I made the normal royal
> copper *ban* and standardised it at 5 *sila*. I standardised the stone
> weights from 1 shekel of silver to 1 mina. I made the bronze *sila* and
> standardised it at 1 mina.[61]

And it seems that such responsibilities were actually carried out in practice. For instance on the accession of Šulgi's son Amar-Suen (r. 2046–2038)
new weights were issued to overseers in the city of Umma:

1 weight of 10 minas
1 weight of 5 minas
1 weight of 2 minas
1 weight of 1 mina
1 weight of ⅓ mina
1 weight of 10 shekels:
Ur-Sakkud received for the Goatskin Bureau.

1 weight of 10 minas
1 weight of 5 minas
1 weight of 4 minas
1 weight of 3 mina
1 weight of 2 minas

1 weight of 1 mina
Ur-Nintud received for the Weavers' Bureau.

Month IV. Year that Amar-Suen became king.[62]

A copy of Hammurabi's famous stele of laws, erected after the unification of Babylonia in 1760, was originally deposited in the temple of Šamaš, god of justice, in Sippar 'in order that the mighty not wrong the weak' (figure 4.10).[63] Its three-hundred-odd laws include some that promote fair measurement and exchange while others punish metrological fraudsters:

> If a merchant gives grain or silver as an interest-bearing loan, he shall take 100 *sila* per *gur* (= 33%) as interest; if he gives silver as an interest-bearing loan, he shall take 36 grains of silver per shekel (= 20%) as interest.

> If a (female) innkeeper refuses to accept grain for the price of beer but accepts only silver measured by the large weight, thus reducing the value of beer in relation to the value of grain, they shall establish the guilt of that (female) innkeeper and they shall throw her into the river.[64]

The arresting image at the top shows Šamaš seated on his divine throne, presenting Hammurabi with an object which is now conventionally known as the 'rod and ring'.[65] The earliest attestation of this exchange shows Ur-Namma accepting the same thing from his dynastic god Nanna-Suen (figure 4.11); the latest are from the early first millennium. The 'rod and ring' was clearly a powerful symbol of kingship, which must have accrued many layers of meaning; debate about its interpretation runs hot amongst Assyriologists.[66]

There are two types of image featuring the 'rod and ring' in Ur III and Old Babylonian times: those depicting a king receiving it from a god; and a naked goddess holding a 'rod and ring' in each hand, for instance on the Burney Relief, also known as the Queen of the Night. Dominique Collon has surveyed various proposals for the identity of this naked goddess and her iconography.[67] Of particular interest is Thorkild Jacobsen's identification of her as Inana holding a 1-rod reed and a coiled-up measuring rope on her way to the Underworld, just as in *Inana's Descent*.[68] On this reading she has already passed through several gates of the Underworld, shedding clothing and divine powers on the way, and is morphing into a feathered Underworld being from the feet up. Just as the Burney Relief parallels the literary image of Inana descending, the royal presentation scenes parallel the passages in the royal hymns in which Nisaba bestows the reed and rope on kings as symbols of literate and numerate justice. The 'rod and ring' features in no other type of visual scene, just as the reed and rope as literary motif is predominantly in the hands of goddesses and kings. But

Figure 4.10 Detail of the law-code stele of king Hammurabi (c. 1760), showing the Šamaš, god of justice, bestowing him with symbols of just kingship. (Scheil 1902a, pl. 3.)

Figure 4.11 Detail of a stele of king Ur-Namma (c. 2100), showing Nanna-Suen, the moon god, bestowing him with a measuring reed and coiled rope. (Canby 2001, pl. 31, courtesy of the University of Pennsylvania Museum of Archaeology and Anthropology.)

whereas in the scribal schools Nisaba reigned supreme, in public images she was replaced by the deity appropriate to the context. In eighteenth-century Mari, for instance (§5.1), a wall painting in the palace throne room shows Ištar (the Akkadian counterpart to Inana) investing the king with rod and ring/reed and rope (figure 4.12).

Figure 4.12 Detail of a wall painting in the palace throne room of Mari (c. 1765), showing the goddess Ištar bestowing the king with symbols of just kingship. (Parrot 1958, pl. XI, courtesy of Librarie Paul Geuthner.)

The reed and rope, then, were public symbols of royal justice as well as scholastic ones, representing the fair mensuration of land amongst the people. For, as the young scribe Enki-manšum asserts in a literary dialogue from House F:

> When I go to divide a plot, I can divide it; when I go to apportion a field, I can apportion the pieces, so that when wronged men have a quarrel I soothe their hearts and [. . .]. Brother will be at peace with brother, their hearts [. . .].[69]

Finally, does the discovery of gendered literacy and numeracy amongst divine actors in the Sumerian literary corpus reveal anything at all about the contemporaneous 'real' world of the scribes? To be sure, we cannot simply infer the existence of numerate female scribes from the prevalence of goddesses in that role in literary Sumerian. But equally, the simplistic assumption that scribes and their students were all male no longer holds water. Records from the palace at Mari document ten anonymous female scribes receiving oil and wool rations in the 'harem' and writing kitchen documentation, including royal menus, while princess Šīmātum received a personal scribe Šīma-ilat as part of her dowry in the 1760s BCE.[70] Several female scribes were active in nineteenth-century Sippar. Inana-amağu, whose father and sisters were also scribes (§4.2), wrote tablets for the judges in Šamaš's temple (where Hammurabi's law stele was later to stand), mostly records of legal cases involving the contested land and property of priestesses.[71] And four school tablets written by female scribal students during the reign of Samsu-iluna, probably also in Sippar, have been identified.[72] In other words, female scribes learned the standard student exercises, administered large households, and assisted in the maintenance of numerate justice: exactly what the images of goddesses in curricular compositions lead us to expect. Female scribes appear to have worked predominantly for female clients, but they existed—and were numerate—nevertheless.

4.5 CONCLUSIONS

While the invention of the sexagesimal place value system was a necessary condition for the creation of mathematics as an intellectual activity, divorced from the mundane necessities of central administration, the SPVS alone is not sufficient to explain that extraordinary development. Nor does it explain the particular form that mathematics took, so clearly focused on the correct calculation of lines and areas, or the sheer volume of extant mathematical tablets. Jens Høyrup has posited an oral 'surveyors' culture' in the third millennium BCE in which professional land administrators entertained and tested each other with mathematical riddles based on their professional interests.[73] But that hypothesis does not adequately account for the turn to literate mathematics in the early second millennium; little remains of any third-millennium-style practical pretext in the examples presented here. Rather, it is necessary to factor in the ideology of the state under which the trainee scribes were preparing to serve.

Piety and militarism were key components of kingship; the third strand was justice. Royal hymns and monumental imagery of the twenty-first to eighteenth centuries give equal weight to all three. The Sumerian word for justice was $niğ_2$-si-sa_2, literally 'straightness, equality, squareness', Akkadian $mīšarum$ 'means of making straight'. The royal regalia of justice were the measuring rod and rope, often presented to the king by a god on royal monuments or, more usually in Sumerian literary works, a goddess. In the light of these passages and images Old Babylonian mathematics, with its twin preoccupations of land and labour management on the one hand and cut-and-paste geometrical algebra on the other, becomes truly comprehensible. This formalised discourse of measurement and rectification was, at one level, simply royal ideology in another form, promoted through practice and through literary representation. Just as Sumerian literature taught not only Sumerian literacy but also what it meant to be professionally literate, Old Babylonian mathematics carried similar messages about the abstract principles of numerate justice as embodied in the correct calculation of lines and areas. And paradoxically, the wealth of written evidence is an artefact of a fundamentally oral, memorised transmission of knowledge and values: Old Babylonian school tablets were essentially ephemera, created to aid and demonstrate recall, destined almost immediately for the recycling bin. Thus, while modern scholars have chosen to portray Old Babylonian literature and mathematics as amongst the world's first truly creative and non-utilitarian writings—and indeed there is much to admire in their beauty—for their producers and consumers they represented above all idealised abstractions of the ordered urban state, with god, king, and scribe at its centre.

CHAPTER FIVE

Assyria

Most evidence for mathematics in ancient Iraq comes from the flat plains of the south, in Sumer and Babylonia. But to the north, on the upper reaches of the Tigris and Euphrates rivers, there was mathematics too, albeit of a rather different flavour (§5.1). Assyria's mathematical history falls into three phases. In the Old Assyrian period, c. 1940–1760 BCE, predominantly decimal arithmetic was used by merchant families from the city of Assur and by palace scribes at Mari on the mid-Euphrates, both in very different ways from their Babylonian contemporaries (§5.2). In both the Middle Assyrian period, c. 1350–1050 BCE, and the Neo-Assyrian period, c. 750–610 BCE, quantifications were an important rhetorical device in royal inscriptions and were often drawn directly from carefully kept records (§5.3); numeracy also played a role in Neo-Assyrian intellectual culture (§5.4).[1]

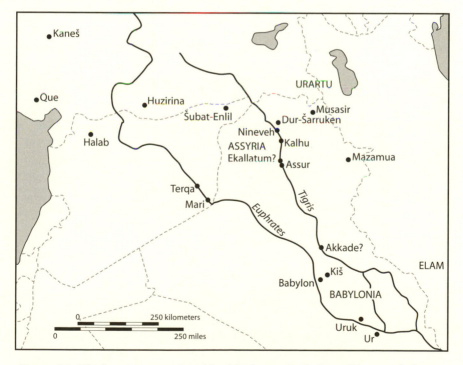

Figure 5.1 The locations mentioned in this chapter, including all known findspots of Assyrian mathematical tablets.

5.1 BACKGROUND AND EVIDENCE

The city of Assur sits on a rocky promontory overlooking the Tigris river, some 100 km south of modern Mosul.[2] Its commanding position over the surrounding plains and easy access to water made it an ideal location for settlement. By the middle of the third millennium Assur must have been a substantial town: it supported at least one temple, where the goddess Ištar was worshipped and whose remains suggest cultural links with Mari (of which more below) and Kiš in northern Babylonia. It later came under the influence of the great territorial empires of Akkad and Ur (see §3.1) but by the early second millennium it was a politically independent city state.

Between c. 1945 and 1835 BCE Assur was home to a large trading community, which transported Afghan tin and Babylonian woollen fabrics to its northwestern neighbours in modern-day eastern Turkey in return for highly valuable silver and gold.[3] This copper-rich area was an important centre for the manufacture of tin bronze, commercially and technologically the most important metal of the age. But tin (which comprised some 10 percent of the alloy) was a rare commodity, originating far to the east, whose value could double or triple between Assur and its final destination. Several hundred Assyrian traders set up residence in the north—six weeks arduous journey by donkey caravan—sometimes settling down with local women, sometimes returning regularly to their families in Assur. They organised themselves into a trade association, or *kārum*, with headquarters at Kaneš, near modern-day Kayseri, and twenty or so further trading posts. This *kārum* negotiated with the local authorities, settled disputes, and oversaw the pooling of resources. Senior family members, women among them, stayed home in Assur to organise finances, supported by the city assembly, the temple of the city god Aššur, and the king. Most, if not all, of the Assyrian traders—male and female—were literate and numerate, using a reduced form of the cuneiform script to strike deals with business partners, write to family members, and record the decisions of the *kārum*, in a local dialect of Akkadian now known as Old Assyrian. They stored their tablets in the ground-floor strong rooms of their Kaneš houses, which were abandoned suddenly during destructive political conflicts in about 1830; over twenty thousand tablets have been archaeologically recovered from the site since the 1920s. There must have been similar archives in Assur but only isolated tablets have so far been found.

The Assyrian tin trade was revived in the late nineteenth century by king Samsi-Addu of Assur (r. 1809–1776 BCE), who by 1796 had united upper Mesopotamia into one triangular kingdom, as far west as the Euphrates and Khabur rivers. He himself made his court at Šubat-Enlil (modern Tell Leilan) at the kingdom's northernmost point and installed his sons to rule from its eastern and western tips: Ekallātum, just south of Assur on the Tigris, and Mari on the Euphrates. Mari had been a major urban centre for a millennium already, controlling the Euphrates trade that linked Babylonia with the Mediterranean. Before Samsi-Addu's conquest it had been the capital of an empire with strong ties to powerful Ḥalab (Aleppo) to the west. Just twenty years later it was restored to its previous ruling family under the kingship of Zimri-Lim—who was to reign for only seven years before Mari was captured, pillaged, and destroyed by his former ally Hammurabi of Babylon in 1760 BCE.

At the heart of Mari was a vast palace, constructed over centuries, comprising nearly three hundred ground-floor rooms over its 2.5 hectares plus an upper storey, with walls up to four metres thick and eight metres high. The palace was formally divided into royal, public space (including throne room and reception areas), sacred space (two chapels), private apartments for the women of the household, and service quarters (kitchens, offices, and storage facilities). Over fifteen thousand excavated tablets from the reigns of both Samsi-Addu's son and Zimri-Lim (c. 1785–1760), left *in situ* by Hammurabi's plundering army, shed fascinating light on the day-to-day running of the palace, the administration of the kingdom, and its relationships with neighbouring polities as well as some further afield. More recently, tablets have been found in the so-called governors' palace some two hundred metres to the east of the main palace, which in Zimri-Lim's time was inhabited by the diviner Asqudum, a senior member of the royal family. Unlike the Assur tablets, the Mari documents were written by professional scribes, trained in the Old Babylonian (more specifically, Ešnunan) tradition.[4]

Altogether twenty mathematical tablets are known from the Old Assyrian period, from Mari, Kaneš, and Assur (tables B.12–13). Those from Kaneš and Assur are predominantly calculations on round tablets, while those that have been published from Mari are for the most part copies of standard Old Babylonian–style elementary lists (see §§4.1–2).

After several centuries of eclipse, Assur finally regained its status as an important political power in the middle of the fourteenth century when, under king Aššur-uballiṭ (r. 1363–1328), it began to expand into a territorial state recognised by Egypt and challenged by Babylonia (see §6.1). Three centuries of prosperity were built upon the protection of trade routes, agricultural production, and military aggression. Archives and libraries from the renovated temples and palaces of Assur, as well as from the homes of senior Assyrian officials, reveal many details of Middle Assyrian economic, social, and intellectual life. Law codes and legal documents show a heavily indebted society whose moral conduct was closely controlled by the crown. The waning of Assyrian power in the early eleventh century is closely related to the migration of Aramean nomads into Assyria from the west—who brought with them alphabetic writing and thus unwittingly initiated the slow death of cuneiform culture over the next thousand years.

The *Epic of Tukulti-Ninurta*, celebrating that Assyrian king's victory over Babylonia in 1225 BCE, describes the capture of tablets, including 'scribal lore' and 'works of the *āšipu* scholars' (see §5.4).[5] Indeed, many Middle Assyrian scholarly tablets end with colophons describing them as copies from Babylonian originals; others are in Babylonian (not Assyrian) script. The sole known Middle Assyrian mathematical tablet is a complete series of multiplication tables, ending with a table of squares. It is clearly based on an Old Babylonian original but dates to c. 1178–1076 (table B.14).

Following a period of decline in the early first millennium, Assyria's fortunes revived in the 880s through efficient militarisation. The king led annual battle campaigns in the name of the god Aššur in which every male subject was liable to serve, and all senior political officials held military rank. First the former city state expanded northward and westward to re-capture the land it had lost as far as the Euphrates in the west and the Zagros mountains to the north and east: this was considered to be the true 'land of Aššur'. By 700 BCE expansion into 'lands under the yoke of Aššur' had taken the borders of empire all the way down the eastern coast of the Mediterranean, the Babylonian gulf coast, and almost to Lakes Van and Urmia in the north. Even Egypt fell briefly to Assyrian rule early in the seventh century. Lands were annexed only rarely through immediate con-quest: more usually, they became semi-autonomous vassal states with an-nual tribute obligations. Only when tribute was not forthcoming were local rulers replaced by puppet regimes or territories placed under the di-rect control of an Assyrian governor. Stability was managed through the displacement of many thousands of people, uprooted from their home-lands on the peripheries of empire and resettled in the heartland or on other peripheries as craftsmen, agricultural labourers, or conscripts.

In about 878 BCE king Aššur-nāṣirpal (r. 883–859) moved the Assyrian capital from Assur to Kalḫu (modern Nimrud) some fifty kilometres to the north. Kalḫu had been an important provincial centre in the Middle Assyr-ian period, but Aššur-nāṣirpal had it completely rebuilt over the next fif-teen years, using labour and resources amassed from recent military cam-paigns. At its heart were an imposing and lavishly decorated royal palace, several temples, and a ziggurat. The city walls ran for eight kilometres, enclosing an area of some 360 hectares. But Assur remained the religious heart of the empire, as well as an economically important city; the kings continued to be buried in the Old Palace there too.

The capital moved again twice: first, briefly in the late eighth century, when king Sargon II (r. 721–705) founded a completely new settlement about forty kilometres to the north, almost as large as Kalḫu, which he named Fort Sargon (Dūr-Šarrukēn, modern Khorsabad). After his unpropitious death in battle the new city was considered ill-omened, so the capital was transferred once again: this time back south to Nineveh, some twenty-five kilometres north of Kalḫu, on a site which had been an important settlement since prehistoric times. The modern city of Mosul now overlooks it from the opposite bank of the Tigris. Once again, temples, palaces, and city walls were constructed anew—this time over seven square kilometres. From Nineveh kings Sennacherib (r. 704–681), Esarhaddon (r. 680–669), and Aššurbanipal (r. 668–627?) ruled almost the entire Middle East, accruing a powerful retinue of generals and scholars to support them. But Assyria's collapse, when it came, was swift: the apparatus of empire began to disintegrate in the 630s and by 612 the Median and Babylonian armies, who had been fighting for independence for a decade, were at the gates of the capital. They laid siege for three months before Nineveh fell and Assyria was destroyed forever.

Some five thousand scholarly tablets from Sennacherib's 'Palace Without Rival' were amongst the first written records to be excavated in modern times, by Austen Henry Layard in the 1840s CE. Now held in the British Museum, they remain amongst the most important sources for Mesopotamian intellectual history today. Just five Neo-Assyrian mathematical tablets are known, however, from three places: Assur and Nineveh, as might be expected, and more tantalisingly the distant provincial settlement of Huzirīna (modern Sultantepe) near Sanlıurfa in southern Turkey (table B.14). Those from Assur and Huzirīna are lists and tables, while two of the three Ninevite fragments appear to be partial duplicates of an Old Babylonian–style word problem.[6]

5.2 PALATIAL AND MERCANTILE NUMERACY IN EARLY ASSYRIA

The arithmetical tables from Mari are no different from those known from Old Babylonian sites—the shape and size of the tablets are the same, the format of the texts, and even the one surviving colophon, on a Type III multiplication table for 22;30. Tablet M 10681 (table B.12) reads, 'Long tablet of [. . .]-zan, completed on Month Dagan, day 5.' The month name is specific to Mari—in fact it is the name of one of the most important deities in the Mari pantheon—proving that that tablet must have been written locally and not imported from the south. Sadly, not enough survives of the student's name to be able to identify him or her with any certainty.

It is very difficult to determine the circumstances in which the mathematical tablets were written. The three metrological lists come from the diviner Asqudum's residence, but the exact circumstances of their find (such as their location in the building, and whether other school tablets were discovered with them) are currently unavailable. The remaining mathematical tablets come from the main palace; similarly, if there were any other school tablets found there, they have not yet been published. The find contexts of the palace tablets tell us very little—those whose exact findspots are known mostly came from the small rooms around the court in front of the throne room, where Hammurabi's soldiers left them as they were pillaging the royal archives during the sack of Mari.[7] But we can be sure that the 'schoolrooms' identified by the excavators, photographs of which are sometimes reproduced in the introductory literature on the history of mathematics, were nothing of the sort (figure 5.2). More recently Jean-Claude Margueron has pointed out that the rooms concerned have no natural light source, making them impossible to read or write in. The 'students' benches' are in fact long tables and the previously unexplained pits in the floor are storage bins: the 'school', in short, is a secure warehousing area with extra-thick walls for insulation.[8] Yet there must have been *some* scribal education within the palace: in particular, it is difficult to imagine why rough calculations should have made their way into the palace from outside, when they were only meant to be ephemerally preserved.

But it is these calculations that reveal the true nature of mathematics at Mari. As Christine Proust has shown, M 7857 in particular reveals the underlying decimal thinking of the Mari scribes (figure 5.3). On one side of the very roughly made tablet the same series of numbers is written out twice, one in purely sexagesimal place value notation and once in predominantly decimal absolute notation, but with some sexagesimal influence. The third column below shows the modern decimal equivalents for convenience:

[1 3]9	99	99
[1]4 51	8 hundred 1 31	981
1 13 39	8 thousand 19	8019
20 02 51	7 great 2 thousand 1 hundred 1 11	72,171
3 00 25ᶦ 39	1 sixty 4 great 9	649,539
	[thousand 5 hundred 39]	

On the other side the series is reversed. The first two lines use a pure centesimal place value notation which is identical to the sexagesimal one except for the addition of six to nine diagonal wedges to write 60, 70, 80, and 90 (also seen on the first line of the other side). The remaining lines revert to the absolute decimal system with sexagesimal influence, as before. The last line gives the total of all the previous lines, namely 730,719:

64 95 [3]9 [(grains?)]
7 21 71 ears of grain
[8 thousand] 19 ants

Figure 5.2 A storeroom (locus 24) of the Old Babylonian palace at Mari, erroneously identified as a schoolroom and sometimes reproduced in the secondary literature as such. (Parrot 1956, pl. XLII, top left, courtesy of Librarie Paul Geuthner.)

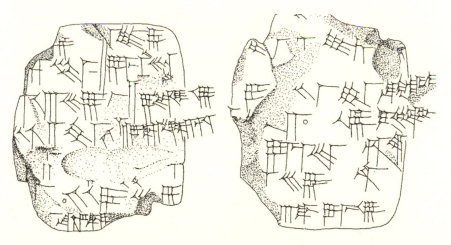

Figure 5.3 Mathematical calculation in decimal notation, from Old Babylonian Mari; note the signs for 90 and 70 on the reverse. (M 7857. Guichard 1997, 315, courtesy of Éditions Recherche sur les Civilisations/ERC (CULTURESFRANCE), ISBN-10: 2-86538-259-1.)

8 hundred 1 31 birds
1 39 men (?)
[1 sixty 1]3 great 7 hundred 19.

This of course is highly reminiscent of the famous Problem 79 in the roughly contemporary Rhind Papyrus from Middle Kingdom Egypt:

7 houses
49 cats
343 mice
2401 [ears of] grain
16,807 *hekat* [of grain]
Total 19,607.[9]

The fact that the Mari version lists the items in descending order and is based on powers of 9, not 7, while the objects themselves are different, suggests that this was not a direct loan in one direction or the other. The first side of M 7857 shows that the series was calculated by starting with 99, and that it was recorded in base 60 first and then converted to decimals afterwards: the ruling down the middle of the tablet separating the sexagesimal and decimal columns is broken in lines three and four, and there are extensive erasures and rewritings.

There are no parallels known from the Old Babylonian mathematical corpus—the eight published OB tables of powers are all for the sexagesimally regular numbers, usually squares, whereas 7, 9, and 11 are all coprime to 60. Further, the OB lists are purely numerical, with no descriptive content.[10] But there is a related tablet from Mari, M 8613 (table B.12), set out in tabular form with a horizontal ruling between each line:

1 grain increases by 1 grain, so that
2 grains on the first day
4 grains, 2nd day
8 grains, 3rd day
16 grains, 4th day
a 6th and 2 grains, 5th day
⅓ shekel 4 grains, 6th day
[⅔ shekel 8 grains,] 7th [day]

The rest of the tablet is missing until:

[6 hundred 54] talents 15 minas [8 shekels 16 grains, 28th] day
1 thousand 3 hundred 48 talents 30 [minas 16] a sixth 2 grains, 29th day
2 thousand 7 (hundred) 37 talents ½ mina 2⅓ shekels 4 grains, 30th day.

And this too—a geometrical progression 2^n—is reminiscent of many other similar mathematical riddles known from all over Asia well into the Middle Ages, often attached to the story of a king giving out grain whose quantity doubles for every day of the month (as perhaps here) or for every square on a chess board. Mathematical riddles, unlike school problems, often require a special trick or insight that will enable them to be solved easily. In the doubling grain riddle, the first trick is on the king who does not realise that he will giving out nearly three thousand metric tonnes by the end of the month; the second trick is to know that the sum of a geometric progression—the total grain that the king gives out over the month—can be found by working out an extra term in the series (in this case the grain for the 31st day) and subtracting 1 from it.[11] Alternatively the trick can be—as in the traditional British 'St Ives' version of the first of the Mari riddles—simply to listen carefully to the question posed. In both cases the solution requires something other than straightforward arithmetical labour.

Scribal administrators at Mari also occasionally wrote numbers with centesimal place value—and the way they did so helps to explain its use in the mathematical calculations too. For instance the total given at the bottom of one account of personnel is given as '1 hundred 1 23', but on the edge it is approximated as '1 85', where '1' represents a hundred and the 80 is written with 8 ten-signs. Other instances of the centesimal place value system occur as rough calculations on the margins of finished administrative tablets, or else on rough drafts of documents that were not meant to survive.[12] Just as the scribes of the Ur III period 250 years before had started to use the sexagesimal place value system as a means of calculating but were prohibited from writing it in their finished documents (§3.4), so too, it seems, the Mari scribes were supposed to use the discrete counting system, with word signs for 'thousand', 'hundred', and some sexagesimalisation—but preferred to use a centesimal adaptation of the sexagesimal for the calculations themselves. Perhaps that was the function of the rough tablet M 7857 too: the student, given the numbers of ants, birds, men, and so forth to sum sexagesimally, preferred to convert them to decimal (on the obverse of the tablet) and then to sum them (on the reverse) in an informal mixture of absolute decimal and centesimal place value sytems. Maybe he converted the answer back to sexagesimal numeration—3 18 15 19—on another rough tablet afterwards or maybe he left it in absolute decimal notation.

The merchants of Assur and Kaneš had no such crises of notation or base. As far as we can tell, they did not attempt to replicate the formal mathematical education of the southern scribes. Not being professional administrators themselves, but wanting literacy and numeracy solely for practical, commercial purposes, they had no need for the sense of professionalism and justice instilled by the supra-utilitarian curricula of the Old Babylonian scribal schools (§4.4). Fewer than twenty identifiable school tablets of any sort are known so far from Kaneš and Assur, about half of which are mathematical or mathematically related (table B.13).[13]

There are no purely arithmetical lists or tables surviving, and just one metrological list of weights (from Kaneš), running originally from 1 shekel to perhaps 100 talents, although much is now lost. Reflecting contemporary Assyrian accounting practice, the weight system is the classic Mesopotamian one, with 180 grains to a shekel and 60 shekels to a mina, 60 minas to a talent (table A.3). The larger values of the talent may have been written sexagesimally or decimally; that part of the tablet is missing so it is impossible to tell. Writing errors, duplicated lines, and an elementary vocabulary of metals and stones on the reverse all show that it is a student exercise and not a reference list.[14] Similarly, the mathematical problems must be school exercises and not a merchant's rough jottings—which, as informal notes, memos, and lists, account for about 20–30 percent of the tablets found in private archives in Kaneš.[15] The calculations are all on round tablets—which is one of the two typical shapes for Old Babylonian rough school work (§4.2) but rarely found in the Kaneš archives—and three of them, one from Kaneš, two from Assur, are exact duplicates of each other:

> 10 minas of pure gold at 5½ shekels (per shekel): its silver is 55 minas.[16]

Like all the Old Assyrian exercises, this problem is about calculating the cost of a precious metal in silver, given the weight of silver that is equivalent in value to a shekel of the metal concerned. Other problems concern choice gold at 15 shekels and red gold at 3½ shekels. In each case the method of solution is simply to multiply the weight of the metal by its value in silver. A further complication is added in Ass 13058k, a problem about copper, whose price is calculated by the 30-mina 'top-load' of a pack donkey (figure 5.4):

> 2 talents 15 minas of copper at ½ mina 5 (shekels) per top-load: its silver is 2 minas 6½ shekels 15 grains.

Both silver and gold were typical high-value means by which the profits of the trading enterprise could be returned to Assur without using large and expensive donkey caravans. Copper, by contrast, was traded extensively in and around Kaneš but was rarely transported back to Assur. Real-life silver exchange rates for copper ranged from 75:1 to 45:1 (here 70:1), while a unique partnership investment contract specifies a silver-gold exchange rate of 4:1. First at least thirteen shareholders are named, with the

Figure 5.4 An exercise in exchange rates from Old Assyrian Assur. (Ass 13058k. Donbaz 1985, 16 top right, courtesy of *Akkadica*.)

weight of each of their investments (between 1 and 4 minas of gold). Before a closing list of seven witnesses the contract states:[17]

> In all 30 minas of gold, Amur-Ištar's partnership capital. Reckoned from the year named after Susaya he will trade for 12 years. He will enjoy a third of the profit. He will be responsible for a third. Whoever gets his money back before the completion of his term must take the silver at the rate of 4 minas of silver for 1 mina of gold. He will not get any of the profit.

Thus the few surviving Old Assyrian witnesses of mathematical training faithfully mirror a primary concern of the merchants themselves: the weights and values of the metals they dealt in. (There are no problems, by contrast, on counting or measuring the cloth they traded too.) The numeration system used looks sexagesimal, as all the values in the problems are small; but in their working lives the merchants, like the Mari scribes, used both the absolute decimal numeration, with sexagesimal notation for figures between 60 and 99, and the decimal place value system, with signs for 60, 70, 80, and 90 just as at Mari.[18]

The extant Old Babylonian problems on prices, by contrast, are so divorced from their original context that they do not even associate commodities with the prices. Rather, they are pretexts for the manipulation of unknown quantities and sexagesimally irregular (that is, co-prime) numbers.

For instance, the third problem on VAT 7530, probably from Uruk, reads in its entirety:[19]

2 23	7 minas each and 11 minas each
1 31	13 minas each and 14 minas each
1 17	1 shekel 11 grains and a 3rd of a grain of silver.
1 11 30	Let the silver rise or fall so that the exchange rates are equal.

Unfortunately the tablet gives no method for solving the problem, or even what the answer was expected to be; nevertheless it is clear that it has little in common with the simple exercises from Assur and Kaneš, which are firmly grounded in the immediate needs of trainee merchants.[20]

5.3 COUNTING HEADS, MARKING TIME: QUANTIFICATIONS IN ROYAL INSCRIPTIONS AND RECORDS

From Early Dynastic times all Mesopotamian kings recorded their upkeep and construction of religious buildings in royal inscriptions, buried in the foundations of the structures they sponsored or displayed prominently in public spaces. In the late twelfth century such building inscriptions began to accrue year-by-year official records of the Assyrian king's military triumphs as well as his domestic building projects. They were revised annually, with the latest year getting the most prominent treatment and previous years being edited down to highlight the most salient events or to exclude embarrassing ones. From the reign of Tiglath-pileser I (r. 1114–1076) to the middle of the reign of Aššurbanipal (c. 640) the annals are the most important sources for Assyrian political history—and are full of exact and approximate quantifications that are revealing of the Assyrian court scribes' political uses of numeracy.[21]

The Middle Assyrian king Shalmaneser I (r. 1273–1244) marked the growing prosperity of his kingdom by extensive renovations to the temple of the city god Aššur, leaving five different versions of a foundation inscription within the temple precinct and elsewhere in the city. One of them starts and ends:

> When Eḫursağ-kurkura, the ancient temple, which Ušpiya, my (fore)father, Aššur's high priest, had previously built and which had become dilapidated, and which Erišum, my (fore)father, Aššur's high priest, rebuilt: 2 sixties 39 years passed and that temple returned to dilapidation, and Samsi-Addu, also my (fore)father, Aššur's high priest rebuilt it. 9 sixties 40 years passed and the temple and its sanctuary were destroyed by fire. . . .

I deposited my monumental inscriptions and foundation documents. He who alters my inscriptions and my name: may Aššur, my lord, overturn his kingship and eradicate his name and his seed from the land.[22]

When, nearly six centuries later, king Esarhaddon commissioned the rebuilding of the temple in 679 BCE, he also acknowledged the reconstruction work of earlier kings:

> Aššur's ancient temple, which Ušpiya my (fore)father, Aššur's high-priest, had previously built, and which had become dilapidated, and which Erišum, son of Ilu-šumma, my (fore)father, Aššur's high priest rebuilt: 2 sixties 6 years passed and it returned to dilapidation, and Samsi-Addu, son of Ilu-kabkabbi, my (fore)father, Aššur's high priest, rebuilt it. 7 sixties 14 years passed and that temple was destroyed by fire. Shalmaneser, son of Adad-nērārī, my (fore)father, Aššur's high priest, rebuilt it. 9 sixties 40 years passed and the interior shrine, dwelling of Aššur my lord, the summit building, the shrine of the *kūbu* images, the astral deities' shrine, the god Ea's shrine, had become worn out, dilapidated, and old.[23]

Esarhaddon's history of the temple is an almost word-for-word copy of Shalmaneser's; one surviving copy even imitates it in deliberately old-fashioned script. Only two significant changes have been made: the addition of his predecessors' parentage and the alteration of the time-spans between them (table 5.1).

Assyrian court scribes were able to assign exact spans of time to the past thanks to their system of naming each year after a *līmu*, or eponym official, who was chosen annually according to his status in the court hierarchy. The date at the bottom of one copy of Esarhaddon's foundation inscription, for instance, reads 'Month IV, day 19. Eponymy of Itti-Adad-annu' (679 BCE). Lists of these *līmus* were maintained for administrative purposes, and in addition *līmu* officials often commemorated their year of office by setting up a stele by the side of the monumental roadway into the city of Assur. *Līmu* lists are completely reconstructible as far back as 910 BCE and fragmentarily attested for two hundred years before that, while the so-called Assyrian King List records the names and lengths of reign of kings of Assur going right back to the early second millennium and the days of '17 kings who lived in tents'.[24] The King List was compiled wherever possible from *līmu* lists but also from other historical sources; for instance, it describes one group of early rulers as 'total of 6 kings [whose names occur on (?)] bricks, whose eponyms are destroyed'.

Esarhaddon's chronology (126 years, 434 years) seems to be based on the Assyrian King List, composed, on present evidence, some three centuries after Shalmaneser's reign. It gives the total time-span between Samsi-Addu

Table 5.1
Time-spans Between Rebuildings of Aššur's Temple According to Assyrian Sources

Time-span	Shalmaneser's Inscription	Esarhaddon's Inscription	Assyrian King List	Modern Maximum	Modern Minimum
Ušpiya (dates unknown)– Erišum (r. 1939–1900)	uncounted	uncounted	uncounted	—	—
Erišum–Samsi-Addu (r. 1813–1781)	159	126	[5 reigns]	158	87
Samsi-Addu– Shalmaneser (r. 1273–1244)	580	434	421 [+ 2 reigns]	569	508
Shalmaneser– Esarhaddon (r. 680–669: 679)	—	580	523 (+ 42)	594	565

I and Shalmaneser I as 421 years and 1 month plus the reigns of two kings whose lengths are missing from the extant exemplars. Esarhaddon's scholars, then, seem to have put more weight on their own evidence-based quantifications of the past than on the authority of ancient writings and were capable of rewriting the historical record accordingly. Ironically, according to modern chronology Shalmaneser's time-spans (159 years, 580 years) are much closer to modern consensus (158 years, 569 years) than Esarhaddon's revisions.

No temporal quantifications are attached to the temple's supposed founder, Ušpiya, who has left no traces whatsoever in the archaeological or historical record except as the penultimate entry in the enumeration of the '17 kings who lived in tents' at the start of the Assyrian King List. There are surviving building inscriptions for both Erišum and Samsi-Addu in Aššur's temple; neither of them mentions any royal builder earlier than Erišum himself.[25] Shalmaneser's scholars appear to have added him to the beginning of the temple's architectural genealogy, as an attested king from the primordial time before the quantifiable past, in order to imbue the temple with deep antiquity.

Descriptions of military campaigns, the other primary focus of Assyrian royal inscriptions, used enumeration with equal prominence. Sargon II (r. 721–705), for instance, in the eighth year of his reign felt it necessary to punish one of his vassals, Urzana, for aiding the enemy Urartians, whom he had just defeated. Urzana was king of Muṣaṣir, now the town of Mirga Sur in Iraqi Kurdistan, sandwiched between Assyria to the southwest and the kingdom of Urarṭu to the northeast. Urarṭu had been a thorn in Assyria's side since the ninth century but the current conflict had been brewing since 716. Muṣaṣir was important both as the cult centre of the Urartian god Ḫaldi and as the site of royal coronations. Narratives of the capture and plunder of Muṣaṣir are known from three different types of inscription. The lengthiest is from a celebratory letter written on Sargon's behalf to the god Aššur, recording the events of his eighth regnal year:

> Because Urzana the king, their prince, did not fear Aššur's name, slandered my lordship and forgot that he was in my servitude, I planned to deport the people of that city and to remove the god Ḫaldi, protector of the land of Urarṭu. Triumphantly I made him dwell outside his city gate and I deported his wife, his sons, his daughters, his people, the offspring of his father's house. I counted with (them) 6 thousand 1 hundred 10 people, 12 mules, 3 hundred 1 20 donkeys, 5 hundred 25 oxen, 1 thousand 2 hundred 35 rams, and I had them enter the walls of my military camp. I triumphantly entered [the city of] Muṣaṣir, Ḫaldi's dwelling, and I lived like a lord in the palace, Urzana's residence. I broke the protective seal on the heaped up [storeroom] which was piled high with abundant treasure. I plundered [. . .] minas of gold, 1 hundred 67 talents 2½ minas of silver, white bronze, tin, carnelian, lapis lazuli, agate, a selection of jewels in vast number, [*there follows a very long itemised list of booty*], the property of his palace as well, and heaped up his belongings. I sent my magnates and soldiers to the temple of Ḫaldi. I plundered his god Ḫaldi and his goddess Bagbartu, as well as numerous belongings of his temple, as much as there was: [. . .]+3 talents 3 minas of gold, 1 hundred 62 talents 20 minas minus 6 shekels of silver, 3 thousand 6 hundred talents of bronze ingots, [*and another long itemised list of treasure*], together with the countless property of his land.[26]

A shorter version is included with the annals carved into the middle band of the reliefs used to decorate Sargon's new palace at Dūr-Šarrukēn ('Fort Sargon') just ten miles north of Nineveh, which was commissioned in 717 BCE, completed in 707, and abandoned on his inauspicious death in 705:[27]

> (Because of) Urzana, of the city of Muṣaṣir, who broke the oath of Aššur and Marduk by writing to Ursaya, of the land of Urarṭu, king Aššur my lord inspired trust in me. With only my 1 chariot and 1 thousand of my very wild cavalrymen and capable infantrymen I traversed the very difficult mountains of Šiak, Ardikši, Ulayu, Alluria, crossing good fields by riding the horses and difficult places on foot. When Urzana of Muṣaṣir heard that I was coming he fled like a bird

TABLE 5.2
Booty from Muṣaṣir According to Neo-Assyrian Inscriptions

	People	Donkeys & Mules	Cattle	Sheep	Silver from the Palace
Letter to Aššur	6110	384	525	1235	167 talents 2½ minas
Annals	6170	692	920	100,235	160 talents 2½ minas

and went up into the inaccessible mountain. I brought out of Muṣaṣir, Ḫaldi's dwelling, Urzana's wife, his sons, his daughters, 6 thousand 1 hundred 1 10 people, 6 hundred 1 32 mules and donkeys, 9 hundred 20 oxen, 1 hundred thousand 2 hundred 35 sheep. I plundered 34 talents 18 minas of gold, 1 hundred 60 talents 2½ minas of [silver], white bronze, tin, a selection of jewels in vast number [. . .] that he [. . .] uncountably many garments of multicoloured fabric and linen [. . .] with [. . .] talents 3 minas of gold 1 [hundred] 62 talents [. . .] minas of silver [. . .] countless bronze and iron equipment [. . .] as well as a bronze bull, bronze cow, and bronze calf. [. . .]

Clearly neither inscription is a direct copy of the other—the circumstantial details are too various. Nevertheless, they agree on the fact that large numbers of people, animals, and precious metals were taken as booty in the capture of Muṣaṣir. But not one set of figures, where they survive in their entirety, is the same from inscription to inscription (table 5.2). Some discrepancies can be accounted for by the simple omission or erroneous inclusion of single wedges or signs: in the people-count an extra vertical wedge bumps up the total by 60 in the annals, while an additional ME 'hundred' sign (just a vertical plus horizontal) gives an astronomically large flock of sheep. The 7 seems to have been omitted from the annalistic account of silver. No such scribal slips can account for the increased numbers of larger animals though. It may well be that the extra 308 donkeys and 395 cattle were added by taking into account livestock captured from surrounding districts too.

Where do figures in the annals and royal inscriptions come from? The first clue is in the depiction of the sacking of Muṣaṣir, from a relief elsewhere in Sargon's palace (figure 5.5).[28] The image is identifiable by a caption reading 'I . . . and conquered the town of Muṣaṣir'. There is fighting on the roof of the temple, decorated with pillars and conical motifs. Outside it are two statues of spear-carriers, a cow and calf, and two cauldrons, as described in the *Letter to Aššur*. To the top left, an Assyrian official seated on a stool and footstool dictates to two standing scribes, on the roof of a fortified building. The scribe in front holds a tablet and stylus, the one

Figure 5.5 The sack of Muṣaṣir, as shown on Flandin's copy of a stone relief from Fort Sargon (now largely lost). (Albenda 1986, pl. 133, courtesy of Éditions Recherche sur les Civilisations/ERC (CULTURESFRANCE), ISBN-10: 2-86538-152-8.)

behind a parchment or papyrus.[29] Further to the bottom left, two soldiers chop up a statue, while others weigh booty in a balance—presumably to be quantified and recorded for posterity.

There are many further images of Neo-Assyrian scribes in the field, always in pairs. Typically a clean-shaven scribe holds a papyrus or parchment, next to a bearded one with a hinged wooden writing board, which opened out to reveal a waxed writing surface that could easily be erased for re-use (though other combinations of scribes are also known). The scribes are more often depicted pointing their styluses at the plunder or bodies they are counting than actually using them to write with (figure 5.6).

Deportees and booty were detailed in both administrative records and letters on cuneiform tablets. Here the governor of the province of Mazamua (now Suleimaniya in northern Iraq) reports to king Sargon on the plunder from a recently captured town:

> To the king, my lord: your servant Šarru-ēmuranni. Good health to the king, my lord!
>
> On the 27th day, at dawn, we opened the treasury of metal scraps at the entrance to the house in the palace on the terrace. [We] weighed 4 hundred 20 talents of bronze scraps and placed it in the cupbearer's storehouse.[30]

An undated document counts and categorises deportees from Que (Cilicia, the easternmost Turkish coast of the Mediterranean) as dispassionately

Figure 5.6 Two scribes accounting for booty on hinged writing boards, on a stone relief from one of king Aššurbanipal's palaces in Nineveh. (Burrell 28.33. Peltenburg 1991, no. 47, courtesy of Edinburgh University Press.)

and anonymously as if they were livestock—although administrative letters reveal that some Assyrian officials were more compassionate to those in their care than lists such as this suggest:

3 hundred 34 healthy men.
38 boys—5 spans
41 boys—4 spans
40 boys—3 spans
28 weaned boys
25 unweaned boys
Total: 1 hundred 1 12 boys.

3 hundred 49 women.
8 girls—5 spans
22 girls—4 spans
49 girls—3 spans
17 weaned girls
25 unweaned girls
Total: 2 hundred 21 girls.

Grand total: 9 hundred 1 17 people deported from the land of Que.[31]

A span was half a cubit, about twenty-five centimetres.

Like the Old Assyrian documentation, Neo-Assyrian administration used a basically decimal system with the addition of a vertical wedge for 60, though six or more 'ten' signs could also be used, as in the weights of precious metals captured from Muşaşir. The few surviving Neo-Assyrian tables (see §6.2) tend rather to be tabular lists, with rare calculations or summary totals.[32] A large, headed tabular account from Huzirīna with fourteen columns and at least three levels of horizontal calculation and two vertical ones shows that Neo-Assyrian scribes could and did use tabulation very effectively;[33] but it is likely that they mostly used waxed writing boards to do so. Administrative tablets and letters also refer to writing boards carrying similar information—very few of which survive and none with their writing surfaces intact—while much must also have been written in alphabetic Aramaic on perishable materials such as papyrus or parchment.[34] Nevertheless enough survives to suggest that Assyrian institutional record-keeping was as numerate as its Babylonian counterpart.

5.4 *ARÛ*: NUMBER MANIPULATION IN NEO-ASSYRIAN SCHOLARSHIP

What of mathematics as an *intellectual* activity in Assyria? Where, if at all, did it belong in the lives and minds of the great scholars who guided the decisions of the Assyrian kings and shaped the destiny of the empire?

The Assyrian court comprised not only political and military officials and advisers but also men of learning, *ummānu*, whose fields of expertise were vast, ranging from religion to science, from medicine to magic. As the rulers were particularly concerned with determining the gods' will by means of terrestrial and celestial omens, there were especially strong motives to improve the predictability of key celestial events. Lunar and solar eclipses were particularly portentous for king and country. The scholars of celestial omens held the title *ţupšar Enūma Anu Ellil* ('scribe of "When the gods Anu and Ellil"'), after the first line of the major compilation of omens, which ran to seventy or more tablets and is so vast that it still has not been published in its entirety. There were scholars, *bārû*, who read omens from the innards of sacrificed rams, and scholars, *asû* and *āšipu*, who advised on the health of the royal family and prescribed medicines and rituals for their well-being. The latter is sometimes translated inappropriately as 'exorcist' or 'sorceror'. There were also scholars, *kalû*, who calmed and propitiated the gods at times of cultic danger and attempted to avert the ill effects of evil omens. One surviving personnel list names seven scribes of *Enūma Anu Ellil*, nine *āšipu*s, five *bārû*s, nine *asû*s, and six *kalû*s, as well as three bird diviners, three Egyptian magicians, and three Egyptian scribes.[35] Hundreds of extant letters and reports from such scholars to the Assyrian kings, especially Esarhaddon and Aššurbanipal, enable us to see how these men actually operated—how they used the scholarly works they knew and owned to advise and influence the king on matters of person and state.[36] These elite literati depended almost entirely, it seems, on royal patronage to keep them in housing and employment, creating a highly competitive atmosphere which drove intellectual innovation as they fought amongst themselves for preferment.

Almost all the surviving Middle and Neo-Assyrian mathematical tablets can be attributed archaeologically or by their colophons to particular owners or social contexts.

In the provincial town of Huzirīna, one Qurdi-Nergal of the Nūr-Šamaš family began his career as a 'junior apprentice scribe' in about 701 BCE, eventually becoming chief temple administrator of the god Zababa and the father of at least one son, Mušallim-Bau. In 1951 British excavators found some four hundred scholarly tablets and fragments, many of which had been written or owned by Qurdi-Nergal and his descendants, in a large pile, protected by wine jars, against the outside wall of a house, presumably his former home. The latest date on the tablets was 619 BCE, just a few years before the final collapse of the Assyrian empire and the destruction of nearby Harran in 610 BCE. The tablets, which must have been the remnants of a scholarly library, were predominantly ritual and medical, but also included myths and epics, prayers and incantations, compendia terrestrial and celestial omens, as well as some school tablets.[37] The one mathematical tablet, a list of reciprocals, stands out because it was the only one amongst all of them to have been baked in antiquity. Other peculiarities include the fact that it was a copy of an older, broken tablet—it is peppered with the standard scribal notation *ḫepi* 'broken'—and that many of the numerals in an otherwise unexceptional table are written not as figures but as syllables that help with the pronunciation of the Sumerian. Such syllabic spellings are commonly found in copies of Sumerian literary works from the edges of the cuneiform-writing world.[38]

Those mathematical and metrological lists and tables from Assur that can be exactly provenanced are all linked in some way to Aššur's temple (table B.14). The Middle Assyrian multiplication series, copied (according to its colophon) in about the eleventh century, was excavated from the southwestern end of Aššur's temple—but from a Neo-Assyrian context, with about sixty other Middle Assyrian tablets and at least three hundred Neo-Assyrian ones.[39] They appear to have constituted an institutional library comprising royal decrees and laws, royal rituals, hymns, myths, and incantations, as well as some scholarly tablets, especially omens from sacrificed rams. The library was housed in the rooms around the southwest courtyard, some of whose holdings (including the mathematical tablet) were over four hundred years old at the time of its destruction in 614 BCE.

A small metrological table summarising the relations between length units was written on a tablet found in the largest-known library from Neo-Assyrian Assur: some eight hundred tablets found in a private house about four hundred metres south of Aššur's temple.[40] Unusually, the table occupies just the left-hand side of an otherwise blank tablet, suggesting either that the scribe planned to write more or that it was an *ad hoc* composition on the nearest tablet that came to hand. (Tablets were usually made or selected to fit the text to be written very closely.) The colophons on many of the other tablets show the library to have been written and owned by one Kişir-Aššur, *āšipu* of Aššur's temple, and his nephew Kişir-Nabû. The two men were primarily interested in incantations, prescriptions, and medical ingredients, as one might expect, as well as a lot of other scholarly material (including Sargon II's *Letter to Aššur* discussed in §5.3). One of Kişir-Aššur's tablets, an incantation ritual, is dated to 658 BCE but the library seems to have been in use right up until 614 BCE. A single leaf of an ivory writing board found in or near the house hints that scholars just as administrators used other writing media, all of which have been lost to posterity.[41]

The three tablets from Nineveh all come, as far as is known, from Sennacherib's 'Palace Without Rival', although their exact findspots are unknown. Two are fragmentary versions (not direct copies) of part of a word problem known from Kassite Nippur (HS 245, table B.15). In its Babylonian formulation it essentially uses the measurement of distances between a group of stars which rise heliacally in sequence as the pretext for a problem about sexagesimally irregular (co-prime) numbers:

> I summed 19 from the moon to the Pleiades, 17 from the Pleiades to Orion, 14 from Orion to the Arrow, 11 from the Arrow to the Bow, 9 from the Bow to the Yoke, 7 from the Yoke to Scorpio, 4 from Scorpio to *Antagub*, so that it was 2(*ğeš*) leagues. How far is a god above (another) god? You, when you proceed: sum 19, 17, 14, 13, 11, 9, 7, (and) 4, so that you see 1 21. The reciprocal of 1 21 is 0;00 44 26 40. (*Unfinished*)

However, the context of the most complete of the Assyrian versions, namely on the reverse of a star map, suggests that its original function—dealing with co-primes—had been lost and it had now become primarily celestial in meaning. That manuscript finishes with a colophon saying that it has been 'extracted for reading'. The following line is probably to be restored as '[Tablet of Nabû-zuqup-kēna, son of Marduk]-šumu-iqīša, scribe'.[42] Nabû-zuqup-kēna (c. 760–680) was one of the most influential court scholars of the Neo-Assyrian empire. He was the senior scribe and scholar at Kalḫu during the reigns of Sargon II and Sennacherib and father and grandfather of Nabû-zēru-lēšir and Ištar-šumu-ēreš, the senior scholars of kings Esarhaddon and Aššurbanipal. Colophons on around seventy of his tablets found at Nineveh (which were probably brought there by his descendants when the court moved from Kalḫu) show that he was interested above all in celestial and terrestrial divination of many different kinds.[43]

In sum, the contextual evidence suggests that while the metrological lists and tables reflect Neo-Assyrian practice, the surviving mathematical works all hark back to the Old Babylonian traditions: some are clearly copies or adaptations of much older originals, while at least one had itself been in circulation for half a millennium. They were written or owned by influential scholarly families who were primarily interested in other types of intellectual activity. So why were mathematical tablets kept and copied at all? Were they simply antiquarian curiosities? Evidence from other types of Neo-Assyrian scholarly work suggests that the answer is not simple.

First, there are works that indicate that mathematics was part of the ideal intellectual training, in theory if not in practice. A bilingual dialogue between a teacher and his student (in Sumerian with Akkadian translation), known in several copies from Nineveh, includes the question:

> Do you know multiplication, reciprocals and their inverses, constants, accounts, how to check allocations, to work out labour rates, to divide inheritance shares and to demarcate fields?[44]

But as this composition may itself be an adaptation of an Old Babylonian work—it is certainly reminiscent (see §4.4)—it tells us little about contemporary Assyrian educational practice. Much more useful are two official compositions for king Aššurbanipal, one of Mesopotamia's few self-proclaimed literate kings, who was educated while still crown prince by the scholar Balasî. First, in a survival or revival of the early Mesopotamian ideal of metrological justice (§4.4), Aššurbanipal's coronation hymn begins:

> May eloquence, understanding, truth, and justice be given to him (Aššurbanipal) as a gift! May the people of Assyria buy 30 *kurru* of grain for 1 shekel of silver! May the people of Assyria buy 3 *sūtu* of oil for 1 shekel of silver! May the people of Assyria buy 30 minas of wool for 1 shekel of silver! May the lesser speak, and the greater listen! May the greater speak, and the lesser listen! May concord and peace be established in Assyria! Aššur is king—indeed Aššur is king! Aššurbanipal is the [representative] of Aššur, the creation of his hand.[45]

Aššurbanipal, then, if not his immediate predecessors, not only saw the link between fair measurement and social equality but felt it important enough to declare at the very moment he came to the throne. Second, he even professed a degree of numeracy himself, as a royal inscription proclaims on his behalf:

> I have learnt the skill of Adapa the sage, secret knowledge of the entire scribal craft; I observe and discuss celestial and terrestrial omens in the meetings of scholars; with expert diviners I interpret the series 'If the liver is the mirror of heaven'; I solve difficult reciprocals and multiplications lacking clear solution; I have read elaborate texts in obscure Sumerian and Akkadian which is difficult to interpret; I examine stone inscriptions from the time before the Flood.[46]

Aššurbanipal thus locates arithmetical ability—multiplication and division—amongst the most esoteric and specialised of scholarly specialisms. That mathematics meant simply calculational facility is reflected in the surviving mathematical documents: even Nabû-zuqup-kēna's version of the Middle Babylonian problem boils down to an exercise in manipulating integers which are co-prime to 60 (§5.1).

The manipulation of numbers even had its own term in Neo-Assyrian: *arû*, an Akkadian derivative of the Sumerian word *a-ra₂* ('times', lit. 'steps'), which occurs in every Old Babylonian school multiplication table in phrases such as 2 *a-ra₂* 2 4 '2 steps of 2 (is) 4' (§4.2). But in first-millennium scholarly contexts it had come to mean '(multiplicative) product' and more generally 'calculation'. The semantic range of *arû* was wide: it encompassed arithmetical tables, ideal periodic schemes for the heavenly bodies, and even what we would now consider numerology.[47] For instance, Assyrian *bārûs*, scholars who read the gods' intentions from the configurations of the entrails of sacrificed rams, used a manual called *nişirti bārûti* 'Secrets of Extispicy' to support their interpretations based on omen compendia. It included calculations to help read ominous marks on the various internal organs, and others to determine the period of time for which the ominous message was valid.[48] Most are very opaque to a modern reader; as for instance this extract on lungs and a protrusion of the liver (the caudate lobe) called the Finger:

> You multiply a *la'bu* mark on the Lungs with the front [. . .]. You multiply [. . . by] 3 10, so that an excess Finger [. . .]. [3] 30 steps of 3 (is) 10 30. The excess Finger is 10 30. 10 30 [. . .] 3 Fingers. See the shape of one place: 10 30 times 3 (is) 31 [30 . . . , so that] you see 35. 35 means 3 fingers and a third [. . .].[49]

Or on the duration of such omens' validity:

> If the top of the right Plain of the Finger has 1 Split, you solve the reciprocal of 9, a *la'bu* mark on the *muštašnintu* (*an unidentified part of the liver*), so that [you see] 6 40. [. . . , so that] you see 6 40.

> Multiply 6 40 by 1, the Split, so that you see 6 40, (which is) 0;20 of the [daytime] watch.[50]

The technical terminology of multiplication, division, and announcing results is identical to that found in other types of mathematical word problem.

More transparently, the short Neo-Assyrian celestial compendium MUL. APIN 'Plough star'—which mostly describes changes in the night sky over the year—gives two short instructions for doing basic calculations for oneself.[51] Simple period schemes, describing the ideal movements of the sun, moon, and five visible planets, had first developed in the Old Babylonian period. Their aim was not so much predictive as portentous: real-life deviations from the ideal were considered to be ominous signals of divine unease. In the earliest attested annual timekeeping scheme, the lengths of day and night are said to change linearly between the winter and summer solstices, when the ratio is 2:1, with equality at the spring and autumn equinoxes.[52] Other schemes are found, for instance, in the lengthy celestial omen series *Enūma Anu Ellil*.[53] Qurdi-Nergal and Kişir-Aššur's libraries in Huzirīna and Assur included copies of 'Plough star' Tablet II. In Kalḫu Nabû-zuqup-kēna had copied most or all of *Enūma Anu Ellil* from Babylonian originals over the period 718–701.[54] Well over a hundred Babylonian tablets and writing boards of *Enūma Anu Ellil* were brought to Nineveh along with much other cultural booty in 647 after Aššurbanipal's conquest of Babylonia.[55]

In 712 Nabû-zuqup-kēna also copied or composed a work called *Inam ğišḫur ankia*, which uses learned wordplay and numerical manipulation to speculate on various aspects of 'the plans of heaven and earth'. It includes a list of OB-style mathematical constants (see §4.1) embedded in a passage about Enlil's temple Ekur at Nippur and straightforward period schemes for moonrise and moonset. But there are also rules for transferring the portent of an omen from one day of the month to another. For instance:

> The 22nd day: the 14th day. You multiply 14 and 10. 14 (×) 10 (=) 2 20, (which is) 22 downwards upwards, upwards downwards.[56]

The final operation, called 'downwards upwards, upwards downwards' involves the left-right swapping of the tens and units. It was not a particularly common method of numerological manipulation in Neo-Assyrian times, but was used most famously and effectively by king Esarhaddon to justify the early rebuilding of Babylon, which had been comprehensively sacked by his father Sennacherib in 689 BCE and was supposed to remain in ruins for seventy years:

> Although he had written 1 10 years as his calculation of abandonment (on the Tablet of Destinies), the merciful one, Marduk, soon became

tired in his heart and turned it upside down to order its rebuilding in the 11th year.[57]

Thus the ominous reversal of digits was given divine sanction by the god Marduk himself, and Babylon was functioning again by 676 (but see §7.1). It is also possible that K 2069, a strange table of reciprocals on a base of 1 10 (table B.14), was drawn up in relation to this event.

5.5 CONCLUSIONS

Assyria has had no place in the history of mathematics until now. It is certainly true that there is much less evidence for mathematical activity in Assyria than for Babylonia in general and the Old Babylonian period in particular. But the superabundance of mathematical tablets from the latter time and place is more an outcome of the pedagogical practice of memorisation through repeated writing, and of the modern archaeological discovery of scribal schools through urban excavation, than any objective proof that Old Babylonian scribes were more, or better, mathematicians than their Assyrian counterparts. In early second-millennium Assur, where literacy and numeracy were, exceptionally, not the exclusive preserve of professionals, there was no need at all for a formal education that enculturated as much as it taught competence. It was presumably young members of merchant families who learned the necessary weights and measures, along with the mathematical skills needed to calculate with them in context.

Given the domestic setting of Old Babylonian scribal schools (§4.2) we should not expect to find mathematical tablets in the palace of Mari; indeed in this light it is remarkable that it has yielded any at all. Although the elementary arithmetical tables found there are firmly within the OB tradition, the lack of metrological exercises may indicate their irrelevance to an administrative culture that depended far less on measurement and quantified prediction than its Babylonian counterpart. (On the other hand, the diviner Asqudum's house has so far produced three metrological tables.) Similarly the calculations attest to a lively local culture of riddles quite divorced from the ideologies of scribal self-identity: neither ants nor birds were the objects of bureaucratic management regimes. The tension between the use of sexagesimal numeration in official documentation and decimals in notes and ephemera suggests, however, that Babylonian numerate culture had more status than the local one.

Some of that status seems to have been sustained through the second millennium and into the first, to judge by the OB-style arithmetical tables occasionally found in Assyrian scholarly libraries. Indeed, it is related to a much bigger phenomenon of Assyrian fascination with Babylonian intellectual culture that occasionally resulted in full-scale appropriation through capture of tablets, deportation of scholars, and/or occupation of the whole region. But the exact nature of that relationship is difficult to pinpoint. After a long evidential hiatus, from the mid-eighth century it is clear that some Babylonian scholars were undertaking systematic observation and quantification of ominous celestial phenomena—planetary events, lunar eclipses (§§8.1–2). Elegantly written arithmetical tables are attested in Babylonian libraries from perhaps the seventh century (§7.1). But amongst the over four thousand scholarly tablets written in Babylonian script that were found amongst the royal libraries of Nineveh, just two have any numerate content: a fragmentary copy of part of the 'Secrets of Extispicy' and a short set of instructions and tables on timekeeping.[58] There are no Babylonian copies of 'Plough star' Tablet II (Tablet I gives primarily qualitative descriptions), and none of any astronomical observations. There is no metrology, no arithmetic (apart from the reciprocal table for 1 10), and no word problems (except for one which had been reinterpreted as primarily astronomical). It is not that numerate methods had no place in Assyrian intellectual culture—Nabû-zuqup-kēna's work and the 'Secrets of Extispicy' show that they did. Further, both Babylonian and Assyrian scholars were concerned to understand the meanings of quantification within the gods' portentous communications. Could it be that the Babylonians were too jealous of their carefully collected observational data to allow the Assyrians access to it, even under extreme duress? Or was it not yet sufficiently useful for Aššurbanipal's scholars to have any interest in it?[59] Nevertheless, Neo-Assyrian mathematical activity was not simply a matter of accountancy or antiquarianism. King Aššurbanipal led the numerate endeavour from the very top, as his official inscriptions claim.

CHAPTER SIX

The Later Second Millennium

The collapse of the Old Babylonian kingdom in c. 1600 BCE apparently precipitated the disappearance of school mathematics from the archaeological record, along with many other textual genres. By c. 1400 BCE, however, new technologies had enabled the establishment of stable, long-term, quasi-national polities across the Middle East, and regular formal contact between them, in the Akkadian language. The concomitant geographical expansion of cuneiform culture entailed some spread of sexagesimal numeracy too, as attested on the eastern Mediterranean coast, in Egypt, and in southwest Iran (§6.1). Although little pedagogical mathematics survives from Babylonia itself, there is a wealth of untapped evidence for professional numeracy there. A northern Babylonian dynasty of land surveyors with high social status can be traced from the fourteenth to the eleventh centuries (§6.2). In the city of Nippur, nearly two millennia of quantitative temple management reached a peak of efficiency and sophistication in complex tabular accounting (§6.3), while the period's greatest literary work, *The Epic of Gilgameš*, uses quantification and measurement to powerful effect (§6.4). This wide range of evidence suggests that mathematics itself was still a vital component of Babylonian intellectual life (§6.5).[1]

6.1 BACKGROUND AND EVIDENCE

The kingdom of Babylon had been shrinking since the mid-eighteenth century, through environmental catastrophes, ineffective government, and external political upheavals. By 1600 BCE almost all textual evidence dries up, leaving historians with a Babylonian 'dark age' of nearly two centuries after which a dynasty of Kassite rulers, originally from beyond the Iranian Zagros mountains, are ruling southern Iraq. People of Kassite ethnicity had been living in northern Babylonia since at least the eighteenth century. Their language is attested for the most part only in their personal names, as they adopted much of Babylonian culture, including the Sumerian and Akkadian languages written in cuneiform script. While Assyriologists have often inferred that Kassite social structure was essentially tribal, with clans or 'houses' (Akkadian *bītu*) named after an ancestor, it has recently been demonstrated that these 'houses' are in fact estates, or areas of land named after a former owner.[2] The most substantial administrative archives are from the city of Nippur (§6.2), but there are also published tablets from Ur, Babylon, Dur-Kurigalzu (the new royal city near the confluence of the Tigris and Diyala rivers, founded in c. 1400), and settlements further north.

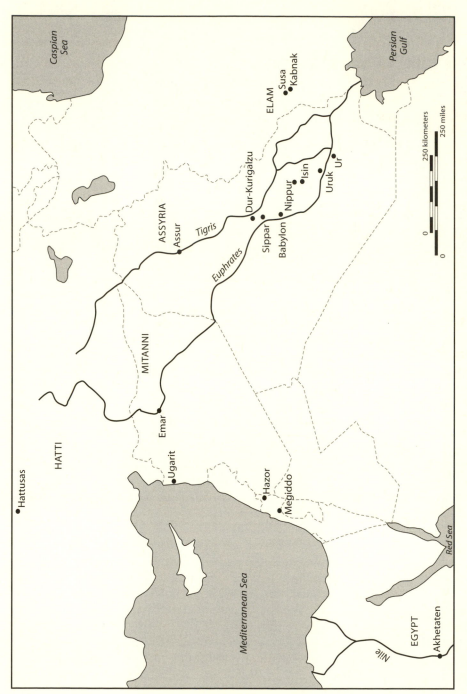

Figure 6.1 The locations mentioned in this chapter, including all known findspots of mathematical tablets from the late second millennium.

Assyria attacked Babylonia in c. 1225 and ruled it for nearly a decade (see §5.1). Eventually the Kassites claimed it back, but a further raid by the Elamite king Šutruk-nahhunte in c. 1160 spelled the end of Kassite dynasty. Both encounters involved the confiscation of large quantities of written material: Babylonian scholarly writings to Assyria (§5.4) and monuments, including records of land grants (*kudurrus*, §6.3) and divine statues, to Elam in southwest Iran. Thereafter, a new Babylonian dynasty, who traced their origins to the city of Isin, took control of the former Kassite empire and even managed to recapture the revered statue of the Babylonian god Marduk from Elam some fifty years later.

The evidence for scholastic mathematics in Kassite Babylonia is sparse to say the least (table B.15), although other scribal exercises are now coming to light.[3] Just four mathematical tablets are thought to date from this time: two erroneously solved or unfinished word problems from Uruk and Nippur; a multi-columnar list of musical and mathematical constants, also from Nippur; and an extract from the standard Ur III–OB reciprocal list (§§3.4, 4.2). This last comes from a private house in late Kassite Babylon which yielded some 150 school exercises in total. The palaeography of scholarly and scholastic writings from this time is still badly understood, while secure archaeological evidence is entirely lacking. Early publications often ascribe a Kassite date to multiplication tables that are now confidently dated to the eighteenth century,[4] while the list of constants just mentioned may be from any time between the fourteenth and the sixth century.[5]

During the mid-second-millennium 'dark age', two major technological innovations occurred which fundamentally reshaped the political landscape. The domestication of the horse and innovations in ship design enabled much faster communications by land and sea. By the time Babylonian scholarship and bureaucracy had reasserted themselves under the new Kassite dynasty, the wider Middle East was a much more accessible place. Large territorial empires, Babylon amongst them, were able to sustain themselves through a network of international trade and diplomacy across the region that lasted, more or less, until the twelfth century. Because the *lingua franca* of this new world order was Akkadian, the concomitant writing technology—clay tablets, cuneiform script, highly standardised pedagogical exercises—spread with it to the eastern coast of the Mediterranean and down into the new Egyptian capital, Akhetaten.

Akhetaten, also known by its modern name Amarna, was founded by the 'heretic' pharaoh Amenophis IV/Akhenaten (r. 1364–1347) to break with Egyptian tradition and institute a religion of the sun. Clandestine digs here in the late nineteenth century uncovered some three hundred cuneiform tablets, which later excavation revealed must have come from a single building now known as the 'Records Office'. Most of those tablets turned out to contain fascinating diplomatic correspondence between Egypt and the other great powers of the region—Babylonia, the Hittite empire of eastern Anatolia, the Hurrian kingdom of northern Syria, and the newly resurgent Assyria. There are also communiqués to and from Egypt's vassals on the eastern Mediterranean coast, as far north as modern Lebanon. Not surprisingly, a good number of scribal training exercises were also found amongst them.[6] A fragment of an Egyptian-Akkadian dictionary, written in cuneiform, was also discovered in a private house in the city (table B.16). Part of it is ruled into columns—and that part correlates Egyptian weights with their Mesopotamian equivalents (now much broken away), running from 1 to 10 shekels in ascending order. It is the only Egyptian-Akkadian bilingual document hitherto known, and contains knowledge that must have been essential to the maintenance of good Egypto-Babylonian diplomatic relations. Almost all of the fourteen surviving letters and inventories of gifts that were exchanged between the Kassite kings Kadašman-Enlil I (r. 1374?–1360) and Burnaburiaš II (r. 1359–1333) and the pharaoh Amenophis IV mention specific weights of gold, silver, and/or lapis lazuli—central commodities in the elaborate ritual of diplomatic gift-giving. For instance, Kadašman-Enlil melodramatically complains, 'The only thing you have sent me as my greeting-gift, in six years, is 30 minas of gold that looked like silver.' When Amenophis presents ebony beds, chairs, and footrests to the Babylonian king to furnish his new palace, he lists the exact weights of the gold (7 minas, 9 shekels) and silver (1 mina, 8½ shekels) with which they are decorated.[7]

One of Egypt's Levantine vassals, the city of Hazor a few miles north of the Sea of Galilee, had been a major centre of Canaanite culture since at least the eighteenth century, when it had been in close commercial and political contact with Mari (§§5.1–2).[8] During excavation of the late second-millennium palace in the mid-1990s a fragment of a four-sided prism was found amongst under-floor rubble, along with thousands of potsherds and a few other fragmentary tablets (table B.16). Its contents—the remains a reciprocal table and the entire standard series of multiplications (§4.2)—and appearance are very Old Babylonian, with spellings in the reciprocal table that link it to the northern OB tradition.[9] However, its clay is identical to that of a diplomatic letter found at Amarna/Akhetaten, sent from Abdi-Tirši, king of Hazor, to the pharaoh in the fourteenth century.[10] It was thus probably the product of a scribal centre that had survived some four centuries since Hazor's links with the Mesopotamian city of Mari.[11]

Ugarit, further north on the Syrian coast, was also under Egyptian control until the mid-fourteenth century, when it became a vassal of the Hittite empire. Nearly three centuries of prosperity, built on trade, ended with a devastating raid on the city in c. 1200. Scribes at Ugarit used a locally invented cuneiform alphabet to write their own Semitic language (Ugaritic) as well as fully fledged Sumero-Akkadian cuneiform. About half a dozen scribal schools have been excavated there.[12] Six surviving exemplars of the so-called 'Table of Measures' (in fact a metrological list) record the units of the traditional Old Babylonian capacity, weight, and area systems (see §4.2) from tiny values to large ones, adapted slightly to fit local notations and without sexagesimal equivalents (table B.16). While the tablets' good archival context shows them to have been part of a conventional Babylonian-style education in Sumerian and Akkadian, their exact curricular position is still uncertain.[13] As in the Old Babylonian cities, the educational tablets were almost all found in private houses (table 4.2), but in Ugarit it appears that a specialist wing of the house was reserved for teaching. Most of the Tables of Measures, along with nearly two hundred other school tablets, come from a house owned by a man called Rap'anu. Three of the metrological manuscripts are signed: a pair of weight and area lists were copied by a student called Šaduya, while a longer sequence was taught to one Yanḫānu by a teacher named Nūr-Mālik. Yanḫānu also learned Sumerian grammar and the standard bilingual vocabulary Ur_5-ra = ḫubullu,[14] but (as usual) none of these people can be identified outside this scholastic context.

The kingdom of Elam was not part of the Egyptian diplomatic network. But southwest Iran, especially the eastern extension of the Tigris river plain and the Zagros mountains behind it, had always had close links with southern Iraq. During the late third millennium, it had been part of the great empire of the Third Dynasty of Ur; in the mid-eighteenth century the powerful city state of Susa was defeated by Hammurabi in his construction of the Old Babylonian kingdom. By around 1500 BCE a native polity, the so-called Middle Elamite kingdom, began to take shape around Susa. Although now free from political interference from Babylonia, the Elamites still chose at first to worship Mesopotamian deities rather than their own, and to write Akkadian rather than the local language, Elamite (which is unrelated to any other surviving language).

Towards the end of the fifteenth century a certain king Tepti-ahar founded a new city, Kabnak, some fifteen kilometres southeast of Susa, which was inhabited for perhaps 150 years before being abandoned. Iranian excavations in the 1960s and 1970s found over four thousand cuneiform tablets in the city, about three hundred of which have been published. Although detailed findspot information is not yet available, one group of school tablets was uncovered in a courtyard a few metres to the east of the ziggurat, just to the north of a large kiln which was used for firing pottery and tablets. Half-recycled tablets and fresh clay were also found in the vicinity.[15] Perhaps this find included or comprised the twenty-four published school tablets from Kabnak, as no other similar find is recorded. For the most part the identifiable exercises are sign lists and word lists, but also three fragments of standard Old Babylonian–style arithmetical and metro-logical tables and lists, as well as a fragment that might be a catalogue of mathematical exercises on the areas of squares (table B.17). There is little to be concluded from these tiny pieces, except for the following general comments: first, OB-style elementary mathematics continued to be taught in Elam at least a century after the last dateable school mathematics from the southern Iraqi cuneiform heartland; and, second, the educational mi-lieu was not the school house but the temple (if the deductions about where they were found are correct)—or at least it was in the workshop area of the temple that the scribes and their students made and recycled their tablets.

But when and how did OB-style mathematics reach Elam? It seems to have been there since at least the nineteenth century—perhaps even since the twenty-first, when Susa was under the direct rule of Ur. The evidence is slight, but convincing: two verbose multiplication tables on Type IV tab-lets (see table 4.5), which omit the multiplicands that are co-prime to 60, just as Ur III reciprocal tables include only sexagesimally regular integers (§3.4). Both were excavated by French teams from the so-called Ville Royale area of Susa. The 3 times table, whose reverse contains a two-line extract from a lexical list, written twice, was found in the 1934–5 season with many other school tablets, including nine further mathematical exercises. They were in a very disturbed archaeological context that is very difficult to interpret. However, not far away a pot was found inscribed with the name of king Adda-ḫušu, who ruled Susa in about 1900 BCE.[16] The 30 times table comes from a house dug in 1966–7, in Level 5 of the so-called Chantier B. In an oven in the courtyard another school exercise was found: a cylindrical tablet bearing a version of the Old Babylonian list of domestic animals.[17] Although this excavation is not yet fully published, the excavators date Level 5 also to the reign of Adda-ḫušu.[18]

A much better-known group of twenty-six mathematical tablets was found in the Ville Royale area in 1933. Although they come from a find-spot close to the 3 times table, the characteristic shape of some of their cuneiform signs suggests a date of about 1650–1600 BCE.[19] The one multi-plication table of this assemblage is very different from the earlier two: rectangular in shape and terse in format (n $25n$), it contains entries for all multiplicands 1–20, 30, 40, 50 in mature OB style (§4.2).[20] Much could also be said of the other tablets of this find—including two bearing math-ematical diagrams, two with long lists of problems, and twenty-odd with model solutions to often complex problems—and their relationship to similar sources from Babylonia, but they are in desperate need of re-edition first.[21]

In sum, then, the mathematical fragments of Kabnak were not a new cultural import into Elam but the end of a long tradition of mathematics in and around Susa, stretching back at least four centuries. They were

components of the standard scribal training package of southern Mesopotamia, perhaps originally imported in the twenty-first century when
Elam was a province of the kingdom of Ur, or very shortly afterwards.[22]
When cultural reforms of the early fourteenth century phased out Sumerian and Akkadian in favour of Elamite cuneiform writing, the traditional
educational system was abandoned, and cuneiform mathematics with it.

6.2 TABULAR ACCOUNTING IN SOUTHERN BABYLONIA

In the high summer of 1237 BCE, the Kassite king Šagarakti-Šuriaš
(r. 1245–1233) commissioned an account of the consumables at the disposal of two of the most powerful women in the city of Nippur. Their
names are unknown, but their title—*entu* priestess—was a prestigious one,
with an ancient history going back deep into the third millennium. The
outcome was a simple table, divided into labelled rows and headed columns, introduced by a sentence describing its function, date, and commis-
sioning authority (figure 6.2, table 6.1).[23] It shows that the senior *entu* had
at her disposal some 3,300,000 litres of grain, 30,000 litres of vegetable
oil, 2900 litres of ghee, over 3000 kg of wool, and nearly 750 kg of goat
hair. Her junior colleague had just 700,000 litres of grain, 16,000 litres of
vegetable oil, less than 150 litres of ghee, about 730 kg of wool, and 340
kg of goat hair. All were managed by five scribes: Bēlu-ana-kala-udammiq
and Addu-šar-ilānī for the senior *entu* priestess, Ḫuzālu and Martuku for
her junior colleague, and Iqīša-Nergal for both women.

A more detailed, balanced account also survives for Ḫuzālu and Martuku's
income and expenditure on behalf of the junior *entu* priestess (table 6.2).[24]
(Iqīša-Nergal's affairs are tabulated in a separate account, probably because he worked for both *entus*.)[25] In each case the balances, shown in
bold, are identical with those in the summary account. In many ways, this
document records much the same sorts of information as the annual balanced accounts of the Ur III period (§3.3). The opening balance, carried
over from the previous account, is given, followed by credits and debits.
The final balance is tallied, and transferred to the summary tablet. But
there are two major differences: in numeration, and in format.

First, absolute decimal numeration is used for large quantities of *gur* and
talents. The words 'hundred' and 'thousand' are written after the numerals
they pertain to, and '1 sixty' is also explicitly noted in cases of ambiguity.
(Compare the notation in the Old Babylon tabular account YBC 4721:
figure 6.4 and table 6.4.) Second, whereas the Ur III accounts mix quantification and description in sentence-like entries organised on chronological
principles, the Kassite accounts use tabulation to separate quantitative
from qualitative information, to partition the quantitative into discrete

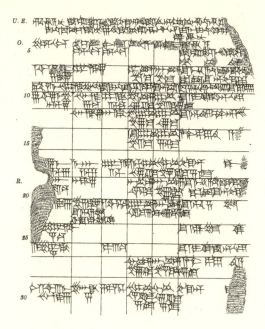

Figure 6.2 Tabular account of the assets of the *entu* priestesses of Nippur for 1237 BCE. (CBS 3359. Clay 1906, no. 136.)

categories, and to calculate up to three levels of totals. When, how, and to what extent did tabular document formatting develop? Why did it happen and what were its effects?[26]

The earliest table in the cuneiform record appears fully formed in the Early Dynastic IIIa period, with headings and a horizontal axis of calculation: the table of square areas from Šuruppag (table B.3, §2.3). The slightly later tablet from Adab which contains similar data, in prosaic layout, shows that tabulation was not the only choice available to scribes when formatting metrological conversions. The Šuruppag table is all the more remarkable, not only for pre-dating any other cuneiform table by some five centuries, but for being the first of just two truly tabular documents in the whole of the cuneiform mathematical tradition—the other being the famous table Plimpton 322, also headed and with a horizontal axis of calculation (§4.3).

There is, to my knowledge, just one published tabular account securely dateable to the Ur III period.[27] HAM 73.0400, probably from the state taxation centre Puzriš-Dagan and written in 2028 BCE, is an account of sheep and goats (figure 6.3).[28] Like later administrative tables, the row labels are at the end of each row, but uniquely the column 'headings' are at

TABLE 6.1
Annual Account of the Junior and Senior *entu* Priestesses of Nippur in 1237 BCE

Balances of the scribes of the *entu* priestesses of Amēl-Marduk, governor of Nippur, in Month V of the 9th year of Šagarakti-Šuriaš, established in accordance with the king.

Grain, by the 10-sila ban	Vegetable Oil, by the 10-sila ban	Ghee, by the 10-sila ban	Wool (minas)	Goat Hair (minas)	Balances of the entu Priestesses' Scribes: Its Name
1t 1s 4.4.5, 8	7.2.4, 8		5g 41m	3g 40m	In the possession of Bēlu-ana-kala-udammiq, son of Irēmšu-Ninurta
From which (subtract) 4h 26.2.5 seed grain that Iqīša-Nergal previously removed from Bēlu-ana-kala-udammiq's possession					
9t 8h 52.1.1, 8	1 22.2.3, 6	6.1.4, 2½ sila	51g 16⅓m	6g 43m 15š	In Iqīša-Nergal's possession
	13.2.4, 8	3.1.5	48g 49m	8g 19m	In Addu-šar-ilani's possession
			1s 8g 42m	6g 5½m	The shepherds' arrears
[Total 10t] 9h [17].1.1, 6	1h 3.3.1, 2	9.3.3, 2½	1h 1 14g 28⅓m	24g 47⅔m 5š	Senior *entu* priestess
[1t 1m] 3.[4].5, 2 sila	2.0.4, 8		2g 14½m	3g 56m 15š	In the possession of Bēlu-ana-kala-udammiq, son of Irēmšu-Ninurta
[5m 14].4.4, 6	50.4.5, 7 in Ḫuzālu and Martuku's possession	3 barig 1½ sila	9g 10⅚m 5š	3g 36m 15	In Ḫuzālu's possession
[4m] 47.3 barig 4 sila					In Martuku's possession
2m 55.0.5, 4		0.1.2, 3			In Iqīša-Nergal's possession
			12g 49½m	3g 44m 15š	The shepherds' arrears
Total 2l 3m 21.2.3, 6	53.0.4, 5	0.4.2, 4½	24g 14⅚m 5š	11g 16⅚m 5š	Junior *entu* priestess

Note: To save space, capacity measures such as 4 *gur*, 4(*barig*), 5(*ban*), 8(*sila*) are represented here as 4.4.5, 8 (see table A.3). Where there are no *ban*, the sign *barig* is written explicitly. Writings of the units of weight are abbreviated š = shekel, m = mina, g = talent (*gun*). In this account *gur* and talents are counted decimally above 60: t = thousand (*līm*), h = hundred (*mē*), s = sixty (*šūši*).

TABLE 6.2
Grain Account of the Junior *Entu* Priestess of Nippur, 1237 BCE

[Account (?)] of the junior *entu* priestess in Month V of the [9th] year of [Šagar]-akti-Šuriaš, made for Amēl-Marduk, governor of Nippur.

[Grain, by the 10-sila ban]	Its Name
6h 1 25.2.5, 2 *sila*	Old grain, balance of the account of the [6th year] of Šagarakti-Šuriaš the king
1m 8.3.1, 4 *sila*	Harvest tax of the junior *entu* priestess in the 8th year
1s 8.0.3	Seed grain that Huzālu received from Martuku's possession in the 8th year
Total 8h 1s 2.2.3, 6 *sila*	Grain, by the 10-*sila ban*, established head of the junior *entu* priestess's account, in Huzālu's possession

Grain, by the 10-sila ban, that was given out from it:

3h 3.1.2	Seed grain, exacted payment that in the 7th year Iqīša-Nergal previously added to the head of Iqīša-Nergal's account from Huzālu's possession
44.1.3	Grain gift for the children of the menial workers, for 7 months, from Month VII of the 8th year to Month I of the 9th year, junior *entu* priestess
Total 3h 47.2.5	Grain, by the 10-*sila ban*, given out, removed from Huzālu's possession
5h 14.4.4, 6	Grain, by the 10-*sila ban*, the balance in Huzālu's possession

Vegetable Oil, by the 10-sila ban	Its Name
42.0.4, 4 *sila*	Old (vegetable oil), balance of the account of the 6th year, in Huzālu and Martuku's possession
8.4.1, 3 *sila*	Harvest tax of the junior *entu* priestess, 8th year
Total **50.4.5, 7**	Vegetable oil, by the 10-sila ban, established head of the junior *entu* priestess's account, in Huzālu's possession
0.2.4, 6½ *sila*	Old (ghee), balance of the account of the 6th year
0.0.1, 5 *sila*	Imposition (?) of the adult cows of the 8th year

Table 6.2 (*Continued*)

[Grain, by the 10-sila ban]	Its Name
Total 3 *barig* 1½ *sila*	Ghee, by the 10-*sila ban*, established head of the junior *entu* priestess's account in Ḫuzālu's possession
9g 10⅔m 5š	Wool, balance of the account of the 6th year in Ḫuzālu's possession
3g 35m 15š	Goat hair, balance of the account of the 6th year
Total of the junior *entu* priestess's account, in Ḫuzālu's possession	
5m 35.2.3, 4 *sila*	Grain, by the 10-*sila ban*, balance of the account of the 6th year that was established concerning Martuku, son of Šamaš-ēriš
Grain, by the 10-sila ban, *that was given out from it:*	
1s 8.0.3	Seed grain, exacted payment that in the 8th year Ḫuzālu previously added to the head of Ḫuzālu's account from Martuku's possession
[9.4.0]	Ditto, that Iqīša-Nergal previously added to the head of Iqīša-Nergal's account from Martuku's possession
[Total 77.4.3]	Grain, by the 10-*sila ban*, given out, removed from Martuku's possession
[447.3.0], 4 *sila*	Grain, by the 10-*sila ban*, the balance in Martuku's possession

Note: To save space, capacity measures such as 4 *gur*, 4(*barig*), 5(*ban*), 8(*sila*) are represented here as 4.4.5, 8 (see table A.3). Where there are no *ban*, the sign *barig* is written explicitly. Writings of the units of weight are abbreviated š = shekel, m = mina, g = talent (*gun*). In this account *gur* and talents are counted decimally above 60: t = thousand (*līm*), h = hundred (*mē*), s = sixty (*šūši*).

the bottom of the obverse below each column total (table 6.3). They appear to be abbreviations, perhaps for personal names or months. Strikingly, the entries are always in the same proportions down the rows (1 lamb : 31 sheep : 3 goats) and across the columns (3 : 3 : 3 : 2 : 1). On the damaged edge and reverse the column formatting is abandoned in order to give the usual sort of summary information: perhaps the names of the official(s) responsible for the record, and the date on which it was drawn up. Contrast the following extract from HAM 73.0639, a more typical monthly

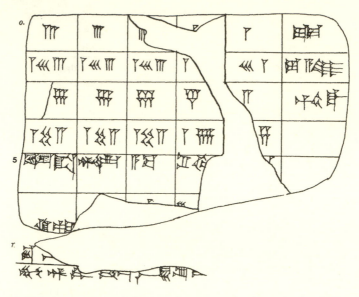

Figure 6.3 Tabular account of sheep and goats received as taxation in Puzriš-Dagan, 2028 BCE. (HAM 73.0400. Sigrist 1984b, no. 56, courtesy of Andrews University Press.)

summary from Puzriš-Dagan, in which the same sorts of data are sorted not by type but by day:

> 7 nanny goats, 3 billy goats: day 10
> 5 sheep, 5 kids, 5 billy goats, 7 nanny goats: day 13
> 4 kids, 1 nanny goat, 1 billy goat: day 14
> 2 sheep, 3 kids, 2 nanny goats, 4 billy goats: day 15
> 3 sheep, 15 kids, 3 billy goats, 9 nanny goats: day 16
> 3 sheep, 5 kids, 1 billy goat, 6 nanny goats: day 18
> 12 sheep, 11 billy goats, 7 nanny goats: day 20.[29]

Although the different types of livestock are always entered in the same order, it is not as easy to read data about a single type of animal—billy goats, say—as it is in this list's tabular counterpart, or to see at a glance which are absent on any one day, or even which days are missing from the dataset.

Why are almost no tabular documents found amongst a published corpus not far short of sixty thousand Ur III administrative records, given the apparently obvious advantages of tables for data display and retrieval? The conspicuous roundness of the entries on HAM 73.0400 hints at a school exercise,[30] but the fact that it is dated suggests otherwise. The use of

TABLE 6.3
Tabular Account of Sheep and Goats Received as Taxation in Puzriš-Dagan, 2028 BCE

3	3	3	2	1	Lambs
1 33	1 33	1 33	1 [02]	31	First-rate sheep
6	6	6	4	2	Billy goats
1 42	1 42	1 42	1 08	[3]4	
puzur$_4$	^{d}*šul*	*a-ba*	*si-sa*$_2$	[. . .]	

abbreviations implies that this was a tablet format designed for rough jottings, estimates, and calculations and that such tables were not intended to be viewed by superiors or archived for posterity—just like the scratch pads for sexagesimal calculations that came into use at the same period (§3.4). This raises the interesting (but untested) possibility that the 'accountant's nightmare' of drawing up annual accounts from daily records might not have been as labour-intensive as it appears if the scribes were able to collect, summarise, and calculate the quantitative data in tabular format—perhaps on erasable writing boards—and then calculate totals using the sexagesimal place value system before transferring them to the final, non-tabular version.

Over the course of the nineteenth century, tables gradually established themselves as an acceptable, but never popular, format for presenting accounts. The earliest and most abundantly attested belong to a single archive from Old Babylonian Nippur. Some 420 administrative accounts record regular *sattukkum*-offerings of bread, flour, and beer made in Ninurta's temple Ešumeša, over about eighty years of the nineteenth century BCE.[31] The archive was excavated in the fifth season of American post–World War II excavations at Nippur. The tablets were all found in the Parthian-period fill near the goddess Inana's temple, but it is unclear whether they had been transported there directly from their original site of use and storage or from some secondary locus such as a rubbish tip. Most exhibit simple linear or list-like structures, with vertical rulings used only as column markings and not as classification separators. Some two hundred, however, are partially tabular. They all follow the same format.

The obverse sides list the offerings made to each of the divine statues and other cult objects in the temple. The first column is always divided into five narrow sub-columns to the left, listing quantities of bread, shortbread, two types of flour, and beer. The wider sub-column to the right lists the divine beneficiaries of the offerings, always in the same order within a single administrative year. The second column is split vertically into three, with the first two narrow sub-columns listing quantities of unknown commodities (the headings, if there were any, do not survive) and the final column again naming the divine recipients. There is no sign of totalling at the bottom of the columns; they are essentially tabular lists. The reverse sides of the tablets, which record the redistribution of these offerings to human functionaries and dependents of the temple, are always in list format, with clear spatial separation of the numbers and the names. As on the obverses, the order of entries remains fixed throughout each administrative year.

There is a clear pattern to the chronological distribution of the tablets in the archive. The tabular records fall into a single phase of fifteen to twenty years from 1855 or 1850 to 1838 BCE. Changes in political rule, either of individual kings or between the rival states of Isin and Larsa, seem to have had little or no effect on bureaucratic style. Rather, it may be that the life-span of this particular tablet format is consonant with the working life of a senior administrator. Tables are so rare in the cuneiform record that we should consider their adoption a matter of individual choice—even if those individuals are anonymous to us—rather than the outcome of large-scale, impersonal forces.

It may simply be the accidents of discovery and publication that explain why just as the Ešumeša archive tails off, accounts in tabular list format begin to appear elsewhere in cities that were also under the political control of Larsa. Within another generation, tabular accounts begin to exhibit a much more sophisticated structure. The tablet YBC 4721 (table 6.4; figure 6.4) is typical.[32] First, it is shaped to fit the data, so that it has a 'landscape' orientation, with the width longer than the height, rather than the more usual 'portrait' format. Second, column headings are used consistently for the first time. The final, qualitative column is labelled MU.BI.IM, literally 'its name'. This became the standard heading for the final column throughout the Old Babylonian period and beyond; it is also found on Plimpton 322 (§4.3). Headings are necessary because a horizontal axis of calculation comes into use: the final column of quantitative data is derived from preceding columns, and (in this case) is used to double-check the values in the first column. Vertical and horizontal additions can be totalled for double-checking. Third, explanatory interpolations can note information outside the categories of the tabular columns. We could almost say that YBC 4721, dating to 1822 BCE, is the world's earliest attested spreadsheet.

Over the following thirty years further formatting innovations took place, so that tabular accounts post-dating Hammurabi's conquest of Larsa involve two or three levels of calculation and add plenty of interlinear comments. Theoretical constants for labouring work rates and harvest yield are also stated explicitly.[33] But the heyday of the cuneiform tabular

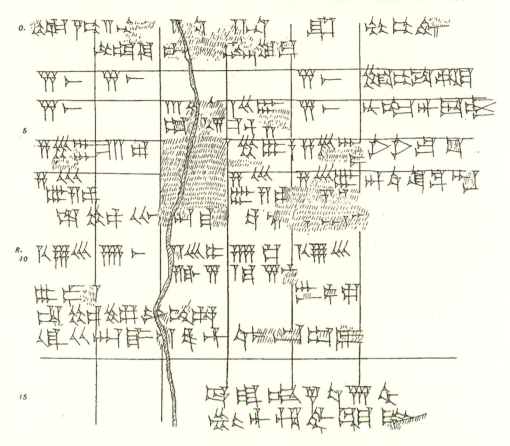

Figure 6.4 Tabular account of grain destined for various towns in the kingdom of Larsa, 1822 BCE. (YBC 4721. Grice 1919, no. 103.)

account was in fourteenth- and thirteenth-century Nippur, where a full third of the six-hundred-odd surviving administrative records are in tabular format: the *entus'* accounts are by no means exceptional. Showing a remarkable structural similarity with tables from the time of Hammurabi and Samsu-iluna, many of them have introductory preambles and column headings, of which the last is MU.BI.IM over the row labels. There may be several levels of calculation, organised horizontally from left to right and vertically from top to bottom. The Kassite accounts are deserving of an in-depth structural study.

For the first millennium the evidence is much scarcer.[34] Tempting though it is to attribute that rarity to some ill-defined decline in bureaucratic ability, a more satisfying explanation is at hand. From the late second millennium onwards, Aramaic became increasingly common as a written language.

TABLE 6.4
Grain Account from the Kingdom of Larsa, 1822 BCE

Grain assets	For Ur	For [. . .]	For Larsa	Total	Its Name
5(ĝeš) 1.0.0	5(ĝeš) 1.0.0			5(ĝeš) 1.0.0	Lipit-Suen
5(ĝeš) 1.0.0		3(ĝeš) 34.4.4, 5	1(ĝeš) 26.0.1, 5	5(ĝeš) 1.0.0	Nūr-Dagan
4(ĝeš) 56.0.0	3(ĝeš) gur	[ ]	56.0.0	4(ĝeš) 56.0.0	Ili-ēriba
4(ĝeš) 37.3.2			4(ĝeš) 37.3.2	[4(ĝeš) 37.3.2]	Šamaš-kīma-ilišu

From 23.[1.4 *sila* of Šamaš-kīma-ilišu's] workers.

| 1(šaru) 9(ĝeš) 36.0.2 *gur* | 8(ĝeš) 1.0.0 | 3(ĝeš) 34.4.4, 5 | 8(ĝeš) *gur* 0.1.3, 5 *sila* | 1(šaru) 9(ĝeš) 36.0.2 *gur* | |

From the grain of Lu-. . .
And from 23.1.4 *sila* of Šamaš-kīma-ilišu's workers.

Month I, 7th day.
Year that Rīm-Sîn became king.

Note: See table A.3 for an explanation of the abbreviated notation for capacity measure used here and table A.2 for the large values of *gur* (taken over from the discrete counting system).

Whereas the Assyrian and Babylonian dialects of the Akkadian language were still recorded with the increasingly cumbersome and recondite cuneiform script on clay tablets, Aramaic used an alphabet of just twenty-two letters, written freehand with ink on parchment or papyrus—neither of which organic media survives at all in the archaeological conditions of Iraq. All three languages, however, were also written on waxed wooden or ivory writing boards, several specimens of which have survived (§5.4), and for which there is ample documentation in the cuneiform record (see already §3.4). The durability, tradition, and tamper-proof qualities of tablets ensured their continuing use for legal documents and religious literature, whereas erasable writing boards and ink-based media were particularly suited to keeping accounts. Cuneiform tablets had to be inscribed while the clay was fresh; the new media, on the other hand, could be drawn up, corrected, and added to over time, like modern-day ledgers. It is highly likely, then, that the vast majority of tabular accounts from the first millennium BCE perished long ago with the objects on which they were written.

6.3 LAND SURVEYORS AND THEIR RECORDS IN NORTHERN BABYLONIA

Amongst the most important sources for the history of the later second millennium are entitlement stelae, sometimes called *kudurru* 'boundary (stone)'.

These monumental copies of legal documents (that were written on clay tablets and which do not survive) record royal grants or acquisitions of land, or other sources of income, to a courtier or other deserving recipient. Some 160 have been published, dating from the fourteenth to the seventh centuries BCE, but mostly from the period c. 1300–1050, and mostly from the areas around Babylon, Sippar, and northeastwards into the Diyala river valley. While they have been studied extensively for their iconography— they are often highly decorated with divine images—and literary form as well as their social, legal, and political import, their significance for the history of land mensuration has often been overlooked.[35]

For instance, BM 90829 records king Meli-Šipak's (r. 1186–1172) grant of some 144 hectares of land in the Diyala valley to a courtier named Hasardu (figure 6.5).[36] It begins:

> The name of this *kudurru* is 'Adad the hero, bestow irrigation ditches of abundance here!'
>
> 40 *gur* of seed grain, at 1 *iku* per 3(*ban*) by the big cubit, meadow of the town of Šaluluni on the bank of the Royal Canal in the province of Bīt-Pir'i-Amurru—
>
>> Upper length, facing north: adjacent to Bīt-Pir'i-Amurru
>> Lower length, facing south: adjacent to the governors of Uštim (?)
>> Upper width, facing west: the bank of the Royal Canal
>> Lower width, facing east: adjacent to Bīt-Pir'i-Amurru—
>
> that Meli-Šipak, king of the world, granted to Hasardu, court official and director, son of Sumû, his servant.

The area of land granted is recorded not in the traditional *bur, iku,* and *sar* but in 'seed grain' metrology, which equates (as the text states) 1 *iku* of land, measured by the 'big cubit' of c. 75 cm, with 3 *ban* of seed for sowing.[37] As this works out at 1 *gur* per 10 *iku,* it may have been a stealthy means of sexagesimalising middle-size area measures by substituting multiples of the *gur* for the messy relationships 1 *bur* = 3 *eše* = 18 *iku* (table A.3).[38] And unlike the third-millennium administrative land surveys (§3.2), the edges of the field are not quantified, but described only in terms of their cardinal directions and their neighbours.[39] This is not a chronological change, however, but a generic difference: land sales from the early third millennium onwards quantify area but not boundaries, presumably because the former was the main determinant of financial value (along with qualitative attributes such as location, soil quality, access to water, and so on).[40] The measurement itself is almost certainly a rounded figure: almost all the extant entitlement stelae record land grants in decimal multiples of *gur*.

Figure 6.5 Stone stele recording king Meli-šipak's grant of land to Hasardu in c. 1180 BCE. (BM 90829. L. W. King 1912a, II pl. XXIII.)

The *kudurru* BM 90829 continues:

Ibni-Marduk, son of Arad-Ea, the field surveyor, Šamaš-muballiṭ, the mayor, Bau-aḫa-iddina, the provincial governor's scribe, and Itti-Marduk-balāṭu, the king's eunuch, measured that field and established it for Hasardu.

The survey could be carried out by a single surveyor or by a team of up to four men, comprising both technically competent professionals as well as representatives of local and royal authority (here the mayor, governor's scribe, and king's eunuch). The recently coined Akkadian verb *mašāḫu* 'to measure' was typically used to describe this process, though the phrase *rēš eqli našû*, literally 'to lift the head of the field', could be added or substituted, or more rarely the verbs *palāku* or *tarāṣu*, both 'to stretch out (a rope)'.[41] Although the stelae never detail the measurement process itself, a fragmentary land sale record from Nippur, probably written in the reign of Burnaburiaš I or II (r. 1359–1333) describes a local governor and a field registrar (Sumerian *saĝ-sug₅ a-šag₄-a*) using a yardstick (Akkadian *wašlum*) and a 1-rod reed, in a manner highly reminiscent of Old Babylonian literary descriptions of goddesses with their mensuration equipment (§4.4).[42]

After the survey, local dignitaries witness a chain of documentation showing previous ownership as well as the new legal tablet recording the royal grant of land.[43] The remainder of the text is typically taken up with a long compilation of curses directed against any future malefactor, which may comprise more than half the total number of lines. Some time after the event the entitlement stele itself was fashioned, combining the text of the legal transaction with images of divine symbols, which together with the textual curses protected the monument from harm.[44] The object was then deposited in a temple, presumably with some ceremonial pomp, for the gods to further protect it. The stele BM 90829, for instance, was discovered in 1881–2 with two other entitlement stelae in a room of the temple of Šamaš, god of justice, during Hormuzd Rassam's archaeological expedition to Sippar for the British Museum.[45]

Very few entitlement stelae include a map or plan.[46] The so-called Hinke *kudurru*, found in the temple precinct of Nippur and named after its early twentieth-century editor, commemorates Nebuchadnezzar I's (r. 1125–1104) benefaction of agricultural land to one Nusku-ibni, who held high political and religious office in the city of Nippur.[47] Under an array of divine symbols at the top of the monument, the inscription begins with the name of the object, 'Ninurta and Nusku are the establishers of the boundary', and a visual record of the granted land (figure 6.6).

The total land area, in seed grain, is stated at the top. Labels around the borders of the field give their lengths (rounded to the nearest 5 rods, c. 60 m), cardinal directions, and neighbours. The short sides of the field both abut water, while the long sides belong to local estates. The vertical line down

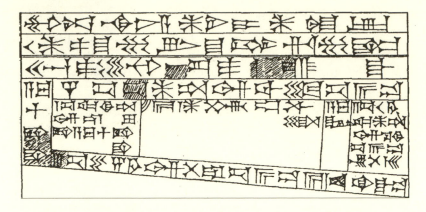

22 (*gur*) 2 (*bariga*) 5 (*ban*) seed grain at 1 *iku* per 3 (*ban*) by the big cubit

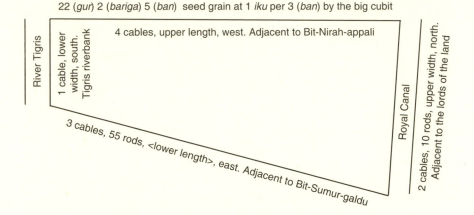

Figure 6.6 Map showing land granted to Nusku-ibni by king Nebuchadnezzar I in c. 1115 BCE. (Drawing: Hinke 1911, no. 5; reconstruction, author's original.)

the middle of the plan does not demarcate an internal boundary, but serves to separate the labels for the west (top) and south sides from each other. Structurally, this map is typical of other Mesopotamian field plans and mathematical diagrams in that its long sides are orientated horizontally and line labels, rather than line lengths, bear the quantitative data (see §3.2). The calculation is not given explicitly, but the most efficient way to use the long-lived 'surveyors' method' (§2.1) to determine the area of the field is to convert the average lengths into decimal multiples of the rod and then multiply: 137½ × 95 = 22,562½ *sar*, which by inspection is just over 22½ *gur* of seed grain (at 1 *gur* per 10 *iku* = 1,000 *sar*). As 1 *barig* (⅕ *gur*) = 200 *sar* and 1 *ban* (⅙ *barig*) = 33⅓ *sar*, it follows that the fractional part is approximately 2 *barig*, 5 *ban*, as stated in the total.

The inscription does not continue, as expected, with a description of the measuring and granting of the land, but rather with a long tripartite account, in elegant literary Akkadian, praising first the great god Enlil (patron of the recipient's home city), then the king, and finally the work and piety of the recipient himself.[48] Eventually, Nebuchadnezzar commissions a local governor to oversee the land grant as reward for Nusku-ibni's faithful service to king and god:

> Bau-šuma-iddina, the governor of Bīt-Sîn-šeme, measured 22(*gur*) 2(*barig*) 5(*ban*) seed grain of field, fallow land that is situated at a channel cutting, the meadow of Ša-Mar-aḫu-attūa town, on the Tigris riverbank in Bīt-Sîn-šeme province; where in previous days no dyke had ever been heaped up, no furrow had ever been prepared, so that it was not (immediately) ready for cultivation but was situated on running water—
>
> > 4 cables the upper length, west, adjacent to Bīt-Nirah-appili and Bīt-Sîn-šeme province;
> > 3 cables 55 rods the lower length, east, adjacent to Bīt-Suḫur-galdu;
> > 2 cables 10 rods the upper width, north, adjacent to Bīt-Ušbula that was given to the lord(s) of the lands (?);
> > 1 cable the lower width, south, the Tigris riverbank:
> > total 22(*gur*) 2(*barig*) 4(*ban*) 8⅔ (*sila*) 5 (shekels) of seed grain at 1 *iku* per 3(*ban*) by the big cubit–
>
> and granted the meadow of Ša-Mar-aḫu-attūa town in Bīt-Sîn-šeme province to Nusku-ibni, son of Upaḫḫir-Nusku, *nešakku* priest of Enlil, *gudapsu* priest of Nusku, mayor of Nippur, his servant, for ever after.
>
> The dividers of that field were Nabû-zēru-lēšir, son of Itti-Marduk-balāṭu, descendant of Arad-Ea, and Nabû-unna, son of Aḫi, administrator of Bīt-Sîn-šeme.

At the beginning of this passage the local governor himself claims agency for the measuring, although it becomes clear at the end that in fact the work itself was carried out jointly by two other men, one of whom is a junior local official. The first part contains unusual detail about the current state of the field, followed by the boundary descriptions from the map (with some minor variations) in descending order of size. At this point the total area is given exactly, instead of the approximate values stated on the map and at the start of this passage. The inscription ends with the usual type of curses and a list of fourteen witnesses, mostly representatives of the crown or local authorities, as well as Nabû-zēru-lēšir of the surveying team.

Nabû-zēru-lēšir, like Ibni-Marduk, the surveyor in BM 90829 discussed above, is described as a descendant or son of one Arad-Ea, as are a dozen more land surveyors whose names are recorded on entitlement stelae spanning the twelfth and eleventh centuries (table 6.5). Although ancestral family names became commonplace in the later first millennium (see §8), Arad-Ea's descendants are the earliest such family by several hundred years.[49] They were clearly of high social standing: several members are called 'royal scribes', while one Marduk-zākir-šumi was himself the recipient of 18 *bur* 2 *eše* of land from king Marduk-apal-iddina I. (Was this rare survival of OB area metrology over 'seed grain' measure a traditionalist signal of numerical virtuosity?) Nabû-zēru-lēšir carried various official titles, including *bēl pīḫāti*, a type of provincial governor—a post which remained in the family until at least the early seventh century.[50] Another family member, Izkur-Nabû, owned fields neighbouring land granted by king Meli-Šipak (surveyed by his kinsman Iqīša-Bau).[51] Indeed, the very fact that these men's individual names and family identities were recorded so prominently on the monumental records of the land they surveyed suggests the high regard in which their professional skills were held: their involvement was an index of authority and accuracy to be acknowledged and valued by posterity.

Land surveying was apparently not the only numerate profession within the Arad-Ea family: an undateable stone weight in the shape of a duck is inscribed '2½ true minas, Nāṣirī, son of Kidin-Gula, descendant of Arad-Ea', while in a legal dispute in the city of Ur, one Ṭāb-ṣilli-Marduk acted on behalf of king Adad-šumu-uṣur (r. 1216–1187) to oversee the correct measurement of 100 *gur* 'by the *5-sila ban*' to settle the case.[52]

Because the descendants of Arad-Ea rarely used patronymics, it is impossible to determine whether these professions stayed within a narrow line of descent or spread widely across a large family network; only Nabû-zēru-lēšir and Šāpiku (table 6.5) can reasonably be assumed to be brothers. The family itself can be traced as far back as the fourteenth century, thanks to BM 114704, an extraordinarily informative Sumerian inscription on a cylinder seal belonging to a courtier of Kurigalzu I (early fourteenth century) or II (r. 1332–1308):

Nin-sumun, exalted lady,
Heir of the great gods (?),
Majestic land registrar of Enlil,
Whose wisdom perfects everything:
May the seeker be happy,
the passer-by joyful,
the follower achieve success.
Uballissu-Marduk,
son of Arad-Ea,
the scholar of accounting,
servant of Kurigalzu, king of the world.[53]

TABLE 6.5
Land Surveyors from the Family of Arad-Ea

King	Name	Patronym(s) and/or Titles	Context	Publication
Meli-Šipak (r. 1186–1172)	Ibni-Marduk	Son of Arad-Ea	With a royal eunuch and a mayor 'they measured'	Sb 22, line i 28 (Scheil 1900, 99)
		Son of Arad-Ea, field surveyor (*šādid eqli*)	With a mayor, a provincial governor's scribe, and a royal eunuch, 'they measured that field'	BM 90829, lines i 13–8 (King 1912a, no. 4)
	Nabû-šuma-iddina	Son of Arad-Ea, royal scribe	With a local governor and one other 'they measured that field'	Sb 26, lines ii 7–10 (Scheil 1905, 31–8)
	Iqīša-Bau	Son of Arad-Ea, scholar of accounting	Witness to land grant	BM 90829, lines i 8–9 (King 1912a, no. 4)
		[. . .] son of Arad-Ea	Last witness to land grant	Sb 169, lines ii 16–9 (Scheil 1905, 42–3)
Marduk-apal-iddina I (r. 1171–1159)		Son of Arad-Ea, local governor	Last witness to land grant	Sb 26, lines iii 21–3 (Scheil 1905, 31–8)
	Bēl-ippašra	Son of Arad-Ea, scribe	With a local governor they 'raised the field's head and measured 30 *gur* of seed grain'	Sb 26, lines iii 1–10 (Scheil 1905, 31–8)
		Son of Arad-Ea, 'that field-measurer'	'In measuring that field' with a local governor	IM 67953, lines ii 14–9 (Page 1967)
	Ina-Esaĝila-zēra-ibni	Son of Arad-Ea	'He measured the field' alone	Sb 33, line i 15 (Scheil 1905, 39–42)

TABLE 6.5 (*Continued*)

King	Name	Patronym(s) and/or Titles	Context	Publication
	Nabû-šākin-šumi	Son of Arad-Ea, charioteer	Last witness to land grant	BM 90850, line ii 36 (King 1912a, no. 5)
	Ša-bāb-Šamši	Son of Arad-Ea	With a royal official, a royal eunuch, and a local governor 'they stretched out (a rope)'	Tehran *kudurru*, lines i 8–19 (Borger 1970)
Nebuchadnezzar I (r. 1125–1104)	Nabû-zēru-lēšir	Son of Itti-Marduk-balāṭu, descendant of Arad-Ea	With a local governor they were 'the surveyors (*pālik*) of that field'	UM 29-20-1, lines iii 13–6 (Hinke 1907)
		Son of Arad-Ea	Last witness to the land grant he had surveyed	UM 29-20-1, line v 24 (Hinke 1907)
Marduk-nādin-aḫḫe (r. 1099–1082)	Šāpiku	Son of Itti-Marduk-balāṭu, son of Arad-Ea, 'that field measurer'	Measures alone	BM 90841, lines i 13–4 (King 1912a, no. 7)
		Son of Arad-Ea, scribe, sealing official	Last witness to private land purchase	YBC 2154, line 29 (Clay 1915, no. 37)
Marduk-šāpik-zēri (r. 1081–1069)	[. . .]-imitti	Son of Arad-Ea	With a local official (?) 'the measured [the field]'	IM 80908, lines iii 1–8 (Reschid 1980)
	Nabû-zēra-iddina	Son of Arad-Ea	With a royal eunuch 'those measurers of that field'	IM 74651, lines i 15–7 (Reschid and Wilcke 1975)
Marduk-aḫḫe-ēriba (r. 1046)	Nabû-ēriš	Son of Arad-Ea, scribe	With a divine and local governor's scribe, a local administrator, and a mayor, 'they lifted the head of the field'	BE 1/2 149, line i 15 (Hilprecht 1896, no. 149)

TABLE 6.5 (*Continued*)

King	Name	Patronym(s) and/or Titles	Context	Publication
Simbar-Šipak (r. 1025–1008)	Esagilaya	Son of Arad-Ea, scribe, seal-holder	Last witness to private land exchange	BM 90937, bottom edge line 2 (King 1912a, no. 27)
Undated	Biraya	Son of Arad-Ea	With three local officials, 'they measured that field'	MDP 6 44, lines i 5–15 (Scheil 1905, 44)
	Ēriba-Marduk	Son of Arad-Ea, royal scribe	With a local official 'they measured the field'	IM 5527, line ii 7 (Sommerfeld 1984)

Note: Based on Lambert 1957, 9–10; Page 1967, 49–50; and Hurowitz 1997, 90.

Uballissu-Marduk also owned another cylinder seal, BM 122696, perhaps commissioned later in his career, in which he lists four generations of ancestors of which Arad-Ea, 'scholar of accounting' (Sumerian um-mi-a nig_2-kas_7) is the first.[54] Thus in both inscriptions 'son' really does mean 'son', and not the more general 'descendant'. Arad-Ea's father had been the governor of Nippur, his grandfather the governor of Dilmun (Bahrain): the family was powerful, well connected, and presumably very wealthy.

The first of Uballissu-Marduk's seals is dedicated to the goddess Nin-sumun; the second describes him as 'the servant of Lugalbanda and Nin-sumun'—the divine parents of the hero Gilgameš (see §6.4). Just three other Kassite 'scholars of accounting' are known, none of whom was apparently a member of the Arad-Ea family. But Dašpī, Kiribti-Marduk, and Qīšti-Marduk all describe themselves in cylinder seal inscriptions as 'the servant of Lugalbanda and Nin-sumun' too.[55] How and why this divine couple came to be revered by the numerate professionals of the late second millennium is not yet clear; there are few clues in the Old Babylonian Sumerian literary tradition. Uballissu-Marduk calls Nin-sumun a majestic 'land registrar'. This professional title (Sumerian sag-sug_5, Akkadian *šassukku*) is first attested in later third-millennium lexical lists and occurs frequently in Ur III administrative documents.[56] By the later second millennium it was rarely applied to humans though: the only published attestations are the thirteenth-century land sale record mentioned above, in which the registrar uses a measuring rod and rope; and a cylinder seal belonging to another

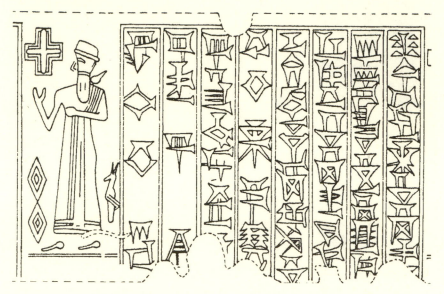

Figure 6.7 Seal of Pir'i-Amurru, chief land registrar and courtier of king Kuri-galzu I or II in the fourteenth century. (D. M. Matthews 1992, no. 54, courtesy of Vandenhoeck and Ruprecht.)

courtier of Kurigalzu I or II, one Pir'i-Amurru, *gal unkina gal sağ-sug₅* 'chief of the assembly, chief land registrar' (figure 6.7).[57] However, in Old Babylonian Sumerian literature the goddesses Bau, Ninisina, and Nisaba were all described on occasion as *sağ-sug₅* or with epithets like Nin-sumun's, such as 'great lady who knows everything'.[58]

Although Nin-sumun is not currently attested amongst the numerate goddesses of literary Sumerian, Uballissu-Marduk's first seal inscription is clearly in that same tradition. And it was not the sole remnant of it either. Bulluṭsa-rābi's hymn in Akkadian to Gula, the goddess of healing, was presumably also composed some time in the late second millennium. It too describes its subject as a numerate land surveyor and equates her with another goddess of similar prowess:[59]

She who handles the yardstick, the cubit made of reeds, the 1-rod reed;
She who carries the reed stylus for her work, the keeper of accounts:
Mother Nanše, lady of *kudurrus*, am I.

It is now clear that numerate goddesses were not simply a literary device of the Old Babylonian scribal curriculum (§4.4), but had real meaning for high-status numerate professionals even centuries later.

6.4 QUANTIFICATION AS LITERARY DEVICE IN THE *EPIC OF GILGAMEŠ*

If the goddess Nin-sumun's numerate attributes have not been generally recognised,[60] that is perhaps because she was and is most frequently identified as the hero Gilgameš's divine mother. The greatest epic of all Babylonian literature, which recounts the eponymous hero's saga of love, friendship, death, and the quest for eternal life, is attested throughout the second and first millennia BCE. In the late second millennium it was certainly a much more successful western cultural export than mathematics or metrology: later second-millennium versions have been found in the Hittite capital Hattusas in central Anatolia, Emar on the Syrian Euphrates, and Megiddo (biblical Armageddon) near the eastern Mediterranean coast. It was considered to have reached its definitive, twelve-chapter form under the editorship of a Babylonian scholar named Sîn-lēqi-unninni some time in the Kassite period. The oldest dateable source of this Standard Babylonian recension is a copy of Tablet XII, made from an older original by the great Assyrian scholar Nabû-zuqup-kēna in 705 BCE (§5.4), the most recent a copy of Tablet X by Itti-Marduk-balāṭu, who became an astronomer at the temple of Marduk in Babylon in 127 BCE (§8.2). Other numerate scholars also feature amongst *Gilgameš*'s many owners and copyists: members of the Šangû-Ninurta family in Uruk owned Tablets II and III in c. 420 BCE, while their successors in the same house, the Ekur-zākirs, had Tablets I, II, and V a century later (§8.3).[61]

Briefly told, Gilgameš is king of the ancient city of Uruk (see §§2.1–2), where he oppresses his subjects, paradoxically by coercing them into too much sex and sport. The gods send the wild man Enkidu to calm and distract him, and together they set off on a great adventure, to the great Cedar Forests of the Lebanon or Amanus area. The formation of the two men's deep friendship forms the first third of the *Epic* (Tablets I–IV). The middle section revolves around death: first the heroes' murder of Humbaba, spirit of the Cedar Forest; then their slaughter of the Bull of Heaven, who is sent to destroy Uruk by the vengeful sex goddess Ištar after Gilgameš has spurned her advances. She retaliates by inflicting upon Enkidu a long, lingering death, bringing Gilgameš face to face with not only his beloved companion's mortality but also his own (Tablets V–VIII). In the final third of the *Epic* Gilgameš searches alone for the secret of everlasting life, eventually discovering Ūta-napištim, the Babylonian Noah, in his retirement home at the ends of the earth. Ūta-napištim recounts his own life story and sets Gilgameš a test to see if he is worthy of immortality too. Failing miserably, the hero returns to Uruk, newly aware that we live forever, if at all, only in others' memories of the good we have done. In a short appendix,

Enkidu returns from the Underworld to confirm that those with the most comfortable afterlives are the ones with plenty of loving descendants who make regular offerings to their dead ancestors (Tablets IX–XII).

Amongst the reasons for the *Epic*'s enduring power and success is the richness and density of its language. Quantifications and measurements are amongst many interesting literary devices put to great effect.[62] At the very beginning and end, the reader is taken on a virtual tour of Gilgameš's city (Tablet I, 18–23; cf. XI, 323–8):

Go up on the wall of Uruk and walk around,
Survey the foundation platform, inspect the brickwork!
See if its brickwork is not baked brick,
And if the Seven Sages did not lay its foundations!
1(*šar*) is city, 1(*šar*) date-grove, 1(*šar*) clay-pit, a half-*šar* the temple of Ištar:
[3(*šar*)] and a half-*šar* is the measurement of Uruk.[63]

The post-OB *šar* was roughly 875 hectares, giving some 30 square kilometres or 12 square miles as the total area of Uruk (table A.4). As *Gilgameš*'s modern editor, Andrew George, points out, this is about six times the size of the space delimited by the archaeologically attested city walls, but that is surely beside the point.[64] In the context of discrete counting the sign *šar* represented 3600 objects (table A.2), and by extension 'an awful lot', rather like modern uses of the words myriad or million. Thus this passage conveys that Uruk was heroically large but quantifiably so, as signalled by the final word *tamšīḫu* 'measurement', derived from the land surveyor's technical term *mašāḫu* 'to measure' (§6.3).

Gilgameš himself—'two-thirds of him god but a third of him human' (Tablet I, 48; IX, 51)—is also built on a heroic scale, as described in a rather damaged and difficult passage (Tablet I, 56–8) that literally reads:

A/the *nikkas* (of) his foot, ½ rod his *purīdu*,
6 cubits between his legs,
[. . .] cubits the foremost of his [. . .].

Even the first word is problematic: does it mean *nikkassu* 'account, calculation', signalling the beginning of a series of quantifications, or the length unit *nikkassu*, 3 cubits? At the end of that same line, *purīdu* means both 'leg' and a length measure of 3 cubits, and in the following line *ammatu* 'cubit' also means 'forearm'. . . Does 'between the legs' here mean the groin area or the length of Gilgameš's stride?[65] The multiple layers of meaning are at once the outcome of metrological nomenclature arising from traditional embodied measuring practice and at the same time a deliberate obfuscation. There are unambiguous ways of quantifying body parts—using sexagesimal numerals and standard length units, for instance—but the scribe/author/editor of *Gilgameš* chose to capture the gigantic stature and masculinity of the hero's body by selecting syllabic writings of words that were both quantitative and corporeal at once.

Heroic size recurs in the door Gilgameš makes from the felled cedars after Humbaba's defeat—'6 rods its height, 2 rods its breadth, 1 cubit its thickness' (Tablet V, 295–6)—and in the offering vessels he fashions from the slain Bull of Heaven (Tablet VI, 162–5):[66]

30 minas each of lapis lazuli was their bulk,
2 minas each their rims (?),
6 *gur* of oil was the capacity of each.
He dedicated them for the anointing of his god, Lugalbanda.

Such large measurements are unambiguously impressive. By contrast, sexagesimally irregular numbers (that are co-prime to 60) signal problematic otherness or liminality: over seven nights of sexual activity Enkidu makes the transition from animal to human (Tablet I, 194);[67] Gilgameš grieves for seven nights over Enkidu's death (Tablet X, 58, 135, 235). Gilgameš accuses Ištar of bringing trouble to her lovers in sevens (Tablet VI, 51–5). For Ūta-napištim the Deluge lasts for seven nights (Tablet XI, 128) and it is a further seven days before he dares to release his birds over the still water (Tablet XI, 147). Ūta-napištim demands that Gilgameš stay awake for seven nights in order earn the immortality he craves (Tablet XI, 209). In his journey to the edge of the world, Gilgameš travels through an uncanny night of eleven double-hours (instead of six) before he sees the sun again (Tablet IX, 139–70). Nin-sumun gives Gilgameš thirteen wild winds to use against Humbaba, the monstrous guardian of the Cedar Forest (Tablet III, 87–93; V, 137–43); Enkidu's death agonies last thirteen days (Tablet VII, 255–60).

Conversely, Gilgameš overcomes difficulty and distance by means of regular, round, whole numbers. Gilgameš and Enkidu undertake the journey to the Cedar Forest in extraordinarily quick time, covering the distance from Uruk to Lebanon in five three-day marches: fifteen days at fifteen times the normal pace (Tablet IV, 1–4, 34–7, 79–82, 120–4, 163–5):[68]

At 20 leagues they broke bread,
At 30 leagues they pitched camp:
50 leagues they travelled in the course of a day,
A journey of a month and a fortnight by the third day.

And in order to cross the Waters of Death to reach the immortal flood-survivor Ūta-napištim, the boatman orders Gilgameš to cut him '3 sixties of punting poles, each 5 rods long' (Tablet X, 160, 166), so that 'by the third day they had travelled (?) a journey of a month and a fortnight' before the poles run out (Tablet X, 171).[69] Similarly, Ūta-napištim's only hope of surviving the Deluge is to follow Ea's instructions and construct a mathematically pleasing Ark (Tablet XI, 28–30):

The boat that you will build: you
Let its measurements be measured against each other;
Let its breadth and length be square to each other.

In both vocabulary and structure, Ea's speech is highly reminiscent of Old Babylonian word problems (§4.1). First he states the subject of the problem, then addresses Ūta-napištim with the emphatic *atta* 'you!' which so often introduces a mathematical procedure. The verbs—*madādu* 'to measure' (not the land surveyors' *mašāḫu*) and *mitḫuru* 'to be square'— are familiar technical terms too. The instructions themselves, however, are not directed at Ūta-napištim in the imperative or present future tense, as in mathematical procedures, but are phrased in an impersonal precative. Ūta-napištim reminisces to Gilgameš about how he carried those orders out, describing first the outer shell and then its interior (Tablet XI, 57–63):[70]

On the fifth day I laid down her form:
1 *iku* was her circumference (*sic*), 10 rods each her walls were high;
10 rods each the edges of her top were square to each other.
I laid down her body, I drew up her design:
I roofed her 6 times—
I split her into 7.
Her interior I split into 9.

In other words, the Ark is a perfect cube with sides of some 60 metres, divided into seven decks, each of which is partitioned into nine (square?) spaces. While the verbs *šaqû* 'to be high' and *mitḫuru* 'to be square' are also mathematical technical terms, the verb *parāsu* 'to split, to divide' is not. Numerical division was performed through multiplication by a recip-rocal or through trial and error; for geometrical division of a line or area into pieces *ḫepû* 'to break' was preferred.[71] In the second line all manu-scripts have *kippassa* 'her circumference/circle' where perhaps *eqelša* 'her area' would be expected. But given that *kippatu* may mean the area con-tained within a circumference as well as the boundary itself (§2.4), its use here may be figurative rather than wrong. The Ark is definitely rectilinear, given the repeated use of *mitḫuru* here and in Ea's instructions, but it would be over-fanciful and anachronistic to imagine that this passage alludes somehow to a squaring of the circle (that is, finding a circle and square which have the same areas: a favourite pre-occupation of classical Greek mathematics, but apparently absent from all Babylonian mathematical sources). In any case, these explicit and (mostly) unambiguous quantifica-tions invite the reader to visualise the Ark most precisely. Finally the boat is waterproofed with three times 3(*šar*) quantities of asphalt, bitumen, and oil (Tablet XI, 66–70). The unit of measurement is not stated, suggesting that—like the land comprising Uruk—the image sought is one of general plenitude rather than exact quantity.

In short, numbers and measurements serve three different functions in the *Epic of Gilgameš*. In conjunction with the body and deeds of Gilgameš himself, they convey the epic scale of his might, speed, and actions, and the trophies of his conquests. Dangerous spaces and times of transition are invariably associated with the sexagesimally irregular numbers 7, 11, and 13—a mathematical preoccupation that extends as far back as the Early Dynastic period (§2.3).[72] Those dangers can be overcome with tools and instruments that are measured or counted in whole, round numbers or which conform to mathematically simple shapes. The editor Sîn-lēqi-unninni (if indeed it was he) seems to have been just as interested in the properties of number and measure as many of his later readers were.

6.5 CONCLUSIONS

Traces of Old Babylonian mathematical learning lingered on long after the political ideology that it supported had disappeared. At Kabnak in Elam and at Hazor in Canaan, the standard elementary lists and tables were being copied until at least the fourteenth century BCE—and as late as the twelfth century at Ugarit and Assur (§5.1). Egyptian scribes learned how to convert Babylonian weights to their local equivalents—essential in the context of international trade and diplomacy. But just like Sumerian literature, that other mainstay of Old Babylonian scribal education, traditional Babylonian metrology and mathematics were apparently peripheral components of late second-millennium cuneiform scholarship. Major scribal and political centres such as Emar on the Syrian Euphrates and the Hittite capital Hattusas in central Anatolia have yielded no traces even of elementary pedagogical arithmetic or metrology.[73] Mathematical word problems seem to have had no life at all outside Babylonia in the later second millennium. Even within Babylonia the evidence is meagre—although that may be more a reflection of the adoption of relatively ephemeral media such as writing boards than of the abandonment of the mathematical tradition.

Certainly, within Babylonia itself continuities in the ideals and (some) practices of numerate professionals can be traced with some confidence, despite the dearth of clearly identifiable pedagogical mathematics. It is noteworthy, however, that the tabular accountants of Nippur preferred absolute decimal notation for larger orders of big capacity and weight units rather than the traditional sexagesimal system. There are also hints of decimalisation in land measurements.

The highly respected Arad-Ea dynasty of surveyors traced their ancestry to a 'scholar of accounting', and retained a numerate goddess—Nin-sumun the 'majestic land registrar'—as patron until at least the fourteenth century. Apparent changes in area metrology were only cosmetic, perhaps designed to avoid the inelegant numerical relationships between mid-range units in the older system. Leaving the borders of land unquantified was equally a continuation of traditional legal documentation; the map and measurements on the exceptional eleventh-century Hinke *kudurru* reveal that the underlying land survey conventions were still much as they had been in the later third millennium (§3.4). The land surveyors and accountants themselves all carry Babylonian, not Kassite, names.[74]

By contrast, the use of tabular formatting in the cuneiform record was fitful and patchy. After two false starts in the Early Dynastic and Ur III periods, tabulation became recognised as an efficient way of recording, storing, and sorting data only in the mid-nineteenth century BCE. Tabulation enabled the horizontal separation of different categories of quantitative information associated with named individuals. This led to easy addition of quantitative data, along a vertical axis. At the same time, data could be sorted by criteria such as destination, or date of transaction. Headings, attested first in the Ešumeša archive and then regularly after about 1822, obviated the necessity to repeat descriptive information. When columns of derived data were introduced, again from at least 1822, they enabled calculations to be performed along both horizontal and vertical axes for the purposes of double-checking. At the same time, the columnar format could be ignored where necessary to provide note-like explanatory interpolations. It was at this point, arguably, that tables became truly powerful information-processing tools, cognitively distinct from well-organised lists. Their evolution, it appears, took no more than thirty years.

Thereafter further refinements were made and uses found. From the 1790s standard calculation constants (for work rates, harvest yields, market rates) were used in tables to make predictions as well as to record events that had already happened. At the same time, introductory preambles added transparency to complex tables, especially when multiple levels of calculations were made. These further developments, including horizontally organised calculations, appear on present evidence to have been exclusive to the south: late OB tabular accounts from the Sippar area show much less complexity and are predominantly vertical in calculational structure. Tantalisingly, the format, terminology, and structure of Kassite tabular accounts from Nippur suggest a direct line of descent from the mid-OB period, but the mechanism for that descent is still entirely unknown.

CHAPTER SEVEN

The Early First Millennium

Very little written evidence survives from Babylonia in the first quarter of the first millennium BCE. The earliest mathematical tablets considered in this chapter are from eighth-century Nippur, at a time when parts of Babylonia were under Assyrian control or influence. They belong to an *ad hoc* scribal education that was tailored to suit the needs of the student. Just a century later, at Kiš, a much more formalised curriculum entailed basic metrological and numerical knowledge in the elementary stages; this style of curriculum is also found at Babylon, Sippar, and Ur from the sixth century onwards (§7.2). From the brief years of Babylonian independence and the first decades of the Persian occupation, together c. 610–480 BCE, a vast range of official and familial documentation enables us to study the uses of numeracy in urban household archives (§7.3), and to analyse the methods of calculation used by two different types of professional land surveyor (§7.1, §7.4). While at first glance the professional methods

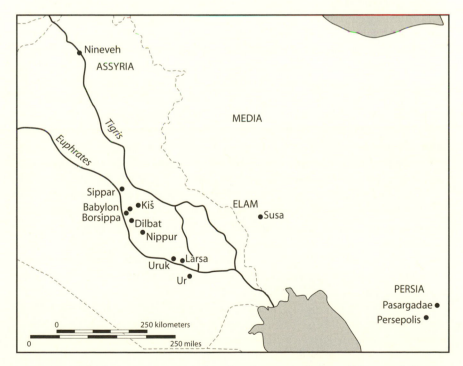

Figure 7.1 The locations mentioned in this chapter, including all known findspots of mathematical tablets from the early first millennium.

seem impressively complex, in fact it turns out that various simple strategies were used to minimise the difficulties involved. Nevertheless, the professional land surveyors show a level of mathematical competence over and above, but not completely divorced from, the basic arithmetical and metrological facts mastered in the contemporary scribal schools (§7.5).[1]

7.1 BACKGROUND AND EVIDENCE

The first quarter of the first millennium BCE was a time of political turmoil for Babylonia. Aramaic-speaking tribes settled in large numbers along the Euphrates and in the southern marshes, often threatening the prosperity and stability of the ancient urban centres around Sippar, Babylon, Nippur, and Ur. Political relationships with Assyria to the north and Elam to the east were often fraught. With the exception of an eighth-century cache of tablets found at Nippur (§7.2) written evidence for the period comes almost exclusively from official monuments, land grant stelae (see §6.3), and Assyrian annals (see §5.3).

Babylonia came under direct Assyrian control in 728 BCE but constantly sought independence, provoking large-scale warfare. In 689 king Sennacherib laid siege to Babylon itself, and deported scholarly tablets and scholars to Nineveh (§5.4). Intellectual life managed to carry on in the devastated city, however: at least two observations of lunar eclipses are recorded during the decade before Sennacherib's son Esarhaddon (r. 680–669) began his great rebuilding programme.[2] He in turn placed one son, Aššurbanipal, on the throne of Nineveh and another, Šamaš-šumu-ukīn, on the throne of Babylon. But the latter was not content to remain subordinate forever. His claim for Babylonian independence in 652 led to a second long siege of the capital and another major destruction in 648. But as before, the astronomers somehow managed to keep observing, despite the famine, pestilence, and even cannibalism that ravaged Babylon.[3]

When the Assyrian empire finally fell to the Medes and Babylonians in the late seventh century, Babylonia formally gained control of its erstwhile oppressor but left the major cities of Assyria in ruins. The newly independent empire took over many of Assyria's former western provinces too and governed them in much the same way as before. Babylon itself was reconstructed by Nabopolassar (r. 626–605) and Nebuchadnezzar II (r. 604–562), so that the palace complex and the nearby temple compound of Marduk, tutelary deity of Babylonia, were even more splendid and awe-inspiring than they had ever been before. The glorious Processional Way of blue-glazed baked brick dates to this period. Meanwhile the Persians, who had ruled western Iran since the seventh century, had been expanding their empire northwestwards (as well as eastwards) so that by the mid-sixth century Babylonia was an island in an otherwise Achaemenid sea stretching from the Indus valley to the Bosphorus. The conquest of Babylon in October 539 made little impact initially on the lives of the urban elite. In fact it brought much-needed stability to Babylonia, which had suffered great political and economic uncertainty under its last indigenous ruler, Nabonidus. Indeed it was not until the reign of Xerxes in the 480s that major social and political changes broke the habits of literacy and numeracy that we are concerned with here.[4]

The largest sources of textual evidence currently exploited by historians come from two major temple archives: Ebabbar, temple of the sun god Šamaš in Sippar, and Eana, temple of the great goddess Ištar in Uruk. The private archives of wealthy, urban families and their businesses also survive, most notably those of the extensive Egibi family in Babylon and the Murašu family in Nippur. This chapter, however, is concerned with a smaller and less influential family, a branch of the descendants of Ea-ilūta-bāni in Borsippa. It also examines land records that were probably drawn up for the central administration in Babylon during the reign of Darius I (r. 521–486 BCE). The few examples with surviving calculations turn out to be particularly useful.

The relationships between capacity measures had altered since Old Babylonian times (table A.4): the size of the *sila*, Akkadian *qû*, had remained constant at about 1 litre, but there were now considered to be just 6 *qû* in a *sūtu* (Sumerian *ban*) where formerly there had been 10. The *qû* itself was now subdivided into 10 *akalu* instead of 60 shekels as earlier. Nevertheless, the basic concept of capacity remained the same. Length measures had shifted in similar ways: the cubit remained fixed at about 0.5 metres but was now composed of just 24 fingers instead of the previous 30. The rod had been lengthened to comprise 14 cubits, but more commonly used now was the reed, at 7 cubits. Area measurement had changed more drastically. The standard unit of area was no longer the *sar*, or square rod, but the square reed, 7 × 7 cubits. It was subdivided into the areal cubit = 7 × 1 linear cubits, and the areal finger = 7 cubits × 1 finger (see §7.4). Alternatively, area could be defined with respect to capacity measures by using seed ratios, as shown in the example below. Weight measures remained almost completely unchanged, except that multiples of the talent (like the *kurru*) were now counted not sexagesimally but decimally.

Strictly mathematical sources from this period are few and far between—partly because of the practical impossibilities of dating tablets without archaeological context. More often than not it is possible only to determine that they were written some time in the first millennium (table B.19). As understanding of palaeography and orthography progresses some of the word problems and less complex tables will doubtless be assigned to the eighth to sixth centuries discussed here. It is equally possible that some of the unprovenanced school tablets from northern Babylonia and Ur analysed in this chapter were actually written in the fifth century or later, but in any case they form part of a curriculum that was established no later than the seventh century (§7.2). The known mathematical tablets comprise two word problems, four tables of squares, and over thirty metrological lists (table B.18). Most are from northern Babylonia—Kiš and Babylon where provenanced—two from Nippur and one from Ur. Most are scrappily written pedagogical exercises, often on the same tablets as other school exercises, but two of the tablets from Kiš are beautifully executed library copies.[5]

Almost no mathematical problems or calculations can be dated with certainty to the first half of the first millennium; without firm chronological evidence I have taken almost all of them to be from later in the millennium (see §8.1). However, two word problems seem to belong to the Neo-Babylonian period, not only because of the shape of the tablets and the style of writing, but also because of the approaches adopted to solving the problems. The obverse of BM 47431 (table B.18) is filled with a numbered diagram of four touching circles inside a square, just like the Old Babylonian drawings on TSŠ 77 (table B.10) and problem (36) of BM 15285 (§2.4, figure 2.10). The reverse names the component figures and states their areas, just as BM 15285 asks the reader to do. A statement at the end asserts that the tablet was 'written in order for the [. . .] scribe to see', suggesting that it is indeed such a student exercise to be shown to a teacher. In all these respects—subject matter, structure, function—it shows a remarkable degree of continuity with the pedagogical mathematics of a millennium before. But there the similarities stop. Although the lengths on the diagram are all stated sexagesimally, the areas on the reverse are given not in *sar* and *iku* but in 'seed measure' and the spelling conventions and technical terminology are much closer to the language of contemporary land surveying and temple administration than they are to Old Babylonian mathematics.[6] By contrast, BM 78822 (figure 7.2) states and solves a question about finding the side of a rectangular field, a problem type that goes back at least as far as the Sargonic period (§4.1).[7]

> The short sides are 50 (cubits). How much should I put as my long sides for 1 seed-*kurru* of plantation and 1 seed-*kurru* of cultivation?
>
> 5 steps of 3 00 00 are 15 00 00. 5 steps of 3 20 00 are 16 40 00. Put them together and (it is) 31 40 00. How many steps of 50 should I go for 31 40 00? Go 38 00 steps of 50 for 31 40 00.
>
> The long sides are 38 00 (cubits).
>
> 50   (erasure) 21 40
> 38
> 45

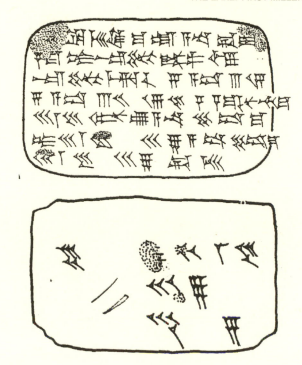

Figure 7.2 A Neo-Babylonian mathematical problem and calculation about fields. (BM 78822. Obverse, Jursa 1993–4, courtesy of Michael Jursa and *Archiv für Orientforschung*; reverse, drawing by the author.)

The problem is to find the length of two fields whose total width is 50 cubits, sown at different rates, one perhaps a date grove in which crops are sown in the cool shade of the trees (the 'plantation') and the other a grain field ('cultivation') (see figure 7.3). Just like its counterparts from earlier periods, it sets out the numerical parameters particular to the problem and asks a question. The solution is given in the form of instructions in which the calculations are made, and the answer is given at the end. So far, so familiar; but the new twist is that area is not measured by the multiplication of linear boundaries ('reed measure', see §7.4) but by 'seed measure', in which areas are considered to be proportional to one of several fixed capacity measures.

As in earlier periods the solver is not told but is expected to know—or to have looked up in a list of constants—three conversion factors, namely:

1 *kurru* = 5 *pānu* seed-measure
3 00 00 = plantation seeding rate (in cubits^2 per *pānu*)
3 20 00 = cultivation seeding rate (in cubits^2 per *pānu*)

Figure 7.3 A modern Iraqi date grove, with crops growing under the trees and an irrigated area in front. (Photo by the author.)

That is, a *pānu* of seed-grain (c. 36 litres) is sufficient to cover 3×3600 square cubits (c. 2700 m²) of plantation or $3;20 \times 3600$ square cubits (c. 3000 m²) of cultivation. Armed with these facts, we can work through the calculation very straightforwardly. The first step in the procedure is to multiply the seed-grain for each field, converted to *pānu*, by the standard seed rates, to give the area of each field in square cubits:

5 × 3 00 00 = 15 00 00
5 × 3 20 00 = 16 40 00.

In the second step, the two results are added to give the total area of the field in square cubits:

15 00 00 + 16 40 00 = 31 40 00.

Third, this total area must be divided by the width of the fields, as given at the start of the problem, to find the length. The problem is phrased in terms of finding a multiplicand ('how many steps of 50 should I go?') rather than using a divisor—or even, as in the Ur III to Kassite periods, just stating the reciprocal of 50, memorised as an entry in the regular sexagesimal reciprocal table (§§3.4, 4.2, 6.1). Traces of this final calculation remain on the reverse, which help us to reconstruct how the calculation was performed (figure 7.2).

The 50 is written to the left of the 31 40 (00), with an erased area in between.[8] The correct answer, 38 (00), is immediately beneath the erasure, itself partially erased, and the figures 40 and 5 immediately beneath them. This layout suggests that this step of the problem was solved by trial-and-error multiplications: the scribe chose a multiplicand of 50 which he knew by experience would result in a product close to the 31 40 (00) required. He wrote the two numbers at the left and lower edges of the tablet. Perhaps he chose 45 first; perhaps those numbers are left over from an earlier calculation.

$$50$$
$$45 \ (00)$$

He wrote the answer, 37 30, in the spot that is now heavily erased, so that the three numbers formed three corners of an imaginary rectangle on the clay:

$$50 \quad 37 \ 30 \ (00)$$
$$45 \ (00)$$

But this product is too large, so the scribe tried again with a smaller estimate, 40 (00), which he wrote immediately above the first, erasing the first answer and writing the second one over it:

$$50 \quad 33 \ 20 \ (00)$$
$$40 \ (00)$$
$$45 \ (00)$$

This of course is still too large, but only by 1 40 (00)—which the scribe must have known was twice 50. He thus erased the final 10 wedge from the 40 (this erasure is still visible on the tablet) and the 33 of the 33 20. This area was now too pitted for re-use and not worth smoothing over, so he took the final 20 as the first digits of the new answer, writing thus:

$$50 \quad \overline{33} \ 21 \ 40 \ (00)$$
$$38 \ (00)$$
$$45$$

There was apparently no need to alter the 20 to 30, for he transferred the answer correctly to the front of the tablet. Thus we can see that the scribe was actually working through the problem as he wrote it: this was no mere copying exercise.

The dimensions of the field—width 50 cubits (c. 25 m), length 2,280 cubits (c. 1.14 km), at a ratio of about 45:1—are commensurate with the

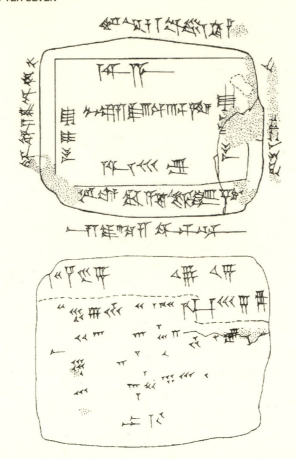

Figure 7.4 A Neo-Babylonian field plan, probably from Babylon. (BM 46719. Nemet-Nejat 1982, no. 2, courtesy of the Pontifical Biblical Institute; plan by the author.)

measurements of nearly fifty fields bought by the Egibi family of Babylon in the period c. 560–500 BCE. The widths of those fields range from 20 cubits to 150, the lengths from 50 cubits to 3,750, and their length-width ratios from 3:2 to 125:1.[9] The Egibis' purchase contracts almost always quantify the areas of their newly acquired fields and date groves, and list the neighbouring properties in a manner similar to that of the Kassite land grants (§6.3). About a third of them explicitly state the measurements of the boundaries, but visual depictions of the Egibis' land are rare.[10]

But Neo-Babylonian field plans do exist. The tablet BM 46719, for instance, is an undated field plan written in northern Babylonia during the reign of Darius (figure 7.4).[11] It has a length-to-width ratio of nearly 13:1

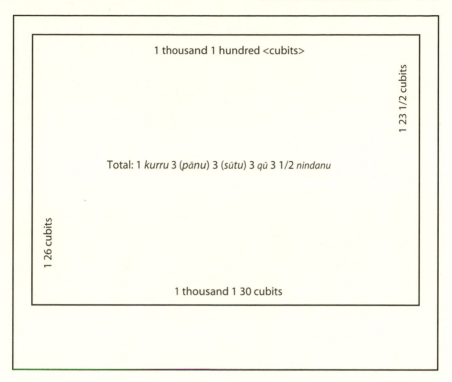

Figure 7.4 (*Continued*).

but as expected it is not drawn to scale (§3.2, §6.3). The cardinal directions of the field boundaries are written on the edges of the tablet, together with damaged details of the neighbouring properties. The dimensions of the boundaries are recorded, along with the area in seed measure. This was calculated by the traditional 'surveyors' method', as shown by the half-erased calculations on the reverse. Above the dotted line (which is not on the original tablet), they read:

1 24 45   18¹ 15
1 thousand 1 35 cubits

The number on the top left is the mean of the two widths: (1 26 + 1 23;30)/2 = 1 24;45 (cubits), while the number on the right is the mean of the two lengths: (18 20 + 18 10)/2 = 18 15 (cubits). Underneath this the average length is written decimally as 1 thousand 1 35, or 1095 (cubits). It is evident from this example that calculations were still carried out in the sexagesimal place value system outside school contexts, even when the preferred recording format used partly decimalised absolute value.

Just as in the mathematical problem, the area of the field is given not in square cubits (1 24;45 × 18 15 = 25 46 41;15, or 92,801.25 cubits²) but in seed measure. As the mathematical problem BM 78822 suggests, and as we shall see again below (§7.4), the scribe probably avoided dividing but rather compared his result in square cubits with integer multiples of the seed ratio per *pānu*, *sūtu*, *qû*, and *akalu*. However, none of the traces of numbers on the reverse of the tablet can currently be identified with any stage of the putative calculation.

7.2 LIBRARIES AND SCHOOLS: THE FORMALISATION OF THE FIRST-MILLENNIUM SCRIBAL CURRICULUM

Although very few mathematical tablets survive from the early and mid-first millennium, almost all of them come from identifiable archaeological contexts. The oldest are two metrological lists, found in American excavations at Nippur in 1973. They were discovered accidentally, in a cache of 128 tablets which had been used as rubble packing around the burial of a ten-year-old child, whose body had been laid to rest in a large terracotta jar in the ruins of a Kassite-period palace in the west of the city. The tablets cannot have been the child's grave goods but were rather being re-used after their working life was over, as all but fifteen of them carried letters addressed to the *šandabakku*, or city governor, of Nippur. The references the letters make to political events and personages enabled their editor, Steve Cole, to date them precisely to the period 755–732 BCE. Almost all the remainder of the tablets in the cache, including the metrological lists, are school exercises. There is no reason to believe that they, like the letters, should not be dated to the third quarter of the eighth century. They are thus amongst the very earliest-known tablets of the Neo-Babylonian period.[12]

The school tablets are extraordinarily important illustrations of scribal education at a time very under-represented in the historical record. The majority of them are *ad hoc* vocabulary lists of Akkadian proper nouns and verbal forms, written either on small tablets similar in size and shape to the letter tablets found with them (and to the Old Babylonian Type III tablets a millennium earlier) or on larger multi-columned tablets (like Old Babylonian Type I) (§4.2, table 4.5). Five of the smaller tablets carry variant lists of professional titles, written in both syllabic Akkadian and occasionally idiosyncratic logographic Sumerian. The social status of those professions ranges from royalty, senior court and temple personnel, and scholars to craftsmen and agricultural labourers.

Only one large tablet contains a standard first-millennium school exercise: the second half of a vocabulary now known as Vocabulary Sᵇ B.[13] In its fullest form each entry would have had three parts: a central column containing the logogram; its reading spelled out syllabically in a column to the left; and its Akkadian translation to the right. This exemplar, however, just has the central logographic column, as so often on first-millennium school tablets: the pronunciation and meaning of each sign would have been committed to memory and perhaps recited as the tablet was written or read.[14]

The two metrological tablets both carry the list of capacities which was to become standard in first-millennium schooling. It runs from 1 *sūtu* up to 1 *kurru* incrementally unit by unit. What makes these exemplars non-standard is the contexts in which they were written. One is on the reverse of a letter that is otherwise indistinguishable in style and content from the other letters in the governor's archive. Perhaps it was set as an exercise just like the metrological list; or perhaps it was a draft, given to the student to re-use while the clay was still workable. The other is disguised as a typical list of rations, with a male personal name set against each entry as follows:[15]

2 *pānu*	Nabû-šar-ilāni
2 *pānu* 1 *sūtu*	Iddina-aha
2 *pānu* 2 *sūtu*	Iddina
2 *pānu* 3 *sūtu*	Iddina-māra
2 *pānu* 4 *sūtu*	Dadiya
2 *pānu* 5 *sūtu*	Babiya
3 *pānu*	Mabiya

As the extract shows, the names are often grouped in threes according to sound (reminiscent of the name lists of the Old Babylonian elementary curriculum). At the end the 'rations' are correctly totalled to 15 *kurru* 2 *pānu* 5 *sūtu* (c. 2800 litres), again as if it were an administrative record of grain distributions. This dressing up of a metrological exercise is almost unique: the only known parallel is a descending list of silver weights, each assigned to a named individual—from the Sargonic period, over 2500 years earlier (§3.1; table B.6).

Together the Nippur school exercises give the impression of one, perhaps two, pupils studying under a teacher who worked as a scribe to the *šandabakku*. He set practical exercises in writing nouns, verbs, names, and measures that his apprentice(s) would need to succeed him in the job, perhaps even letting his own draft documents be re-used. There was little place for institutionalised, abstract learning in this educational setting.

The archaeologically contextualised evidence from later centuries paints a different picture. A further thirty-odd metrological lists can be placed squarely within a formal elementary curriculum that seems to have varied little from city to city. Petra Gesche has studied several hundred very fragmentary school tablets from the seventh to perhaps first centuries BCE, primarily from the northern Babylonian cities of Sippar, Borsippa, Dilbat, Kiš, and Babylon itself, as well as Ur in the south. She identifies two stages

Figure 7.5 A typical Type 1b Neo-Babylonian school tablet from Kiš. (Ashmolean 1924.1242. Robson 2004a, no. 40; photos by the author.)

of scribal education, in each of which three different types of tablet were used. The metrological lists belong to the first stage, in which exercises were written, often very incompetently, on the following tablet types:[16]

Type 1a multi-columned tablet with a long extract from one of seven 'canonical' exercises:

- the syllabary now called S^a;
- two vocabularies whose modern names are S^b A and S^b B;
- a monolingual list of deities, now known as the Weidner god list; and
- Tablets I–III of the bilingual thematic noun list called *Ur_5-ra* = *ḫubullu* 'interest-bearing loan', after its first line.

The extract runs from obverse to reverse; the exercise may end with a colophon and/or a metrological list in the final column.[17]

Type 1b obverse with a long 'canonical' extract like Type 1a; reverse contains extracts from *ad hoc* 'non-canonical' lists (for instance metrology, personal names, place names, professional designations, lexical lists), a literary work, proverbs, or administrative formulae, occasionally with a colophon in the final column (figure 7.5; table 7.1).

Figure 7.5 (*Continued*).

TABLE 7.1
Ashmolean 1924.1242, a Typical Neo-Babylonian School Exercise from Kiš

Obverse	Reverse i'	Reverse ii'	Reverse iii'	Reverse iv'
(The four surviving columns contain Syllabary S^a—a long, standard list of Akkadian syllables)		[. . .] 20 minas 30 minas 40 minas	[. . .] I [suck] I suck	(traces of words)
	[. . .] 10 [shekels] 11 [shekels] 12 [shekels] 13 [shekels] 14 [shekels] 15 [shekels] 16 [shekels] 17 [shekels] 18 [shekels] [. . .]	50 minas 1 talent 2 talents 3 talents 4 talents 5 talents 6 talents 7 talents 8 talents 9 talents 10 talents 11 talents ——— [. . .]	I reach I reach ——— I reach [I] reach [. . .] (The second member of each pair is spelled differently from the first)	

Type 1c   standard extracts from all of the 'canonical' lists in turn and no 'non-canonical' material; thus by definition no metrological content.

Metrological lists occupied a marginal position in this curriculum, being found on around 5 percent of published school tablets. Lists of capacities, weights, and lengths are attested in roughly equal numbers. Most are long extracts or entire lists (e.g., figure 7.5), very like those from eighth-century Nippur, but some consist of two or three lines repeated over and over again. There are no tables, either giving sexagesimal values of metrological units as in the Old Babylonian period (§4.2), or describing relations between metrological systems as later in the first millennium (§8.3). However, three small tablets that fall outside this categorisation do show that at least some students formally learned metrological conversions. Two speedily written exercises from Nippur give the sexagesimal values and names of fractions of the shekel—mostly unit fractions—in ascending order, while a compilation of historical and astronomical exercises from Babylon tabulates sexagesimal units of weight.[18] Just one other type of mathematical exercise is attested on Neo-Babylonian school tablets: lists of squares of integers. One Type 1b tablet contains the line '[8] steps of 8 (is) 1 04', repeated at least three times.[19] Another of the same type, from Ur, contains a table of squares from 1 to 12.[20] A single Type 2a tablet, containing sequential exercises written by a more advanced student, has a literary extract and a list of plants on the obverse, and a table of squares (originally from 1 to 30?) on the reverse.[21]

Many of these tablets have useful archaeological context. Six were excavated by an Iraqi team in the late 1970s, amongst some 260 school tablets that had been re-used as packing materials under the floors of two temples near the northeast corner of the precinct of Etemenanki, Marduk's ziggurat in Babylon. The larger temple was dedicated to Nabû of Harû (a name for his main temple in Borsippa) and the tablets found there were associated with building work carried out for Nebuchadnezzar II (r. 604–562); the smaller temple belonged to the minor goddess Ašratum 'Lady of the steppe'.[22] Almost all of them originally bore colophons naming their student authors and dedicating them to the god Nabû, either in his primary aspect or as 'Nabû of accounts'. The most complete colophon on a tablet with a metrological list reads:[23]

[To Nabû, his lord: for the life] of his breath, [the length of] his days, the health of his seed, [the well-being] of his heart, [the well-being] of his flesh, the non-existence of his illness, Nabû-ēṣir-napišti, son of Sîn-ahu, wrote the tablet and deposited it in Eğišla-anki ('House of the auditor of heaven and earth'), the temple of his lordship, in the *gunnu* receptacle. [He placed (?)] the tablet [in front of] Nabû [. . .].

Like the school tablets from House F in Old Babylonian Nippur (§4.2), the tablets had been re-used when their primary purpose had been served; but unlike their eighteenth-century counterparts they were probably not re-used *in situ*. Although the temples are not yet fully published, the excavators have given no description of the sort of furniture and fittings, such as recycling boxes, that one would expect to find in an area of scribal activity. The tablets may well have been brought to the temple as votive offerings, having been written somewhere else. A further Type 1a tablet carrying a metrological exercise, that entered the British Museum in 1892 from excavations in northern Babylonia, has a very similar colophon dedicated to the sun-god Šamaš and is thus probably from the Ebabbar temple in Sippar.[24]

Finally, about a dozen of these tablets probably come from a badly excavated seventh-century building in Kiš, whose date has since been established on stratigraphic evidence, in which large jars full of tablets were found lining the walls.[25] Frustratingly, no record was kept of the disposition of the tablets in the jars, or even exactly which tablets were found in the building, but close study of the excavation notes and museum records suggests that several hundred school tablets were found in the building, along with at least sixty library tablets. These library tablets are very different in appearance from the school tablets: they are long and elegant, with beautiful writing, and often with colophons and decorative 'firing holes'. Their contents range from myths, epics, and hymns to sequences from the standard series of medical recipes, omens, and astronomy. Cuneiform libraries are a post–Old Babylonian phenomenon, as the intellectual tradition of person-to-person memorisation (whose compositions thereby had a certain fluidity) gave way to one in which tablet-to-tablet copying prevailed. That in turn is probably the outcome of collecting, organising, and editing intellectual writings into compendia and series undertaken by scholars such as Sîn-lēqi-unninni in the later second millennium (see §6.4).[26]

Amongst the Kiš library tablets are a tiny list of capacities, from 1 *akalu* to 100,000 *kurru* (see table A.4), and a beautifully written table of squares and inverse squares of integers and half-integers from 1 to 60. In other words, the mathematical contents of the school and library tablets closely reflect each other.[27] The table of squares, however, is particularly important as the earliest attestation of a medial zero sign to represent a completely empty sexagesimal place.[28] It is frequently attested in Late Babylonian arithmetical tables.

The evidence taken together thus suggests that the mathematical elements of formal elementary scribal education in the mid-first millennium BCE consisted *only* of metrological lists and tables of squares. The following sections ask if and how this relates to the mathematics used by the professionally numerate (§7.4) and by wealthy families in their legal transactions (§7.3).

7.3 HOME ECONOMICS: NUMERACY IN A MID-FIRST-MILLENNIUM URBAN HOUSEHOLD

Which mathematical skills might affluent urban householders have needed in the sixth or fifth century? The legal documents of a couple named Aḫušunu and Lurindu, who lived in the city of Borsippa near Babylon in around 500 BCE, provide some interesting answers. Several hundred cuneiform tablets belonging to their household were clandestinely excavated and sold to museums around the world in the last decades of the nineteenth century; the archive was pieced together again on the basis of internal evidence by Francis Joannès in the 1980s.[29] Most record the affairs of Lurindu's substantial family going back six generations to the early seventh century, but twenty-seven are directly concerned with Aḫušunu and Lurindu's own financial and legal dealings. Eighteen of them contain some sort of quantitative data (table 7.2). They show three main areas of numerate activity: handling money and interest; the capacity measure of dry goods; and the measurement of land.

TABLE 7.2
Aḫušunu and Lurindu's Legal and Financial Documents

No.	Description	Date BCE	Publication
1	Aḫušunu is to receive 30 *kurru* of dates from his plantation in Ṭābānu as rent from his tenant farmer at harvest time, delivered to his house and measured with his own capacity measure.	521	*TuM* 2/3 150 (Joannès 1989, 89, 207)
2	Aḫušunu is to receive 12 *kurru* of dates from his plantation in Ṭābānu as rent from his tenant farmer at harvest time, delivered to his house and measured with his own capacity measure.	519	*TuM* 2/3 165 (Joannès 1989, 89, 219)
3	Aḫušunu receives 3 shekels of silver in partial payment of *ilku* service obligation.	513	Ist L 1643 (Joannès 1989, 155, 241)
4	Aḫušunu is owed 1 *kurru* of barley as a loan and 1 *kurru* of dates as tenancy dues.	508	Bod A 133 (Joannès 1989, 89, 286)
5	Aḫušunu and his partner Iddin-Bēl are to receive from their tenant farmer at harvest time 3 *kurru* of barley as rent on the field they co-own, as well as ⅔ of a bundle of straw.	508	*TuM* 2/3 168 (Joannès 1989, 90, 221)
6	Aḫušunu owes ½ mina of silver at interest of 1 shekel per mina per month, not including an earlier debt for whom he has pledged his servant woman Šaḫarratu.	506	Bod A 119 (Joannès 1989, 62, 280)
7	Aḫušunu owes 3 *kurru* of dates without interest, to be paid back in four months' time.	506	Bod A 169 (Joannès 1989, 292)
8	Aḫušunu must repay a debt of 1 mina 52 shekels, or his creditor will claim Aḫušunu's servant woman Nanaya-rēṣūa.	506/5	*TuM* 2/3 121 (Joannès 1989, 110, 199)
9	Aḫušunu pays a debt of 2 minas of silver, for which he had pledged his land in Ṭābānu.	505	Bod A 120 (Joannès 1989, 62, 281)
10	Aḫušunu owes 1 mina of silver at interest of 12 shekels per mina, and 15 shekels of silver to his debtor's husband.	504	NBC 8405 (Joannès 1989, 62, 353)

TABLE 7.2 (*Continued*)

No.	Description	Date BCE	Publication
11	Aḫušunu and his partner Nabû-zēru-ukīn are paid in silver by their tenant farmer for 3 years' rental arrears on an orchard.	504	*TuM* 2/3 170 (Joannès 1989, 90, 233)
12	Aḫušunu and Kāṣirtu get married; she brings a dowry of 5 minas of silver plus jewellery, furniture, and kitchen equipment.	503	NBC 8410 (Joannès 1989, 60, 355)
13	Aḫušunu pays 2¼ shekels of silver as guarantor for another man's debt.	495	NBC 8339 (Joannès 1989, 63, 338)
14	Aḫušunu and Lurindu get married; she brings a dowry of 2 minas of silver plus a house of area 5 reeds, furniture, and kitchen equipment.	493	*TuM* 2/3 2 (Joannès 1989, 3, 166)
15	Aḫušunu sells his servant woman Nanaya-ittiya and her son to Lurindu in exchange for the 2 minas of silver from her dowry.	492	*TCL* 12-13 200 (Joannès 1989, 46, 319)
16	Lurindu rents out her dowry house for an annual rent of 2 shekels of silver.	491	Ist L 1652 (Joannès 1989, 95, 246)
17	Aḫušunu must pay the full price of 7 minas 17½ shekels for a field he wishes to buy.	491	Ist L 1663 (Joannès 1989, 91, 251)
18	Aḫušunu witnesses the marriage of his old friend Nidintu-Bēl to the widow Tablūṭu.	486	Ist L 1634 (Joannès 1989, 64, 236)

Lurindu was Aḫušunu's second or perhaps even third wife. He had been economically active for about thirty years by the time of their marriage in 493 (table 7.2: 14), by which time he must have been at least fifty. Almost nothing is known of his parents' generation or earlier family members. A previous marriage, to Kāṣirtu in 503, seems to have ended in her death and produced no children (table 7.2: 12). This was Lurindu's first marriage, though: in the marriage contract she is described as an adolescent, and Aḫušunu had to ask her widowed mother Amti-Sutīti's permission to marry. A widow or divorcée, on the other hand, such as Tablūṭu (table 7.2: 18) was free to marry as she wished. If Lurindu and Aḫušunu's marriage produced any children who survived to adulthood, no evidence remains of them in the historical record.

Kāṣirtu and Lurindu's marriage contracts list the dowries they brought to their new homes. Kāṣirtu's father supplied her with:

5 minas of white, standard-quality silver; 1 silver *karallu* (jewellery) and 2 silver rings, whose weight is ⅓ mina; 1 covered bed of *musukannu* wood; 1 woman's chair of *musukannu* wood; 1 bronze cauldron for 1 *pānu* 4 *sūtu* of liquid; 1 bronze oil-bowl for [. . .] of liquid; 1 bronze censer; a bronze cup; a bronze bowl; a bronze grate.

Ten years later, after Kāṣirtu's possessions had presumably been returned to her family, Lurindu was equipped for her new life by her mother Amti-Sutīti, with the following items:

2 minas of white, standard-quality silver; 5 reeds of the south (wing of the) house, until it reaches a full 5 reeds, within Amti-Sutīti's reeds, adjacent to the house of Rīmūt-Bēl, son of Iddina, descendant of Marduk-šākin-abušu, and adjacent to the house of Iddina, son of Marduk-šākin-šumi, descendant of Dannu-Papsukkal—they will deny each other neither the right of way nor the descent [to (an amenity)], and Amti-Sutīti will pay the rent of the small house; 1 bed of willow; 1 chest of *musukannu* wood decorated with the features of a gazelle; 1 lamp of *musukannu* wood; 2 bronze bowls; 1 bronze platter; 2 overgarments; 2 KUR.RA garments of painted (?) wool; 4 tunics; 1 table; 3 chairs of willow.

Both women have substantial sums of money: 2½ kg silver plus at least another 150 g of silver jewellery in Kāṣirtu's case, 1 kg in Lurindu's. We get some idea of its purchasing power from the fact that Lurindu realises her land assets by renting out her dowry house for just 2 shekels, or about 17 g, of silver a year (table 7.2: 16).

But is it appropriate to refer to silver as 'money'? Throughout Mesopotamia silver and copper had both been used as media of exchange for millennia already. Silver began to circulate as coinage in Babylonia in the decade 490–480 BCE—that is, in the early years of Aḫušunu and Lurindu's marriage—as shown by finds of coin hoards and the evidence of documents which refer to 'stamped silver', but coins did not become the only form of currency until the Seleucid period (late fourth century BCE). Meanwhile silver was classified according to its purity, and weighed. This practice is made more explicit in a sale contract between Aḫušunu and Lurindu in which her dowry silver is exchanged for two servants, giving him full control of the money, perhaps to invest in another property. Lurindu had presumably decided that help around the house was more valuable to her than monetary capital.[30]

Aḫušunu has voluntarily sold Nanaya-ittiya his servant woman and Mār-bīti-iddin, Nanaya-ittiya's son, for 2 minas of white, standard-quality

silver alloyed at ⅛, in place of Lurindu's dowry. Aḫušunu has received the 2 minas of white, standard-quality silver from Lurindu's
dowry from Amti-Sutīti, her mother: it is paid. Nanaya-ittiya and
Mār-bīti-iddin are Lurindu's property forever. (table 7.2: 15)

Here we see that Lurindu's 2 minas of 'standard quality' silver are 87.5
percent pure—which was the normal degree of fineness at this time. Under
Artaxerxes I in the 450s the standard of purity was increased to 91.667
percent, or ¹⁄₁₂ parts alloy. Even coins, it seems, were weighed and not
counted, and could even be cut up into little bits: for the inhabitants of
Mesopotamia, so long used to valuing by weight, coinage was simply a
means of marking the purity of silver currency, not a revolutionary new
method of economic exchange.[31]

Whenever Aḫušunu borrowed money—and he seems to have been regularly in debt in the years before his marriage to Kāṣirtu (table 7.2: 6–10)—
his creditor could choose whether or not to charge him interest, and
whether that interest should accrue monthly or simply be calculated at
the end of the term. The creditor could also demand some security for the
loan: in one case Aḫušunu pledges his servant woman Šaḫarratu to his
creditor Turinnītu. She gets the financial benefit of Šaḫarratu's labour until
such time as Aḫušunu reclaims Šaḫarratu by paying back the debt:

½ mina of white, standard-quality silver, the capital of Turinnītu,
daughter of [Nabû-šumu-uṣur], descendant of Ahiya'ūtu, is the responsibility of Aḫušunu. Every month interest of 1 shekel of white
silver will grow per mina. Not including the earlier debt [of . . . minas]
for which Šaḫarratu, Aḫušunu's servant woman, is pledged to Turinnītu.
[Until] Turinnītu has been paid the [silver] with its arrears and interest, Šaḫarratu will be pledged to Turinnītu; she will serve Turrinītu.
(table 7.2: 6)

In this case, Aḫušunu seems to have decided that it was economically in his
favour to renege on the debt and allow Turrinītu to keep Šaḫarratu, who
was presumably worth less than the debt plus interest that had built up.
Otherwise, if the debt had been paid the tablet would have been destroyed.
He made a similar judgement just a few months later, defaulting on a tax
payment of '1 mina 52 shekels of cut-off silver, at ⅛ alloy' and letting another of his servant women, Nanaya-rēṣūa, remain in his creditor's service
(table 7.2: 8).

Land, though, was worth reclaiming. Just a year later Aḫušunu paid
back another 2 minas that he owed to Turrinītu, in order to get back the
arable land he had pledged against the loan:

(Concerning the) debt-note for 2 minas of white, standard-quality silver of Turinnītu, which is the responsibility of Aḫušunu, and for which

Nabû-aḫḫe-iddina, son of Kalbaya, is guarantor, and for which his arable land in the Ṭābānu district, which is shared with Bēl-kāṣir, is pledged. Bēl-iddin, Turrinītu's husband, has been paid by Aḫušunu. She (= Turrinītu) has given back the debt-note that Bēl-iddin drew up for Aḫušunu and Nabû-aḫḫe-iddina; any future document concerning the debt, wherever it appears, will belong to Aḫušunu and Nabû-aḫḫe-iddina. (table 7.2: 9)

This land in Ṭābānu was presumably the date orchard that Aḫušunu had owned and let out since at least 521 (table 7.2: 1–2). The annual rent depended on the success of the date harvest, and could be assessed only a few months before it was due. Some years were more productive than others, depending on the weather and crop conditions. In any event, the tenant was obliged to bring the contracted quantity of dates to Aḫušunu's house to be measured and assessed, as well as a requisite quantity of palm-tree by-products:

12 *kurru* of dates, tenancy dues of the date plantation, at 5 measuring vessels per *kurru*, the harvest of the field in the Ṭābānu district, belonging to Aḫušunu, are the responsibility of Nabû-šum-ukīn, son of Mušēzib, descendant of Mudammiq-Nabû. In Month VIII, he will deliver all the dates to Aḫušunu's house, according to Aḫušunu's measuring vessel; for every 1 *kurru* of dates (he will also deliver) a load of wood and a palm-leaf basket. The work service and the (payment to the) irrigation official are not paid off. (table 7.2: 2)

The contract specifies not only the capacity of dates to be brought (12 *kurru*, or over 2000 litres) but also the measuring vessel by which the quantity will be counted. As is usual in these circumstances, it belongs to the landlord, Aḫušunu, not the tenant. This particular *kurru*, then, was not necessarily a capacity measure that had been defined, standardised, and controlled by a central authority but one agreed locally between the two contracting parties. We can assume that it did not differ markedly from other *kurru* measures in use locally but neither was it rigorously identical.

On other occasions, royal standards were used, as when Aḫušunu's tenant borrowed grain from him. Here the debt contract scrupulously distinguishes between the loan, in grain, and the overdue rent, in dates:

1 *kurru* of barley, capital, and 1 *kurru* of dates, belonging to Aḫušunu, are the responsibility of (Nabû)-šum-ukīn, son of Mušēzib, descendant of [Mudammiq-Nabû]. In Month II, he will deliver all the barley, according to the royal measuring vessel, to Borsippa, at Aḫušunu's house. The dates are the remains of the tenancy dues for year 14. (table 7.2: 4)

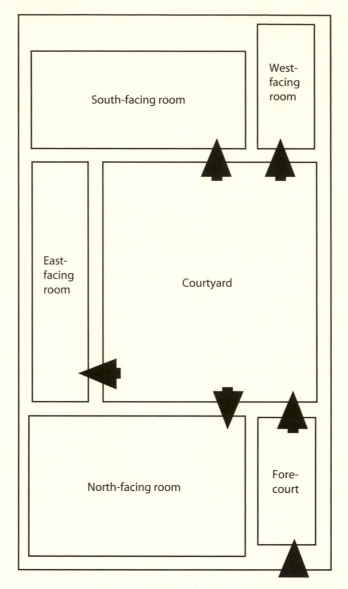

Figure 7.6 Reconstruction of a Neo-Babylonian house from its description, *left* (AO 17648: author's reconstruction), compared with a contemporaneous house plan, *right*. (Louvre A 139: Parrot 1968, 157 courtesy of Librarie Paul Geuthner.)

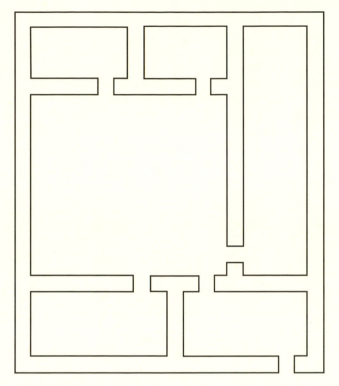

Figure 7.6 (*Continued*).

The dates, perhaps, were to be paid as usual according to Aḫušunu's measure. In any case, the measurement and storage of large quantities of agricultural products was domestic business: it happened at home, not in the fields. The women of the family were also familiar with capacity measures, as Kāširtu's dowry implies. She owned '1 bronze cauldron for 1 *pānu* 4 *sūtu* of liquid', that is about 60 litres, and '1 bronze oil-bowl for [a specified capacity] of liquid' (table 7.2: 12). It is difficult to imagine why capacity measures should be specified for vessels destined solely for kitchen use: compare Lurindu's dowry, which simply lists '2 bronze bowls' (table 7.2: 14). It is suggestive that the large cauldron—perhaps too big for everyday family culinary activity—has a capacity equivalent to ⅓ *kurru*, possibly implying that it could be used for measuring or storing harvested crops such as dates and grain.[32]

The areas of Aḫušunu's agricultural and urban properties are not generally spelled out in his documents. Only in Lurindu's dowry are we told that her house measures 5 (areal) reeds (table 7.2: 14). As the (linear) reed was about 3.5 metres in the mid-first millennium the house would have had a total floor area of $5 \times (3.5)^2$, over 60 square metres. The dowry gives both the situation of the property in relation to the rest of Amti-Sutīti's house and the names of the owners of neighbouring properties, as well as their common rights of access.

Aḫušunu and Lurindu's legal documents, then, show them to have been both numerate and financially astute. They could weigh the relative economic values of land, silver, and servants and make judgements about indebtedness. They were also capable of handling large capacity measures in legally binding contexts. There is less evidence, though, of either of them being much involved in area measures; as land and house mensuration was complex, this may well have been considered a professional domain.

7.4 MEASURING HOUSES, MAINTAINING PROFESSIONALISM

The archaeology of Neo-Babylonian cities has typically focussed on prestigious city-centre houses with a floor area of 200 square metres or more and many rooms.[33] However, contemporaneous household documents like Aḫušunu's and Lurindu's show that most urban, literate families lived in much smaller dwellings. One anonymous document—unique in that it seems to have been written for the author's own personal use—gives a sense of typical living space:

> This is the *taprīstu* ('dividing up?') of our house, which I live in:
> North-facing room: 11 cubits, 8 fingers length by 7 cubits, 8 fingers width;
> Forecourt: 6⅔ cubits length by 3 cubits, 8 fingers width;
> Courtyard: 12⅔ cubits length by 11 cubits 8 fingers width;
> South-facing room: 11 cubits, 8 fingers length by 5 cubits, 8 fingers width;
> East-facing room: 12⅔ cubits length by 3 cubits, 8 fingers width;
> West-facing room: 6⅔ cubits length by 3 cubits, 8 fingers width.[34]

Taking a cubit to be roughly 0.5 metres, composed of 24 fingers, the total living space is just under 94 square metres. Maybe it is significant that the inhabitant scribe has not calculated the area for himself. Although the description does not explain how the rooms interconnect, it can usefully be compared to a roughly contemporaneous house plan from Larsa (which, however, does not give any measurements) (figure 7.6). Although the two houses have different proportions, their layouts are very similar and it is a reasonable assumption that the doorways were likewise similarly positioned.[35]

These plans and descriptions give a sense of the interior, private space of mid-first-millennium houses, from the inhabitant's point of view. Another type of house plan, drawn up by professional surveyors, is concerned with the external appearance and location of dwellings. The majority were made during the reign of Darius, perhaps for taxation purposes, and seem to have been housed in a central archive with agricultural land surveys (§7.1), although their original archaeological context is now lost.[36] The tablet BM 47437 (figure 7.7) for instance is the plan of a derelict house in Babylon, made in 499, just six years before Aḫušunu and Lurindu's marriage.[37]

1 reed 1 cubit 7 fingers
4 reeds 6 cubits 9 fingers
Total: 6 reeds [16] fingers
Derelict property. New Town, of Holy Place, of Šahû, son of Nabû-kilanni, of the property of Marduk-šumu-uṣur, chief administrator of the Esağila temple.
Month VI, day 26, year 23 of Darius the king.
Measured.

On the front, or obverse, is the plan itself, showing a variety of qualitative information—the cardinal directions, the neighbours, the access routes—as well as primary and secondary quantitative data: the linear measurements of the two wings and the areas calculated from them. On the reverse the areas of the two wings are totalled, and the location, state of repair, and ownership of the house described. The derelict building is in New Town, in the northeast of the walled city centre of Babylon, and belongs to the most senior official of Marduk's main temple Esağila, a very grand and important person indeed.[38] Perhaps he considered this plot as an investment property.

The construction of this house plan entailed a higher level of mathematical skill than any of Aḫušunu and Lurindu's documents. Here the surveyor, having measured the house, had a whole set of complex calculations to carry out. The fact that he recorded those measurements in cubits, rather than converting to reeds where he could (for instance in the south wing) hints that the cubit was his basic unit of calculation as well as of measurement.

The tablet BM 46703, a half-finished plan that probably comes from the same archive, gives vital clues about they way that the arithmetical operations and metrological conversions were performed (figure 7.8).[39] Here the surveyor left his scribbled numbers and scrappy calculations on the back of the tablet, instead of replacing them with the ownership and location information recorded on Marduk-šumu-uṣur's plan (figure 7.7). The surveyor uses the so-called 'surveyors' method' to find the area of the irregularly quadrilateral house, a method which had been in regular Mesopotamian

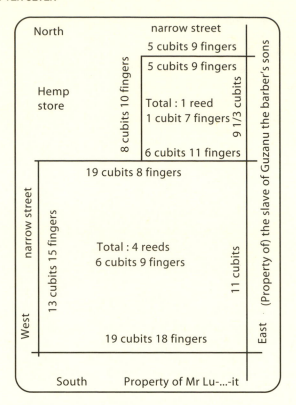

Figure 7.7 A house plan from Babylon, drawn up in 499 BCE. (BM 47437. Drawing by the author after Nemet-Nejat 1982, no. 24.)

use since the late fourth millennium (§2.1). First the opposite sides are averaged, and the results written on the top corners of the reverse:

20½ (cubits)   22 (cubits) 18 fingers

These average sides are multiplied to give the area in square cubits. But what the tablet shows is not a purely sexagesimal calculation—20;30 × 22;45 = 7 46;22 30—but rather one in which only the integers are sexagesimalised and the fractional parts kept in fingers (or rather cubits × fingers). Here for clarity cubits are denoted by c and fingers by f, although no such notation occurs on the tablet:

20½c × 22c 18f = 7 46c^2 9cf

And this is exactly what we see—7 46 09—written very untidily under the 22 18 on the right-hand side of the tablet. In fact, the pattern of erasures

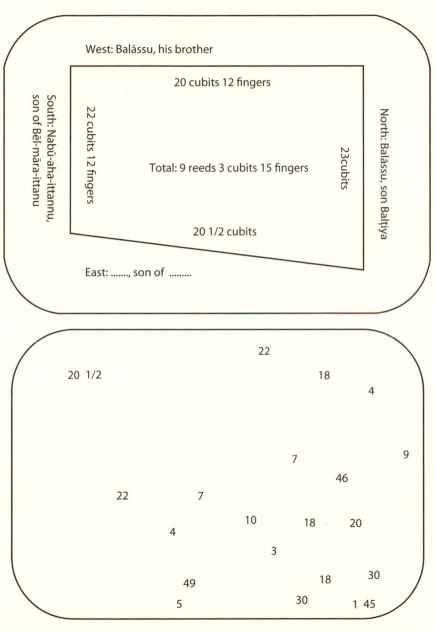

Figure 7.8 A draft house plan from mid-first-millennium Babylonia. (BM 46703. Drawings by the author after Nemet-Nejat 1982, no. 3.)

around this number suggests that the scribe dealt with the multiplication in several stages:

$(20c \times 22c =) \; 7 \; 20c^2$
$(+ \{20c \times 18f = 15c^2\} =) \; 7 \; 35c^2$ (the remains of the 5 are just
  visible)
$(+ \{½c \times 22c = 11c^2\} =) \; 7 \; 46c^2$
$(+ \{½c \times 18f = 9cf\} = 7 \; 46c^2$ $9cf$

Then he had to find the number of areal reeds in this figure. Now, 1 areal reed = $7 \times 7 = 49$ cubits. Dividing by 49 in the sexagesimal system is not a pleasant prospect. Instead, the scribe probably used his knowledge of the multiples of 49, with which as a professional land surveyor he must have been very familiar. He could ignore the fractional values for the moment. The number 49 is written at the bottom left of the tablet.

$9 \times 49 = 7 \; 21 < 7 \; 46 < 10 \times 49 = 8 \; 10$

Thus the house has an area of between 9 and 10 reeds. Taking the remainder, he does the same for areal cubits, where 1 areal cubit = 7×1 linear cubit:

$7 \; 46 - 7 \; 21 = 25$
$3 \times 7 = 21 < 25 < 4 \times 7 = 28$

He now has a more accurate value of 9 reeds, 3–4 cubits. Taking the remainder for the last time, he now finds the number of areal fingers, where 1 areal finger = $\frac{7}{24}$. This time he has to take the fractional value $\frac{7}{24}$ into account:

$25\frac{9}{24} - 21 = 4\frac{9}{24}$

At this point it is most useful for him to work in areal fingers, not square cubits, so he multiplies out by 24:

$4 \times 24 + 9 = 1 \; 45$

and compares this once more to his 7 times table:

$15 \times 7 = 1 \; 45$

This is the very last number written at the bottom of the tablet. Luckily he does not have to round up or down: the area of the house is exactly 9 reeds, 3 cubits, 15 fingers.

Returning to the plan of the derelict house owned by the senior temple administrator (figure 7.7), it is now possible to reconstruct each stage of the calculations involved, sometimes abbreviating cubits to c and fingers to f for convenience:

- Find the average lengths and widths of the north and south wings of the house:

$$(5c\ 9f + 6c\ 11f)/2 = (11c\ 20f)/2 = 5 \text{ cubits } 22 \text{ fingers}$$
$$(8c\ 10f + 9c\ 8f)/2 = (17c\ 18f)/2 = 8 \text{ cubits } 21 \text{ fingers}$$
$$(19c\ 8f + 19c\ 18f)/2 = (39c\ 2f)/2 = 19 \text{ cubits } 13 \text{ fingers}$$
$$(13c\ 15f + 11c)/2 = (24c\ 15f)/2 = 12 \text{ cubits } 7\tfrac{1}{2} \text{ fingers}$$

- Multiply the average sides to find the area of each wing in square cubits, leaving the fractional parts expressed in cubits $\times$ fingers:

$$5c\ 22f \times 8c\ 21f = 5c \times 8c = 40c^2\ 0cf$$
$$+ 5c \times 21f = 4c^2\ 9cf$$
$$+ 22f \times 8c = 7c^2\ 8cf$$
$$+ 22f \times 21f \times 19cf$$
$$= 52c^2\ 12cf$$

$$19c\ 13f \times 12c\ 7\tfrac{1}{2}f = 19c \times 12c = 3\ 48c^2$$
$$+ 19c \times 7\tfrac{1}{2}f = 5c^2\ 22\tfrac{1}{2}cf$$
$$+ 13f \times 12c = 6c^2\ 12cf$$
$$+ 13f \times 7\tfrac{1}{2}f \times 4cf$$
$$= 4\ 00c^2\ 12\tfrac{1}{2}cf$$

- Convert into area measure:

$52c^2\ 12cf$:
$49 < 52 \Rightarrow 1$ reed
$52 - 49 = 3 < 7 \Rightarrow$ no cubits
$3\tfrac{12}{24} \times 24 = 84 = 12 \times 7 \Rightarrow 12$ fingers
$\Rightarrow 1$ reed 12 fingers [cf. plan: 1 reed 1 cubit 7 fingers]

$4\ 00c^2\ 12\tfrac{1}{2}cf$:
$49 \times 4 = 3\ 16 < 4\ 00 < 9 \times 5 = 4\ 05 \Rightarrow 4$ reeds
$4\ 00 - 3\ 16 = 44$
$7 \times 6 = 42 < 44 < 7 \times 7 = 49 \Rightarrow 6$ cubits
$44 - 42 = 2$
$2\tfrac{12}{24} \times 24 = 60 \times 7 \times 9 \Rightarrow 9$ fingers
$\Rightarrow 4$ reeds 6 cubits 9 fingers [cf. plan: 4 reeds 6 cubits 9 fingers]

In the first case the surveyor's result is 19 fingers too large, an overestimate of about 1.4 square metres. Such errors are fairly common, and often arise through misreading the numbers on the plan, or through difficulties in averaging opposite sides. However, it is not clear to me in this particular instance how the error came about.

In sum, professional Achaemenid surveyors used a complex and confusingly defined metrological system that did not sit well with sexagesimal numeration. As Cornelia Wunsch suggests, this may even have been a deliberate move to professionalise and restrict access to urban land measurement—as we have seen (§7.1), agricultural land was measured and calculated with another areal system again, conceptualised in a completely different manner. Yet analysis of the actual calculations involved in house mensuration shows that the surveyor used several simple strategies to lessen the burden of calculation and conversion between sexagesimal and metrological systems. First, he recorded his linear measurements only in cubits and fingers, not converting longer lengths to reeds. Second, in his multiplications, he used the sexagesimal system only for integer multiples of square cubits, leaving fractional values in base 24, with dimensions cubits × fingers. Third, in his conversions from square cubits to area measure he avoided dividing by 49s and 7s, which are sexagesimally irregular numbers. Instead he compared multiples of these numbers with the value in square cubits in order to establish integer values of areal reeds, cubits, and fingers. His use of base 24 fractions was particularly useful at the final stage of the calculation as he simply had to compare this value (plus any remaining integers, multiplied by 24) with multiples of 7 in order to approximate the integer value of areal fingers. Nevertheless, it was an arithmetically fiddly operation which must have been learned on the job: as we have seen (§7.2), institutionalised schooling would have prepared him only to measure and write the numerals and metrological systems, not to convert between and multiply with them.

7.5 CONCLUSIONS

In the first half of the first millennium we find a low level of mathematical sophistication in school, consumer, and professional contexts. The sexagesimal place value system was used only for calculating with integers and simple fractions; the absolute decimal system was preferred for recording, and unit fractions were often used for metrologies. Reciprocals and division were avoided by using trial-and-error multiplication even where, in other periods, reciprocals of sexagesimally regular numbers were easily found.

The two methods of area calculation—urban 'reed measure' and agricultural 'seed measure'—were perhaps used by professional surveyors only. The scribe who measured the dimensions of all the rooms in his own house did not calculate their areas. The two published word problems attributable to this period cannot be situated within a context of institutionalised scribal training and may well have been part of a professional apprenticeship: the tablets on which they are written are even the same size and shape as the house and field plans of the surveyors.

Yet we should not assume that this evidence necessarily represents *all* mathematical activity at this period. Numerate scholarly works such as *Inam ǧišḫur ankia* (§5.4) and cultic topographies (§8.1) are attested in Neo-Babylonian copies. The enormous archives of the temples and family businesses will doubtless reveal a whole range of numerate practices of which we are currently hardly aware. There is also the real possibility of being able to re-date, or date more accurately, already published material as the chronology of handwriting becomes better understood. Further, a significant quantity of unpublished first-millennium mathematics (currently similarly difficult to date), especially in the British Museum, will inevitably enrich our comprehension of the period when it is edited. And we must always remember that other writing media have perished wholesale. Perhaps we will never know whether sophisticated mathematics was ever written in Aramaic but, as the following chapter shows, waxed writing boards were certainly used to record mathematics in cuneiform later in the first millennium. Perhaps we should be surprised not at the dearth of mathematical evidence from the Neo-Babylonian and Persian periods but rather that any survives at all.

CHAPTER EIGHT

The Later First Millennium

The latter half of the first millennium BCE is characterised by the end of local rule in Mesopotamia and the gradual erosion of indigenous culture in favour of Greek and Iranian practices. The locus of traditional intellectual endeavour remained the age-old temples, despite the abandonment of royal patronage. A significant quantity of mathematics survives, with interesting links to, yet developments from, the Old Babylonian period (§8.1). While firm conclusions are still frustratingly elusive it appears that mathematics and computational astronomy were closely connected under the auspices of Marduk's temple Esaǧila, where astronomical 'almanacs' were still being computed as late as the first century CE (§8.2). At Uruk individuals within three families were involved in the production and transmission of mathematical knowledge. Two generations of the Šangû-Ninurta family, active c. 400 BCE, were associated with Inana's temple Eana (§8.3), while the Ḫunzûs and Sîn-lēqi-unninnis were closely linked with Anu's

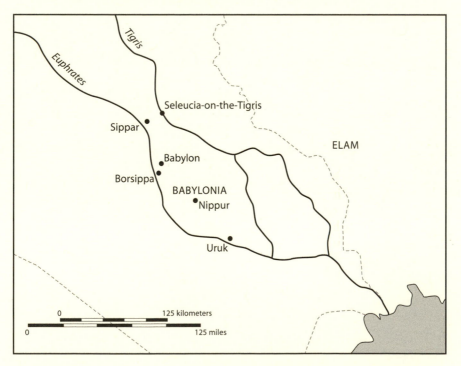

Figure 8.1 The locations mentioned in this chapter, including all known findspots of mathematical tablets from the late first millennium.

temple Rēš, c. 200 BCE. But only members of the latter family also made astronomical calculations, which suggests that mathematics was not as dependent on astronomy for its continued
existence as might have been thought (§8.4).[1]

8.1 BACKGROUND AND EVIDENCE

The later first millennium BCE, or Late Babylonian period, splits conveniently into three roughly equal phases, although confusingly each is commonly known by two different names: the late Persian or Achaemenid
period (c. 485–330), the Hellenistic or Seleucid period (c. 330–125), and
the Parthian or Arsacid period (c. 125 BCE–225 CE). The indigenous system of dating to the Seleucid Era (SE), however, was introduced only in 305
BCE with a retrospective start date of 311 BCE. For much of the later first
millennium Babylonia was a relatively peaceful province of vast empires
whose non-indigenous rulers tolerated and occasionally supported the traditional life-ways while promoting Greek or Iranian culture among the
local elites. Antiochus I (r. 280–261), for instance, funded the rebuilding of
Marduk's great temple Esagila in Babylon while moving the political capital
northwards to the newly founded Seleucia-on-the-Tigris.[2]

Anti-Persian revolts early in the reign of Xerxes (r. 485–465) led to the
execution of many leading figures in northern Babylonian society and the
resultant winding down of their archives, both private and institutional.[3]
Although some smaller archives continue uninterrupted throughout the rupture, it marks the end of large-scale production of cuneiform tablets in Babylonia. By the third century BCE only Uruk, Babylon, and Borsippa are known
as major centres of cuneiform literacy; by the first century just Babylon remained as the last bastion of cuneiform culture, with tablets being produced
there at least until the late first century CE if not later.[4] The latest securely
dateable cuneiform tablet is an astronomical almanac from Babylon for
75 CE.[5] The majority of documentation, including all institutional administration, continued to be in alphabetic Aramaic, written on papyrus, leather
rolls, ostraca (potsherds), and wooden writing boards.[6] Nothing of this survives from Babylonia; all that remains are references to these media on the
relatively indestructible clay tablets. Cuneiform became restricted to legal,
religious, and scholarly functions, for an ever-shrinking number of users.
But cuneiform was often written on writing boards too, so that even this
portion of the written record can never be recovered in its entirety.

As of late 2006 some sixty mathematical tablets have been published
from the later first millennium BCE, nearly two-thirds of which are from
Babylon (table B.21), about a quarter from Uruk (table B.20), and the rest
from Nippur, Sippar, or unknown locations (table B.19).[7] Many of them,
especially from Uruk, close with colophons stating the owner, scribe, date

of writing, and other invaluable evidence about the circumstances of their creation (§§8.3–4). Like their less numerous Neo-Babylonian counterparts (§7.1), they are broadly similar in content to their Old Babylonian precursors but their presentation tends to be very different.[8] Reciprocal tables, often for sexagesimally regular numbers up to six places long, run in ascending rather than descending order;[9] weights and measures (table A.4) are grouped and presented differently in metrological tables;[10] some multiplication tables are for hitherto unattested multiplicands; and new methods are used to solve old problems, with new terminology.

The use of mathematics for religious purposes also flourished in the Late Babylonian period: a whole genre of compositions records the dimensions of the great cultic complexes of the major deities including Marduk's temple Esağila in Babylon and its associated sanctuaries.[11] Conceptually these compositions are the descendants of the linear descriptions of space first encountered in the later third millennium (§3.2). Archaeological findings and information from colophons indicate that at least some of them had a life outside the architect's office, being carefully copied and stored in scholarly libraries. The so-called 'Esağila tablet', copied in Uruk as well as Babylon (tables B.20, B.21), even uses the measurements of the great courts and the ziggurat Etemenanki to the north of Esağila as a pretext for some simple metrological exercises in different kinds of reed and seed-measure (§§7.1, 7.4). This extract uses both reed and seed measure:

> For [you] to see the measurements of the base of Etemenanki, long side and short side:
>
> the long side is 3 chains, the short side 3 chains, by the cable-cubit. To multiply their calculation 3 [steps of 3] is 9; 9 steps of 2 is 18 (*sūtu*).
> If you do not understand 18 it is 3 *pānu* by the small cubit.
> The base of Etemenanki. The height is in accordance with the long side [and short side].[12]

The author assumes basic metrological data as background knowledge:[13]

> 1 chain = 60 cubits (c. 30 m) and 1 cable = 2 chains in the small or cable-cubit system
> 1 square chain = 2 *sūtu* (approx. 900 m²) and 6 *sūtu* = 1 *pānu*

The first multiplication thus gives the base area of the ziggurat as 9 square chains. This reed-measure area is multiplied by the factor 2 to give an equivalent in seed-measure, namely 18 *sūtu*. One last conversion is made, in order to end up at the smallest integer answer in the same metrology. The final statement appears to mean that the overall height of the multi-storey

EINE WIEDERGABE
VON BABYLON
zeigt den Turm von Babel
und die Brücke über den Euphrat

Figure 8.2 The Etemenanki ziggurat in Babylon. (Courtesy of Bildarchiv
Preußischer Kulturbesitz.)

ziggurat is the same as the square base; and it is on this basis (amongst others) that the real ziggurat, which has deteriorated greatly over the millennia, has been conjecturally restored (figure 8.2).[14] The base is indeed about 90 metres square.

By casting the net more widely it is possible to more than double the Late Babylonian mathematical corpus. Anything which has been 'written for the purpose of communicating or recording a mathematical technique or aiding a mathematical procedure to be carried out' is by this (I hope) unexceptional definition a mathematical document, whatever the application of that mathematics may be.[15] With that definition in mind, a further category of Late Babylonian mathematical writings immediately commands attention, and a very well-published and edited category it is too: the so-called astronomical procedure texts, or precepts.

Babylonian scholars began to develop period relations to describe and predict key celestial events began no later than 500 BCE and perhaps as early as the eighth and seventh centuries, driven, at least initially, by the needs of celestial divination.[16] The most sophisticated mathematical astronomy comes from the Seleucid Era: on the one hand several hundred extant ephemerides, or tabulated predictions of the positions of the sun, moon, and five visible planets at key moments in their paths across the sky; and on the other hand about seventy precepts and associated auxiliary tables, which provided instructions and reference data for calculating the ephemerides.[17] Roughly speaking, there were two means by which celestial positions were calculated, which historians now call Systems A and B. System A, fully developed by 320 BCE and mostly attested at Babylon, is based on so-called step functions (as we would describe them), in which 'fixed zones of constant amplitude . . . vary discontinuously at their boundaries' and 'the variable quantity—typically a synodic arc—is expressed strictly as a function of longitude'. System B, perfected by about 260 BCE, and apparently used mostly at Uruk, was based on linear zigzag functions, in which a variable 'increases or decreases by a constant amount in successive intervals, reflecting off some fixed maximum and minimum' (figure 8.3).[18] Simple linear zigzag functions are much older than the Seleucid Era: they are first attested in the Old Babylonian period.[19] These 'functions', of course, appear solely as patterns in the numerical values of the variables given in the tables of ephemerides, or through the instructions given in the precepts, not in any graphical, abstract, or symbolic formulation.

Attempts were also made to find patterns in other apparently random natural phenomena. Scholars kept and copied six-monthly diaries of celestial, atmospheric, and terrestrial observations, from at least the mid-seventh century, and perhaps even a century earlier. As well as lunar, solar, and planetary events, their compilers recorded the weather, the occasional

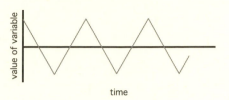

Figure 8.3 Modern graphical depictions of a step function (*left*) and a linear zigzag function (*right*).

appearance of comets and meteors, the level of the Euphrates at Babylon, notable political events, and the market prices of grain, wool, dates, and two spices. They appear to have been looking for periodicities in these long-term phenomena that were not controllable by human intervention. Two tablets from Seleucid Uruk relate the weather to planetary periodicities, while another predicts the fortunes of business from the movement of the planets. One badly understood tablet, copied in about 200 BCE, mixes observational and computational rules for predicting and interpreting eclipses and other lunar phenomena. Another scholarly genre now (unsatisfactorily) known as 'calendar texts', with about a dozen published exemplars from Uruk and Babylon, uses a simple scheme for lunar motion to associate days of the year and zodiacal with medical ingredients and incantations.[20] As shown below (§§8.3, 8.4), many of these scholarly genres are found in the same family libraries as mathematical writings.

Is it really justifiable to consider cult and astronomy as domains of mathematics? The Late Babylonian scribes appear to have done so. One compilation of otherwise unexceptional metrological tables from Uruk begins with a list of integers associated with the major deities and ends with a month-by-month table of shadow-clock shadow lengths (table 8.2:10):

[1]   Anu

[2]   Ellil

[3]   Ea

[4]   Sîn

[5]   Marduk (?)

[6]   Enki (?)

[7]   The Seven gods

8   The Igigi gods

9   The Anuna gods

10	Bēl
20	Šamaš
30	Sîn
40	Ea
50	Ellil.

A metrological table was incorporated into the colophon at the end of the 'Esaǧila tablet', the most famous of the cultic topographical collections, quoted above (and see further §8.4). Two compilations of astronomical precepts from Babylon have buried within them mathematical problems on finding the area of a trapezoid (§8.2). The intellectual context of Late Babylonian mathematics, and the social position of its practitioners, is the subject of the rest of this chapter.

8.2 BABYLON: MATHEMATICS IN THE SERVICE OF ASTRONOMY?

The majority of published Late Babylonian mathematical tablets are from the city of Babylon (table B.21), which over the years has been excavated by several official academic expeditions with recording techniques ranging from minimal to substantial. In theory, then, one could hope for good archaeological context for at least some mathematics. But although the tablets excavated from Babylon by the Deutsche Orientgesellschaft in the early twentieth century were recorded meticulously, a contextual catalogue has only recently been published—and in the meantime many of the tablets have been lost or destroyed. Just three Late Babylonian mathematical tablets from those excavations have provisionally been identified, none with any archival context.[21] The published scholarly material from Babylon is almost without exception from the British Museum's digs of the late nineteenth century, which predate the ability to recognise and record mudbrick architecture and whose aims were limited to the efficient, undocumented recovery of objects for the museum's collections. At the same time, museum staff were buying large batches of illicitly excavated tablets from antiquities dealers in Baghdad and London, which naturally had no documentation at all. However, museology—the study of museum records, in particular the circumstances of acquisition and accession—can shed some light on those tablets' find circumstances.

The British Museum assigns accession numbers to all its incoming tablets, comprising the date of registration and a running number within the lot of tablets registered that day. Museum numbers are then allocated, which run from 1 to (now) over 130,000. In recent decades the tablets acquired in the late nineteenth century, including much Late Babylonian mathematics, have been systematically catalogued and much has been discovered about the circumstances of their arrival at the museum via excavators and/or dealers. For the most part they arrived in coherent lots, in which tablets from larger archives remained more or less clustered. Three-quarters of the published mathematical tablets (and many more unpublished ones) are from just two accession lots, Sp II and 1880-6-17—which also contain large numbers of astronomical pieces, both theoretical and observational. The 1880-6-17 collection, comprising some 1800 tablets and fragments, was excavated in Babylon as part of an ongoing series of British Museum digs all over Babylonia and Assyria under the supervision of museum employee Hormuzd Rassam and his foreman Daud Thoma. Most of this particular lot was found between October 1979 and January 1880 while Rassam was in London, and accessioned on 17 June 1880. The pieces come from the Amran mound in the southern corner of Babylon, in or around Marduk's temple Esağila (figure 8.4), and from the temple of Nabû, god of writing and scholarship, in nearby Borsippa. No detailed records of findspots were made.[22] The Sp II collection, comprising nearly a thousand pieces, was acquired from the London dealers Messrs. Spartali & Co. and reached the museum in August 1879.[23] As many of the astronomical pieces in the Sp II collection join tablets from other groups acquired through excavation or purchase in the late 1870s and throughout the 1880s, it is clearly not a discrete archive. By contrast there are just half a dozen direct joins between the astronomical tablets of the 1880-6-17 collection and those in other batches. Roughly speaking then, 1880-6-17 comprises a reasonably intact find group with minimal leakage to and from other groups of tablets (although, as will become clear, the situation is a little more complicated than stated here).

The 1880-6-17 collection contains about seventy published astronomical tablets, with the latest dateable diary from 130 BCE and the latest dateable ephemeris from 105 BCE.[24] Four colophons on those tablets mention members of the Mušēzib ('Saviour') family, for whom Joachim Oelsner has constructed a family tree, using data from a variety of scholarly colophons and legal data (table 8.1).[25] Their tablets paint a fascinating picture of mathematical astronomy in flux. System A theory is thought, from the dates of the earliest ephemerides, to have been fully developed by around 320 BCE. Bēl-apla-iddina, who wrote his observational diary in 322 BCE (table 8.1: 1), thus belongs to the first generation of System A astronomers. He owned a copy of a System A precept (table 8.1: 3), as well as a precept for Venus which uses a non-System A ('atypical') step function scheme (table 8.1: 2). His son Marduk-šāpik-zēri also owned copies of both System A (table 8.1: 7) and atypical (table 8.1: 5, 6) precepts, all copied by his own son Iddin-Bēl. The latter copied further atypical precepts (table 8.1: 8, 9) and owned old

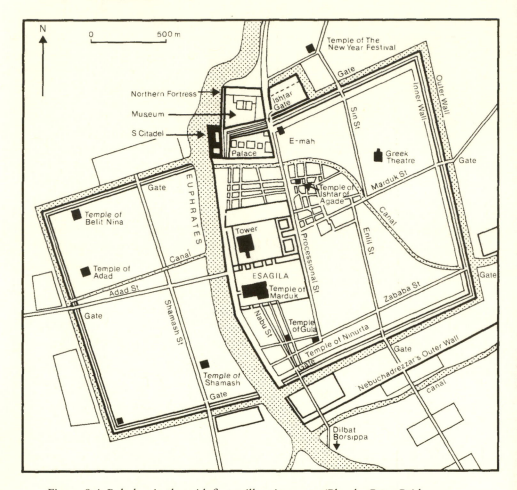

Figure 8.4 Babylon in the mid–first millennium BCE. (Plan by Peter Bridgewater in Oates 1986, 148 fig. 100, courtesy of Thames & Hudson Ltd., London.)

observational records (table 8.1: 10) as well as the standard celestial omen series *Enūma Anu Ellil* and an incantation (table 8.1: 11, 12).[26] Seven of the remaining eight known atypical astronomical precepts and tables are also from the 1880-6-17 collection. As none of those eight contain extant colophon information, it is tempting to associate them with the Mušēzib men too.[27]

But it is not simply a matter of associating the 1880-6-17 collection with the Mušēzib family of astronomers: until the British Musums' Babylon catalogue is fully published, much remains unknown about the contents of the rest of the collection—and, as table 8.1 shows, some of the Mušēzibs'

TABLE 8.1
Scholarly Tablets of the Mušēzib Family of Babylon

No.	Description	Colophon	Publication
Bēl-apla-iddina and [. . .] -Bēl, sons of Mušallim-Bēl			
1	Diary for 322 BCE	From [the tablet of?] Bēl-apla-iddina, son of Mušallim-Bēl, descendant of Mušēzib, which he wrote for his good health; which was in the archive of [. . .] written and (blank).	BM 34093+ = Sp, 192 + 544 + Sp III, 281 (Sachs and Hunger 1988–, I no. −321)
2	Atypical precept for Venus	Long tablet of Bēl-apla-iddina, son of Mušallim-Bēl, descendant of Mušēzib.	BM 33552 = Rm IV, 108 (Britton and Walker 1991)
3	System A3 precept for Mercury	[Tablet of?] Bēl-apla-iddina, son of Mušallim-Bēl, descendant of Mušēzib. (Must be a copy of an older tablet because of the phrase "new break", rev. 8.)	BM 36321 = 1880-6-17, 47 (Neugebauer 1955, no. 816 Zu)
4	Diary for 325 BCE	From [the tablet of? . . .]-Bēl, son of Mušallim-Bēl [. . .].	BM 35024+ = Sp II, 551 + 597 + 625 + Sp III, 129 + 144 + 1881-7-6, 394 (Sachs and Hunger 1988–, I no. −324)
Marduk-šāpik-zēri, son of Bēl-apla-iddina, and his son Iddin-Bēl I			
5	Atypical precept for lunar latitude and planetary periods	Written and checked from a wax writing board. Tablet of Marduk-šāpik-zēri, son of Bēl-apla-iddina, descendant of Mušēzib. Hand of Iddin-Bēl, son of Marduk-šāpik-[zēri, descendant] of Mušēzib.	BM 41004 = 1881-4-28, 551 (Neugebauer and Sachs 1967, text E)
6	Atypical precept for planetary motion and latitudes	[Tablet of] Marduk-šāpik-zēri, son of Bēl-apla-iddina, descendant of Mušēzib. Hand of Iddin-Bēl, his son.	BM 37266 = 1880-6-17, 1022 (Neugebauer and Sachs 1967, text F)

TABLE 8.1 (*Continued*)

No.	Description	Colophon	Publication
7	System A precepts for Jupiter, Saturn, and Mars	According to an original on a wooden writing-board, of [. . . . . .]. Not to be lost. Long tablet of Marduk-šāpik-zēri, [son of Bēl-apla-iddina], descendant of Mušēzib. Hand of Iddin-Bēl, his son.	BM 33801 = Rm IV, 361 (Neugebauer 1955, no. 811 Zs)
8	Atypical table and precepts for first and last visibility of the moon and lunar velocity	[The knowing may show the knowing; he shall] not show it [to the unknowing.] Iddin-Bēl [. . . . . . wrote it according to] its original. By the command of Bēl and Bēltī, lords of the gods of heaven [and earth, may it go well].	BM 36772+ = 1880-6-17, 445 (+) 1881-2-1, 47 (Neugebauer and Sachs 1969, text K)
9	Atypical table for lunar and solar motion; atypical precept for Mars, dated 319 BCE	This is for the 5th year of Philip. Hand of Iddin-Bēl, son of Marduk-šāpik-zēri.	MNB 1865 (Neugebauer and Sachs 1969, text H; Durand 1981, pl. 82)
10	Observations of Jupiter for 387–346 BCE	Copy of tablets and writing boards [of] diaries belonging to [Bēl-apla-iddina (?)], son of Mušallim-Bēl, [descendant of Mušēzib]. Tablet of [Iddin-bēl (?)], son of Marduk-šāpik-zēri, [descendant of Mušēzib]. Hand of Bēlšunu, son of [. . . . . .].	BM 34750+ = Sp II, 241 + 332 + 662 (+) 901 (Sachs and Hunger 1988–, V no. 60)

Iddin-Bēl, son of Marduk-šāpik-zēri, and Nergal-ina-tēšê-ēṭir, son of Gugua

No.	Description	Colophon	Publication
11	From the standard celestial omen series *Enūma Anu Ellil* Tablet 24	[Written, checked and] made good [according to its original]. Tablet of Iddin-Bēl, son of Marduk-šāpik-zēri, descendant of Mušēzib. Hand of Nergal-<ina>-tēšê-ēṭir, son of Gugua. [Whoever fears Nabû] and Tašmētu shall not secretly remove it; shall not deliberately remove it from there. [. . . . . .]	BM 36319 = 1880-6-17, 43 (van Soldt 1995b, text G)

TABLE 8.1 (*Continued*)

No.	Description	Colophon	Publication
12	Bilingual incantation from the series SAG.GIG	Written, checked, and made good according to its original. Tablet of Iddin-Bēl, son of [Marduk-šāpik]-zēri, descendant of Mušēzib. Hand of Nergal-<ina>-tēšê-ēṭir, son of Gugua.	BM 33534 = Rm IV, 90 (Thompson 1903, no. 33; Hunger 1968, no. 417)

Itti-Marduk-balāṭu, son of Iddin-Bēl II, and his son Bēl-aḫḫe-uṣur

No.	Description	Colophon	Publication
13	*Epic of Gilgameš*, Tablet X	Tablet 10, series of Gilgameš, not finished. Written, checked, and made good according to its original. Tablet of Itti-Marduk-balāṭu, son of Iddin-Bēl, descendant of Mušēzib. Hand of Bēl-aḫḫe-uṣur, his son. Whoever fears Bēl and Bēltiya shall return it [and shall not take it away. Babylon (?)], Month IX, day 15, year 1 hundred [. . . Arsaces, king of kings . . .].	BM 34160+ = Sp, 265 (George 2003, MS X b, pls. 114–5)
14	Astronomical compilation *Mul-Apin*, Tablet I	[. . . . . .] Itti-Marduk-balāṭu, descendant of Mušēzib, [. . .] Tašmētu and Kizaza (?) of the Ezida temple [. . . . . .] Seleucus the king.	BM 32311 = 1876-11-17, 2040 (Hunger and Pingree 1989, text K)

Uncertain attribution

No.	Description	Colophon	Publication
15	*Epic of Gilgameš*, Tablet XII	[. . . . . .]-Bēl, descendant of Mušēzib, Hand of [. . . . . .], junior [. . .]. Whoever fears Bēl [must not] steal (?) [it].	BM 30559 + 32418 = 1876-11-17, 286 + 1876-11-17, 2152 (George 2003, MS XII a, pl. 147)
16	Precept for the moon, System A	[. . .] Jupiter [. . . . . .] Marduk-šāpik-zēri, son of [. . . . . .].	BM 36004 = Sp III, 547 + 1882-7-4, 107 (Neugebauer 1955, no. 207ca Zrb)

tablets were found in other collections too. It is not even certain exactly where or how the tablets were found. In the early twentieth century, German archaeologists found contemporaneous libraries both in the Esağila temple and in neighbouring houses, at least one of which included astronomical material. But there are not enough data in the recently published catalogue to enable the authors of those tablets to be identified.[28] The 1880-6-17 collection thus yields nothing but a circumstantial association between mathematics and theoretical astronomy. To determine whether they really were written by the same people the internal, textual connections and similarities between astronomy and mathematics need to be examined.

As already noted, two compilations of System A astronomical precepts for Jupiter have mathematical problems buried within them, both on finding the area of a regular trapezoid (§8.1, table B.21). In their structure and terminology the astronomical precepts are generally similar to the mathematical problems, but there is one striking difference: the precepts are not always worked numerical examples, but instead they may name the parameters in the calculations. Here, for instance, is an extract from a System A precept from Babylon for finding the length of time between moonset and sunrise on the morning before a full moon, whose technical name is 'Disappearance':

> For your calculation of Disappearance: if the time after sunset exceeds the duration of night, [lift] the duration of night from [the time after sunset]. The day of opposition falls in its day. If the time after sunset is less than the duration of night, you increase (*sic, for 'pile up'?*) the time after sunset [and the duration of night] together and you decrease by one day. You multiply what remains by 0;10 and it is the distance travelled by the sun. You put it down. You multiply it by [the velocity of the moon] of its month and it is the distance travelled by the moon. You put it down. You lift the distance travelled by sun from the distance travelled by the moon and [you put it down] <as> the elongation. . . . [29]

Compare the language of this Late Babylonian problem, taken from a mathematical compilation from Babylon:[30]

> I have piled up the length and width: 14. And the surface is 48. Since you do not know, 14 steps of 14 is 3 16. 48 steps of 4 is 3 12. You lift 3 12 from 3 16: 4 is left over. How many steps of how many should I go for 4? 2 steps of 2 is 4. You lift 2 from 14: 12 is left over. 12 steps of 0;30 is 6. The width is 6. You join 2 to 6: 8. The length is 8.

In the both of these extracts, as in Old Babylonian mathematics, a distinction is made between physical addition and subtraction ('piling up', 'joining', and 'lifting from') and their numerical counterparts ('increasing' and 'decreasing') (see §4.1). Instructions are introduced by the standardised phrases *ana epēšika* ('for your working') and *aššu lā tīdû* ('since you do not know') and given in the second person of the present-future tense. A close study of the language and orthography of the astronomical precepts and Late Babylonian mathematics, such as has now been done for Old Babylonian mathematics, would enable scribal writing habits to be distinguished and perhaps linked between the two genres.

This Late Babylonian problem-type is also very well attested in Old Babylonian times: given the area of a rectangle, and the sum of its length and width, the length and width must be found (see §4.1). But the method of solution is not the usual completing the square, as still used in Seleucid Uruk (§8.4).[31] Instead it involves an entirely new procedure, in which a square is made from the sum of the two unknowns (see §9.3, figure 9.2). The known area of the rectangle is subtracted four times, leaving a square whose side is the difference between the length and width. This side is subtracted from the original sum, to give twice the width, then added to the width to find the length. Although, as already noted, geometrical terms for addition and subtraction are employed, arithmetical multiplication is used throughout.

To sum up, Late Babylonian mathematics and astronomy are linked in several ways: the language and structure of the word problems and the precepts are similar; some of the mathematical tables may have been constructed specifically with astronomical needs in mind; and there is circumstantial museological evidence from the British Museum that they come from the same archaeological contexts, albeit in the absence of directly recorded evidence. However, the southern city of Uruk has yielded excellent, archaeologically grounded data that present a more nuanced picture.

8.3 ACHAEMENID URUK: THE ŠANGÛ-NINURTA AND EKUR-ZĀKIR FAMILIES

Between 1969 and 1972 the German excavators of Uruk uncovered a house to the east of the city dating to the fifth and fourth centuries BCE (figure 8.5). Two different scribal families, the descendants of Šangû-Ninurta and of Ekur-zākir, had successively occupied it, both maintaining scholarly libraries there.[32] Some 500 tablets and fragments were found in the house, around 180 of which can be associated with the Šangû-Ninurta family and about 240 with the Ekur-zākirs.[33] The eighteen dateable legal and scholarly tablets belonging to the Šangû-Ninurtas span the sixth and fifth centuries BCE, the latest nine of which are from early in the reign of the Persian king Darius II (r. 423–405).[34] They suggest that the Šangû-Ninurtas left the house some time after 412 BCE. Some time later, perhaps immediately afterwards, the Ekur-zākirs moved in. The latest nine of their

Figure 8.5 Archaeological plan of Uruk, with the Rēš temple in the centre and the scholars' house to the southeast. (Schmidt et al. 1979, pl. 65, courtesy of WHOM.)

sixteen dateable legal and scholarly tablets cover the period 322–300 BCE; that is, from the accession of Alexander the Great's successor Philip Arrhidaeus (r. 323–316) to early in the reign of Seleucus I Nicator (r. 305–281).[35] The higher archaeological levels of the immediate area were eroded away, so it is not known what happened to the house and its inhabitants after that.

Three rooms and a courtyard have survived of the Šangû-Ninurtas' house (figure 8.6). Before they left, they carefully buried much of their household library, and whatever archival tablets they did not want to take with them, in clay jars in a rather strange room within a room (locus 4), which may

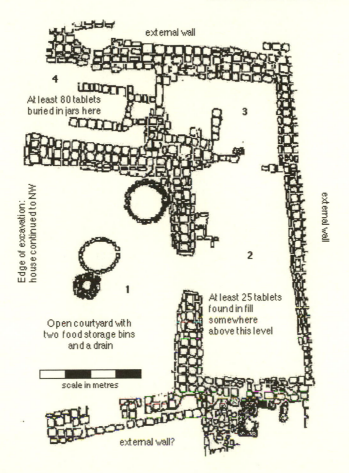

Figure 8.6 The scholars' house in Uruk Ue XVIII/1: Level IV, belonging to the Šangû-Ninurta family. (Drawing by the author after Schmidt et al. 1979, pl. 70.)

have originally been some sort of food store. (There were large gaps between the bricks, allowing air to circulate.) Excavators found other tablets scattered over other rooms, where they had been disturbed by later human burials dug down into the house, perhaps by the Ekur-zākirs.[36] The approximate proportions of the library's scholarly contents were as follows:

- 30 percent medical (physiognomic and diagnostic omens; medical prescriptions and incantations);
- 20 percent other incantations, rituals, and magic;
- 19 percent hymns, literature, and lexical lists;
- 12 percent observed and induced omens (*Enūma Anu Ellil*, terrestrial omens, extispicy, etc.);

- 12 percent astronomy, astrology, and mathematics;
- 7 percent unidentified.

The library included fifty tablets with colophons, recording information such as the owner and scribe of the tablet, its source, and its place within a scholarly series. Forty of those tablets belonged to three generations of the family: one Šamaš-iddin; his sons Rīmūt-Anu and Anu-ikṣur; and the latter's son Anu-ušallim (table 8.2). All describe themselves as *āšipus*, or incantation priests; their professional interests are reflected in the heavy preponderance of prescriptions, incantations, and rituals in the library. However, three of the Šangû-Ninurtas' tablets, owned by Šamaš-iddin (table 8.2: 9) and Rīmūt-Anu (table 8.2: 10, 11), are mathematical. Four others were also found in their level of the house (see table B.20). Together they comprise a coherent corpus.

Two large tablets contain a series of around fifty mathematical rules and problems about different kinds of area measure; one tablet appears to be a continuation of the other. The second (table 8.2: 10) has a catchline before the colophon, '*Seed and reeds*. Finished', which suggests that the problems comprised a standard series. A typical problem on this second tablet is about finding the square area paveable with standard square baked bricks ⅔ cubit (c. 35 cm) long:

> 9 hundred baked bricks of ⅔ cubits each. I enlarged a courtyard. What is the square side of the courtyard? You go 0;40 steps for each brick. You take 30 each of 15 00. You go 30 steps of 0;40: 20. The square side of the courtyard is 20 cubits.[37]

The solution is simply to find the number of bricks that would make up the edge of the square area, by finding the square root of the total number (happily a square integer), then to multiply by the length of each brick.[38]

Three tablets bear metrological tables and reciprocal tables. One (table 8.2: 10) contains tables of length and area measures (reed measure) with their sexagesimal equivalents. It begins with a short list of numerical writings for the major gods (§8.1) and ends with two short and damaged tables which seem to contain simple monthly timekeeping schemes. At the top of the reverse is a little table relating linear units to their nearest equivalent (reed) area unit:[39]

Fingers to grains	(1 finger × 1 finger = 0;50 grain)
cubits to shekels	(1 cubit × 1 cubit = 0;25 shekel)
reeds to sar	(1 reed × 1 reed = 0;15 sar)
10 rods to 1 *iku*	(10 rods × 10 rods = 1 *iku*)
1 cable to 1 *bur*	(1 cable × 1 cable = 2 *bur*)
1 league to 1 *šar*	(1 league × 1 league = 30 *šar*)
6 cable to 1 *šar*	(6 cables × 6 cables = 1;12 *šar*)

TABLE 8.2
Scholarly Tablets of the Šangû-Ninurta Family in Fifth-Century Uruk

No.	Contents	Colophon	Publication
Šamaš-iddin, son of Nādin			
1	Section 9 of the standard series of medical omens, *If a man clutches his head thus*	Tablet of Šamaš-iddin, son of Nādin, descendant of Šangû-[Ninurta].	W 22315 (Hunger 1976b, no. 44)
2	Tablet 45 of the standard series of medical omens, *If a man clutches his head thus*	Written and checked against its original. Tablet of Šamaš-iddin, junior incantation priest, son of [Nādin], descendant of Šangû-Ninurta, Urukean. Whoever fears Anu and Ištar shall not [take it away].	W 22307/23 (Hunger 1976b, no. 48)
3	Fragmentary commentary to medical (?) omens	[From the mouths of] experts. Reading out of Šamaš-iddin, son of Nādin, descendant of Šangû-Ninurta.	W 22660/4 (von Weiher 1983–98, III no. 100)
4	Fragmentary commentary to medical (?) omens	Copy of a writing-board [. . .] Šamaš-iddin, son of [Nādin], descendant of Šangû-Ninurta, incantation priest. Whoever fears Anu and Antu shall not take it away.	W 23268 (von Weiher 1983–98, V no. 254)
5	Incantations against the demoness of childbirth, Lamaštu	Copy of a writing-board, written and checked. Tablet of Šamaš-iddin, incantation priest, son of Nādin, descendant of Šangû-Ninurta, Urukean. Whoever fears Anu and Ištar shall not take it away.	W 23287 (von Weiher 1983–98, III no. 84; Farber 1989)
6	Third House of the standard series of bilingual incantations, *Bath-house*	Copy of a writing-board, property of the Eana temple, written and checked. Tablet of Šamaš-iddin, junior incantation priest, descendant of Šangû-Ninurta.	W 23263 (von Weiher 1983–98, III no. 66)
7	Sixth House of the standard series of bilingual incantations, *Bath-house*	[Copy of] a writing-board, property of the Eana temple, written and checked. [Tablet of Šamaš]-iddin, junior incantation priest, descendant of Šangû-Ninurta.	W 23299/1 (von Weiher 1983–98, IV no. 127)

TABLE 8.2 (*Continued*)

No.	Contents	Colophon	Publication
8	Bilingual incantations	Written and checked against its original. Tablet of Šamaš-iddin, incantation priest, son of Šangû-[Ninurta].	W 23290 (von Weiher 1983–98, IV no. 128)
9	Mathematical problems, *Seed and reeds*	Copy of a writing board, written and checked against its original. Tablet of Šamaš-iddin, son of Nādin, descendant of Šangû-Ninurta.	W 23291x (see table B.20)

Rīmūt-Anu, son of Šamaš-iddin

No.	Contents	Colophon	Publication
10	Various metrological tables	According to an old tablet, Urukean copy. Rīmūt-Anu, son of Šamaš-iddin, descendant of Šangû-Ninurta, wrote and checked it.	W 23273 (see table B.20)
11	Table of many-place reciprocals	According to an old original, Rīmūt-Anu, son of Šamaš-iddin, descendant of Šangû-[Ninurta]. . . . wrote and checked it.	W 23283+ (see table B.20)
12	List of diseases, ordered by body-part	Written and checked against its original. Long tablet of Rīmūt-Anu son of Šamaš-iddin, descendant of Šangû-Ninurta, incantation priest. Hand of Bēlu-kāṣir, son of Balāṭu.	W 22307/11 (Hunger 1976b, no. 43)
13	Descriptions of diseases	Written and checked [against] its [original]. Tablet of Rīmūt-[Anu son of Šamaš-iddin, descendant of] Šangû-Ninurta, incantation priest, Urukean. [Whoever fears Anu and Ištar (?)] shall not take it away.	W 23292 (von Weiher 1983–98, IV no. 152)
14	The standard composition now known as the *Introduction to incantations*	Written, checked and made good against its original tablet. [. . .] of Rīmūt-Anu [son of] Šamaš-iddin, descendant of Šangû-Ninurta. Uruk, Month VII, [day . . ., year . . . of] Darius the king (r. 423–405 BCE).	W 23293/4 (von Weiher 1983–98, V no. 231)
15	Tablet 41 of the standard series of medical omens, *If a man clutches his head thus*	Urukean copy. Anu-ikṣur, son of [Šamaš-iddin . . .] wrote and checked it [against its original]. Tablet of Šamaš-iddin [. . .]. Whoever fears Anu and Ištar shall not [take it away].	W 22307/1 (Hunger 1976b, no. 59)

TABLE 8.2 (*Continued*)

No.	Contents	Colophon	Publication
Anu-ikṣur, son of Šamaš-iddin			
16	Commentary to the standard series of diagnostic omens, *When an* āšipu *goes to a patient's house*, Tablet I	[. . . of Anu]-ikṣur, descendant of Šangû-Ninurta [. . .].	W 22307/15 (Hunger 1976b, no. 28)
17	Commentary to the standard series of diagnostic omens, *When an* āšipu *goes to a patient's house*, Tablet V	Reading out of Anu-ikṣur, descendant of Šangû-Ninurta, incantation priest.	W 22307/16 (Hunger 1976b, no. 31)
18	Commentary to the standard series of diagnostic omens, *When an* āšipu *goes to a patient's house*, Tablet VII	Reading out of Anu-ikṣur, son of Šamaš-iddin, descendant of Šangû-Ninurta.	W 22307/2 (Hunger 1976b, no. 32)
19	Commentary to the standard series of diagnostic omens, *When an* āšipu *goes to a patient's house*, Tablet VII	Reading out of Anu-ikṣur, son of Šamaš-iddin, descendant of Šangû-Ninurta, junior incantation priest, Urukean. Whoever fears Anu and Antu shall not take it away.	W 22307/10 (Hunger 1976b, no. 33)
20	Commentary to the standard series of diagnostic omens, *When an* āšipu *goes to a patient's house*, Tablet XIX	[. . . Anu]-ikṣur, junior incantation priest [. . .]	W 22307/32 (Hunger 1976b, no. 38)
21	Prescriptions against nasal diseases	[Tablet (?) of] Anu ikṣur, son of Šamaš-iddin, incantation priest, [descendant of Šangû]-Ninurta, Urukean. Whoever fears Anu and Ištar shall not take it away.	W 22307/21 (Hunger 1976b, no. 45)

Table 8.2 (*Continued*)

No.	Contents	Colophon	Publication
22	Commentary to the standard series of medical omens, *If a man clutches his head thus*, Section 10	[. . . Anu]-ikṣur, incantation priest [. . . descendant of Šangû]-Ninurta. Whoever fears Gula shall treasure it.	W 22307/35 (Hunger 1976b, no. 47)
23	Commentary to prescriptions for illnesses caused by ghosts, *If a man is seized by the hand of a ghost*	Reading out of Anu-ikṣur, junior incantation priest.	W 22307/17 (Hunger 1976b, no. 49)
24	Commentary to prescriptions for epilepsy, *If a man is struck down by epilepsy*	From the mouths of experts, reading out of Anu-ikṣur, junior incantation priest.	W 22307/3 (Hunger 1976b, no. 50)
25	Commentary to medical prescriptions, *If a man clutches his head thus*	From the mouths of experts, prescriptions from . . . 's house. Reading out of Anu-ikṣur, son of Šamaš-[iddin], incantation priest, descendant of Šangû-Ninurta.	W 22307/15 (Hunger 1976b, no. 51)
26	List of ingredients for magical purposes	Excerpted from a writing-board. Long tablet of Anu-ikṣur, [. . .], incantation priest.	W 22307/1 (Hunger 1976b, no. 56)
27	Incantations and rituals against female infertility	Written and [checked] against its original. [Tablet of Anu]-ikṣur, son of Šamaš-iddin, descendant of Šangû-Ninurga. Whoever fears Anu and Antu shall not take it away, shall not deliberately let it be stolen.	W 23262 (von Weiher 1983–98, V no. 248)
28	Commentary to the standard series of physiognomic omens, section *If there is the head of a chameleon*	Reading out of Anu-ikṣur, son of Šamaš-iddin, descendant of Šangû-Ninurta.	W 22312a (Hunger 1976b, no. 83)

TABLE 8.2 (*Continued*)

No.	Contents	Colophon	Publication
29	Commentary to the standard series of terrestrial omens, *If a city is situated on a hill*: section on bird omens, *If a francolin sits on a man's head*	Reading out of Anu-ikšur, son of Šamaš-iddin, [descendant of] Šangû-Ninurta, [junior incantation priest], Urukean.	W 22659 (von Weiher 1983–98, III no. 99)
30	'These are the rituals, figurines, signs, and incantations (of the standard series) *House of confinement*	Anu-ikšur, incantation priest, son of Šamaš-iddin, descendant of Šangû-Ninurta, wrote and checked it against its original. Whoever fears Anu and Antu shall not take it away, shall not deliberately [. . .] it.	W 22762/2 (von Weiher 1983–98, II no. 8)
31	Tabular overview of the standard series of incantations, *House of confinement*	[. . .] Anu-ikšur, son of Šamaš-iddin, descendant of Šangû-Ninurta, junior incantation priest, wrote and checked it against its original. [Tablet of Šamaš-iddin], junior [incantation priest], son of Nādin, descendant of Šangû-Ninurta, Urukean. Whoever fears Anu and Ištar [. . .]	W 23266 (von Weiher 1983–98, III no. 69)

Anu-ušallim, son of Anu-ikṣur

No.	Contents	Colophon	Publication
32	The standard series of birth omens, *If an anomaly*, Tablet I	Urukean copy, written and checked against its original. Tablet of Anu-ikšur, son of Šamaš-iddin, descendant of Šangû-Ninurta, junior incantation priest. Hand of Anu-ušallim, his son. Whoever fears Anu and Antu shall not take it away.	W 23272 (von Weiher 1983–98, III no. 90)
33	From the standard series of physiognomic omens, section *If his tongue is shiny*	Copy of a writing-board [. . .]. Tablet of Anu-ikšur [. . .]. Hand of Anu-[ušallim, his son].	W 22660/7a+ (von Weiher 1983–98, IV no. 151)

TABLE 8.2 (*Continued*)

No.	Contents	Colophon	Publication
34	Tablets 3–4 of the standard magical series *Burning*	Copy [of a . . .], written, checked, and made good. Tablet of Šamaš-iddin, descendant of Šangû-Ninurta. [Hand of Anu-ušallim], son of Anu-ikṣur [. . . . Whoever fears Anu] and Antu shall not [take it away].	W 23215/1 (von Weiher 1983–98, V no. 242)
Uncertain attribution			
35	Medical prescriptions	[. . ., son of Šangû]-Ninurta, incantation priest.	W 22307/31 (Hunger 1976b, no. 60)
36	Non-canonical material from the standard series of birth omens, *If an anomaly*	Reading out of [. . .], junior incantation priest, descendant of Šangû-[Ninurta].	W 22307/12 (Hunger 1976b, no. 72; Farber 1989)
37	Tablet 3 of the standard list of gods *An = Anum*	[Written and checked] against its original. [Hand of . . .], junior incantation priest, son of Šangû-[Ninurta]. Tablet of Šamaš-iddin, junior incantation priest [. . .]. Whoever fears Anu and Antu shall not take it away.	W 22288/0 (Hunger 1976b, no. 126)

The obverse of the second tablet is also about lengths and areas. Its first column contains tables of the numerical relationships between the different units in each system, and between reed and seed area measures, while the second column relates lengths to various units for measuring time. The reverse of this second tablet has a unique reciprocal table in which the first sexagesimal place of each entry increases by 1. The remaining two to four places can contain any digits, as long as the number is regular. The table runs from 1 to 60 and then from 1 to 30 again with a different set of pairs. Here are two extracts, in which *igi* is the standard term for reciprocal:[40]

igi 10 40	5 37 30	[igi 10 . . .	. . .]
igi 11 06 40	5 20	igi 11 31 12	5 12 30
igi 12 48	4 41 15	igi 12 30	4 48
igi 13 30	4 26 40	igi 13 53 20	4 18(*sic*, for 19) 12

| igi 14 24 | 4 10 | [igi 14] 03 45 | 4 16 |
| igi 15 37 30 | 3 50 24 | [igi 15 1]1 15 | 3 57 02 13 20 |

The third of the tablets is a table of many-place reciprocal pairs from 1 to 4, with the idiosyncratic logogram IB_2.SI 'square (side)' at the end of an occasional line and with some entries not in numerical order (see §8.4). It appears to have been calculated, at least in part, not simply copied from another exemplar, because two rough tablets contain nine calculations of regular reciprocals taken from near the beginning of the table. The method is to multiply repeatedly by simple factors of 60 until the original number is reduced to 1, and then to multiply up from 1 again by those same simple factors in reverse:[41]

↓	49 22 57 46 40	(× 3 =)	1 12 54	
↓	2 28 08 53 20	(× 3 =)	24 18	↑
↓	7 24 26 40	(× 3 =)	8 06	↑
↓	22 13 20	(× 3 =)	2 42	↑
↓	1 06 40	(× 3 =)	54	↑
↓	3 20	(× 3 =)	18	↑
↓	10	(× 6 =)	6	↑
	1	→	igi 1	↑

Thus once again, a favourite Old Babylonian school problem is solved using a new method: not cut-and-paste procedures (§4.3) but repeated factorisation.

In short, the predominant concerns of Šamaš-iddin's and Rīmūt-Anu's mathematics are reciprocals, lengths, and areas, with a secondary interest in timekeeping. But the only astronomical texts securely dateable to the same library are three fragments of planetary diaries, dating from the reigns of Nabonidus to Darius I (late sixth century) and Artaxerxes I (mid-fifth century).[42] There is no evidence of any interest in theoretical or mathematical astronomy.

After the Šangû-Ninurta family left the house some time in the last decade of the fifth century, it was rebuilt to a similar plan to accommodate the Ekur-zākirs (figure 8.7). When, after a few decades, the mud-brick walls were again in need of replacement the family cleared out their belongings, leaving behind a few dozen tablets they no longer needed, mostly out-of-date legal contracts and elementary school exercises. Many of the tablets had been dumped, along with unused tablet clay and bone writing styluses, in two areas on the periphery of the house whose floors had been waterproofed with bitumen, especially in order to provide an area for making and re-using tablets.[43] The elementary education of the son of the house must have been over, for in the succeeding house only a tiny number of school exercises were found.

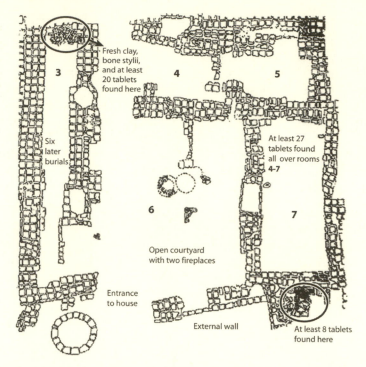

Figure 8.7 The scholars' house in Uruk Ue XVIII/1: Levels III and II, belonging to the Ekur-zākir family. (Drawing by the author after Schmidt et al. 1979, pls. 69, 68.)

Amongst the discarded tablets were a 45 times multiplication table and two fragments of calculations. Two fragmentary tablets of metrological tables probably belong to this phase too, as although their content is similar to those found in the Šangû-Ninurtas' house, their spelling conventions suggest a slightly later date (see table B.20). Perhaps these scribal exercises were the work of the young Ištar-šuma-ēreš, son of Iqīšâ, who wrote some ten scholarly tablets, including seven written for his father, in around 318. Iqīšâ left behind at least another thirty-two. Four of the scholarly tablets are dated, to 322–318.[44] Like the Šangû-Ninurtas, Iqīšâ was an incantation priest, and like them he earned his income from the temples in the centre of Uruk. Legal documents recording the sale of prebends, or rights to temple income, were found in both houses. While the Šangû-Ninurta family invested in Building Inspectors' prebends, Iqīšâ bought into the Brewers' prebends in 316:

Ina-qibīt-Anu, son of Ištar-Uruk-ēriba spoke freely to Iqīšâ, son of Ištar-šuma-ēreš, thus: "Give me 2 *kurru* of dates, 2 [barrels], a sack and (= 'of'?) mustard as the purchase price and I will carry out the

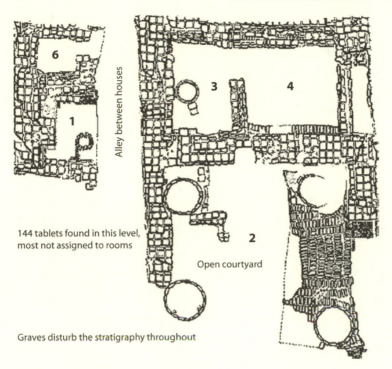

Figure 8.7 (*Continued*).

Brewers' prebend in the Rēš temple." Iqīšâ, son of Ištar-šuma-ēreš, heard him and gave him 2 *kurru* of dates, 2 barrels, a sack, and mustard as his purchase price in Month IV, year 6 of Philip the king. [. . .] In Uruk Ina-qibīt-[Anu, son of] Ištar-Uruk-ēriba received the dates, [the barrels], the sack, and the mustard from Iqīšâ, [son of] Ištar-šuma-ēreš, and was paid off. . . .

Nidinti-Anu the scribe, son of Anu-uballiṭ. Uruk, day 7, Month XI, year 6 of Philip the king.[45]

Apart from the discards mentioned above, none of the Ekur-zākir family's tablets is narrowly mathematical. Otherwise their library was very similar in make-up to the Šangû-Ninurtas', including a great deal of the standard series of omens; magical, medical, and ritual works; and half a dozen astronomical fragments. Iqīšâ owned one of the calculation tablets of 'Secrets of Extispicy' (see §5.4), at least two zodiacal-medical 'calendars' (§8.1), and an astrological scheme for predicting financial success.[46] The library also includes three other simple calendrical schemes: one table of (amongst other things) the monthly change in the length of daylight over the year in a traditional linear zigzag, written by Iqīšâ; prose descriptions of the movements and timings of heavenly bodies over the year; and a series of instructions for finding the date of the summer solstice for thirty-seven years from 361 BCE onwards:[47]

> Year 44 of Artaxerxes the king. Solstice on Month III, 16. For each year [you add] 11 days.
>
> > 1 times 11. 11 and 16 is 27. In year 45 solstice on Month III, 27.
> > 2 times 11. 22 and 16 is 38. You release 30. The remainder is 8. Solstice on Month IV, 8. (Intercalary) Month XII.
> > 3 times 11. 33 and 16 is 49. You release 30. The remainder is 19. Solstice on Month III, 19.
> > (etc., to:)
> > 37 times 11. 6 47 [and] 16 is 7 03. You release 7 00. [The remainder is 3.] Solstice on Month IV, 3. (Intercalary) Month XII.

In other words, the algorithm is to add 11 modulo 30 to the date of the previous year's solstice. In each year that the date of the solstice falls in Month IV instead of Month III you add an intercalary month at the end of the year to bring the solstice back into the correct month.[48]

A System B ephemeris for Saturn during the years 312–291 BCE is among the latest of the scholarly tablets in the Ekur-zākirs' collection. It is the oldest ephemeris known for this planet by some eighty years, and predates the previously assumed development of System B by about half a century (see §8.1).[49]

Thus there is a strong contrast between the scholarly interests of the two families of incantation priests, separated by almost exactly a hundred years. In the late fifth century Šamaš-iddin and Rīmūt-Anu of the Šangû-Ninurta family were copying mathematical problems and calculating many-place reciprocals a century before the advent of mathematical astronomy. The Ekur-zākir family, on the other hand, who wrote or owned a very early System B ephemeris as well as some much simpler applications of linear zigzag functions, have left almost no trace of any non-astronomical interest in mathematics.[50]

8.4 SELEUCID URUK: THE ḪUNZÛ AND SÎN-LĒQI-UNNINNI FAMILIES

Moving on another century, and into the Seleucid or Hellenistic era, the latest dateable mathematical cuneiform tablets were written by members of two other scholarly families in Uruk: the Ḫunzûs and the Sîn-lēqi-unninnis. There is no archaeological information about their homes but, for the Sîn-lēqi-unninni family at least, there are a lot of other pertinent data. About the Ḫunzûs rather less can be said (table 8.3).

TABLE 8.3
Scholarly Tablets of the Ḫunzû Family of Seleucid Uruk

No.	Date	Contents	Colophon	Publication
1	—	Table of many-place reciprocals from 1 to 3	Tablet of Nidinti-Anu son of Ina-qibīt-Anu, descendant of Ḫunzû, incantation priest of Anu and Antu, Urukean. Hand of Ina-qibīt-Anu, his son.	AO 6456 (see table B.20)
2	—	Tablet from the standard sacrificial omen series *Bārûtu* ('Extispicy'), tablet 20 of the section 'Path' (a part of the liver)	Tablet of Nidinti-Anu, son of Anu-bēlšunu, descendant of Ekur-zākir, incantation priest of Anu and Antu, Urukean. Hand of Ina-qibīt-Anu, son of Nidinti-Anu, descendant of Ḫunzû.	AO 6454 (Thureau-Dangin 1922, no. 5)
3	—	Compilation of mathematical procedures for calculating celestial data	Tablet of Nidinti-Anu son of Ina-qibīt-Anu, descendant of Ḫunzû, incantation priest of Anu and Antu, Urukean. Hand of Anu-uballiṭ, his son.	AO 6455 (Thureau-Dangin 1922, no. 11; Brack-Bernsen and Hunger 2002)
4	SE 97 (214 BCE)	Tablet 56 of the standard celestial omen series *Enūma Anu Ellil*	Written and checked against and old writing board from Babylon. Tablet of Nidinti-Anu, son of Anu-bēlšunu, descendant of Ekur-zākir, incantation priest of Anu and Antu, Urukean. Hand of Anu-uballiṭ, son of Nidinti-Anu, descendant of Ḫunzû, incantation priest of Anu and Antu, Urukean. Uruk, Month VII, day 30, year 97. Antiochus was king.	AO 6450 (Thureau-Dangin 1922, no. 16; Largement 1957)

TABLE 8.3 (*Continued*)

No.	Date	Contents	Colophon	Publication
5	SE 98 (213 BCE)	Tablet 55 of the standard sacrificial omen series *Bārûtu* ('Extispicy'), tablet 6 of the section 'Finger' (gall bladder)	Tablet of Nidinti-Anu, son of Anu-bēlšunu, descendant of Ekur-zākir, incantation priest of Anu and Antu, Urukean. Hand of Anu-uballiṭ, son of Nidinti-Anu, descendant of Ḫunzû, incantation priest of Anu and Antu, Urukean. Uruk, Month II, day 25, year 98. Antiochus was king.	MLC 1865 (Clay 1923, no. 12)
6	—	Equinoctial *akītu* rituals of the god Anu for Month VII	Urukean copy written and checked from an old writing board. Tablet of Anu-uballiṭ son of Nidinti-Anu, descendant of Ḫunzû, incantation priest of Anu and Antu, Urukean. Hand of Šamaš-ēṭir, son of Ina-qibīt-Anu, son of Šipqat-Anu, descendant of Ekur-zākir.	AO 6459 + 6465 (Thureau-Dangin 1922, nos. 39–40; Linssen 2004, 184–96)

The many-place table of reciprocals (table 8.3: 1) is one of only two known scholarly tablets written by Ina-qibīt-Anu. As he bears his paternal grandfather's name, he must be Nidinti-Anu's eldest son and so his table presumably antedates his younger brother Anu-uballiṭ's work—roughly the 90s SE, say 225–215 BCE. It differs from Rīmūt-Anu's table of two hundred years earlier: not only has the writing style changed (as might be expected) but different entries have been chosen for inclusion and a cuneiform 'colon' (two corner wedges one above the other) is used to separate the members of each pair. Where Rīmūt-Anu occasionally used the idiosyncratic logogram IB$_2$.SI 'square (side)', Ina-qibīt-Anu writes the equally unconventional ŠA$_3$.BI 'from it' to fill the line in short entries. Compare, for example, the six reciprocal pairs starting from 1 04 (table 8.4). In fact, the later table has thirty-three entries in the range 1–2 that are not found in the earlier one, while the earlier only has four that are not found in the later. But the situation is reversed in the range 2–3, where the earlier table has twenty-three extra entries compared to the later table's two.

TABLE 8.4
Comparison of Two Reciprocal Tables from Late Babylonian Uruk

Rīmūt-Anu's Reciprocal Table (c. 420)		Ina-qibīt-Anu's Reciprocal Table (c. 220)
Lines i 17–20		*Lines i 13–18*
igi 1 04	56 15	[igi] 1 04 : from it 56 15
—		[igi] 1 04 17 01 28 53 20 : 55 59 13 55 12
igi 1 04 48	55 33 20	[igi] 1 04 48 : 45 (*sic*, for 55) 33 20
igi 1 05 [06 15]	55 17 45 36	igi 1 05 06 15 : 55 17 45 36
—		[igi] 1 05 32 09 36 : 54 55 53 54 22 30
igi 1 05 [36 36]	54 52 10 51 51 06 40	[igi] 1 05 36 36 : 54 52 10 51 51 06 40
Lines iii 2–7		*Lines iv 9–12*
igi 2 08	2[8 07 30]	igi 2 08 : from it 28 07 30
igi 2 09 36	2[7 46 40]	igi 2 09 36 : 27 46 40
igi 2 10 12 30	[27 38 52 48]	igi 2 10 12 30 : 27 38 52 48
igi 2 11 13 12	[27 26 05 25 55 33 20]	—
igi 2 11 41 14 04 26 40	27 20 [15]	igi 2 11 41 14 04 26 40 : 27 20 : 15
igi 2 11 50: 09 22 30	27 1[8 24]	—

Thus, although Ina-qibīt-Anu's table contains several copying errors, it is clearly not a direct descendant of Rīmūt-Anu's table of two centuries earlier. But like the Ḥunzûs' other scholarly tablets, it shows great continuity of content with the works owned by the Šangû-Ninurta and Ekur-zākir families (§8.3). Four of them (table 8.3: 2, 4–6) even show the Ḥunzû brothers working closely with members of the Ekur-zākir family—whether direct or collateral descendants of Iqīšâ and Ištar-šuma-ēreš it is impossible to say. But while some also show a concern with timekeeping and celestial omens (table 8.3: 3–4, 6), there is no evidence of any family involvement with mathematical astronomy, or with the medical works which predominated the earlier incantation priests' libraries. This is particularly striking as Anu-uballiṭ's associate Šamaš-ēṭir of the Ekur-zākir family (table 8.3: 6) owned two planetary ephemerides, dating to SE 118 (= 193 BCE), and two lunar auxiliary tables (table 8.5: 16–9). He was also the scribe of a fragmentary precept for calculating the retrogradations of Mars.[51]

Even without the direct evidence of household remains it is clear that the Ḥunzû family were members of respectable Uruk society. Like the earlier Urukean owners of mathematical tablets they were *āšipu* incantation priests, and their extended family counted scribes, tax assessors, financial officers, and even city governors amongst their number.[52] They witnessed legal documents, endorsing them with their personal stamp seals (figure 8.8).[53] Nidinti-Anu's and Ina-qibīt-Anu's sphinxes were very popular sealing motifs in Seleucid Uruk, but the image on Anu-uballiṭ's seal is more unusual. The apotropaic motif of a bearded *apkallu* or sage holding a bucket and a sprinkler is well known in Babylonian and Assyrian art but is strongly associated at this time and place with the incantation priests, particularly the Ekur-zākir family—with whom, as table 8.3 shows, the Ḥunzû brothers were closely connected too.[54]

Another family with scholarly links to Šamaš-ēṭir of the Ekur-zākir family copied both mathematical and astronomical tablets. The Sîn-lēqi-unninnis are the best known of all the scribal families of Late Babylonian Uruk, partly because their eponymous ancestor was considered to be the author of the famous *Epic of Gilgameš* (§6.4), and partly because they have left a vast amount of documentary evidence.[55] Their reconstructed family tree covers two centuries and six or seven generations. Two members of this family owned or wrote no fewer than sixteen extant mathematical and astronomical tablets (table 8.5), while several were parties or witnesses to a whole variety of legal documents (figure 8.9). By profession they were *kalû*, or lamentation priests, of the great god Anu and his consort Antu.

Just one of the Sîn-lēqi-unninnis' thirty tablets is narrowly mathematical: Anu-aba-utēr's compilation of word problems (table 8.5: 26), which is the latest dated mathematical cuneiform tablet known. Suggestively, the colophon does not describe it as a copy of an older original, but that is not to say that it contains all new material. Few of its seventeen problems, allowing for changes in writing habits, would look out of place within the Old Babylonian mathematical corpus: the problems set and the methods of solution seem to have changed little over the millennia. Problems (1)–(2), for instance, concern the sum of arithmetical series, (9)–(13) the capacity of a cube whose sides are known, and (14)–(17) regular reciprocal pairs whose sums are known, solved by a standard Old Babylonian cut-and-paste method.[56] The only innovatory methods of solution are for problems (3)–(8), about triangles, squares, and the diagonals of rectangles—which are not exactly innovatory mathematical subjects in themselves. Just problem (6)—to find the area of a rectangle given the length, width, and diagonal—is not attested at all in the OB corpus but is also found in three

TABLE 8.5
Scholarly Tablets of the Sîn-lēqi-unninni Family of Seleucid Uruk

No.	Date	Contents	Colophon	Publication
\multicolumn{5}{l}{Anu-bēlšunu I, son of Nidinti-Anu I}				
1	SE 63	Birth horoscope for Anu-bēlšunu, giving his birth date as 29 December 249 BCE	[None].	NCBT 1231 (Beaulieu and Rochberg 1996)
2	SE 81	From the standard ritual series *Kalûtu*, section "When the door sockets are installed"	Tablet of Nidinti-Anu. Hand of Anu-bēlšunu his son, junior lamentation priest, Urukean. Month IV, day 28, year 1 21, Seleucus the king.	Ist O 174 (Thureau-Dangin 1922, no. 46; Linssen 2004, 293–8)
3	SE 83	The 'Esağila tablet', a compilation of word problems about Marduk's temple in Babylon	The knowing may show the knowing; the unknowing shall not see it. Written, made good, and checked against an old tablet from Borsippa. Tablet of Anu-bēlšunu, son of Anu-balāssu-iqbi, descendant of Aḫi'ūtu, Urukean. Hand of Anu-bēlšunu, son of Nidinti-Anu, descendant of Sîn-lēqi-unninni. Uruk, Month IX, day 26, year 1 23, Seleucus the king.	AO 6555 (see table B.20)
4	SE 84	Tablet 26 of the standard terrestrial omen series *If a city stands on a hill*	[Tablet of] Nidinti-Anu, son of Anu-uballiṭ, descendant of Sîn-lēqi-unninni, Urukean. [Hand of Anu]-bēlšunu his son. Uruk, Month XI, day 2, year 1 24, Seleucus the king.	MLC 1867 (Clay 1923, no. 21)
5	—	Auxiliary table for the ephemeris of Mercury	[. . . . Anu-bēl]šunu descendant of Sîn-lēqi-unninni [. . .].	A 4309 (Neugebauer 1955, no. 800a R)
6	—	Ephemeris for Jupiter, System B, for at least SE 127–194	[. . .] Anu-[bēl]šunu [. . .] Antiochus [the king.].	AO 6480 (Thureau-Dangin 1922, no. 29; Neugebauer 1955, no. 620 Zb)

TABLE 8.5 (*Continued*)

No.	Date	Contents	Colophon	Publication
Anu-bēlšunu I and his son Nidinti-Anu II				
7	SE 108	Lamentation to Enlil from the standard lamentation series *Kalûtu*	[Tablet of] Anu-bēlšunu, son of Nidinti-Anu, descendant of Sîn-lēqi-unninni, lamentation priest of Anu and Antu, Urukean. Hand of Nidinti-Anu his son. Uruk, Month XI, day 18, [year 1] hundred 8. [. . . the king].	MLC 1857 (Clay 1923, no. 11)
Anu-bēlšunu I and his son Anu-aba-utēr				
8	SE 112	Bilingual building ritual, for the foundation of a temple, from the standard lamentation series *Kalûtu*	Written and [checked] against its original. Tablet of Anu-bēlšunu, son of [Nidinti-Anu, descendant of Sîn-lēqi-unninni], lamentation priest of Anu and Antu, the Urukean. Hand of Anu-[aba-utēr, his son (?)]. Uruk, Month II, day 30, year 1 hundred 12, Antiochus and Antiochus his son, the kings.	W 20030/6 (Mayer 1978; Van Dijk and Mayer 1980, no. 10; Linssen 2004, 299–300)
9	SE 119	Ephemeris and procedure for Jupiter, System B', for at least SE 131–161	Tablet of Anu-bēlšunu son of Nidinti-Anu, lamentation priest of Anu, descendant of Sîn-lēqi-unninni. Hand of Anu-aba-[utēr, his son. Uruk, Month . . . , day . . . ,] year 1 hundred 19, Antiochus and Antiochus his son, [the kings].	A 3426 (Neugebauer 1955, no. 640 + 820 Q)
10	SE 120	Lunar ephemeris, System B, for SE 121	Tablet of Anu-bēlšunu [descendant of Sîn-lēqi-unninni. Hand of Ana-aba-uter, his son, scribe of] *Enūma Anu Ellil*. Uruk, Month XII, day 12, year 1 hundred 20, Antiochus [the king].[a]	A 3432 + AO 6491 (Thureau-Dangin 1922, no. 23; Neugebauer 1955, no. 102 T)

TABLE 8.5 (*Continued*)

No.	Date	Contents	Colophon	Publication
11	SE 120	Tabular zodiacal-medical 'calendar', in which lunar eclipse omens, zodiacal signs, cities, temples, stones, and plants are systematically related	Urukean copy, written and checked against an old writing-board. Tablet of Anu-bēlšunu, lamentation priest of Anu, son of Nidinti-Anu, descendant of Sîn-lēqi-unninni. Hand of Anu-aba-utēr, his son, scribe of *Enūma Anu Ellil*, Urukean. Uruk, Month X, day 14, year 1 hundred 20, Antiochus the king.	VAT 7815 (Weidner 1967; Brack-Bernsen and Steele 2004)
12	SE 121	Table of planetary phases for SE 60–70	Tablet of Anu-bēlšunu, son of Nidinti-Anu, lamentation priest of Anu, descendant of Sîn-lēqi-unninni. Hand of Anu-aba-utēr, his son, scribe of *Enūma Anu Ellil*. Uruk, Month IX, day 14, year 1 hundred 21, Antiochus the king.	A 3405 (J. M. Steele 2000)
13	SE 121	Ephemeris and procedure for lunar eclipses, System B, for SE 113–130	Tablet of Anu-bēlšunu, lamentation priest of Anu, son of Nidinti-Anu, descendant of Sîn-lēqi-unninni, Urukean. Hand of Anu-[aba-utēr, his son, scribe of *Enūma*] *Anu Ellil*, Urukean. Uruk, Month I, year 1 hundred 21, Antiochus [the king].	AO 6485 + (Thureau-Dangin 1922, no. 24; Neugebauer 1955, no. 135+220 U)
14	—	Illustrated tabulation of a scheme for lunar motion, in which each zodiacal sign is divided into 12 micro-signs (so-called *Dodekatemoria*)	Urukean copy, written and checked against an old writing-board. Tablet of Anu-bēlšunu, lamentation priest of Anu, son of Nidinti-Anu, descendant of Sîn-lēqi-unninni, Urukean. Hand of Anu-aba-utēr, his son, scribe of *Enūma Anu Ellil*. [Uruk, Month . . . , day . . . , year . . .], Antiochus the king. Whoever fears Anu, Ellil, and Ea shall not deliberately take it away.	VAT 7847 + AO 6448 (Thureau-Dangin 1922, no. 12; Weidner 1967; cf. Brack-Bernsen and Steele 2004)
15	—	Ephemeris for Venus, System A0, for SE 111–135	Tablet of Anu-bēlšunu, lamentation priest of Anu, son of Nidinti-Anu, descendant of Sîn-lēqi-unninni. Hand of Anu-aba-utēr, his son.	A 3415 (Neugebauer 1995, no. 400 D)

Table 8.5 (*Continued*)

No.	Date	Contents	Colophon	Publication
Šamaš-ēṭir of the Ekur-zākir family and Anu-aba-utēr, son of Anu-bēlšunu I				
16	SE 118	Ephemeris for Jupiter, System A, for SE 113–173	Tablet of Šamaš-ēṭir, son of Ina-qibīt-Anu, son of Šipqat-Anu, descendant of Ekur-zākir, <incantation priest> of Anu and Antu, Urukean. Hand of Anu-aba-utēr, son of Anu-bēlšunu, descendant of Sîn-lēqi-unninni, lamentation priest of Anu and Antu, Urukean. Uruk, Month VII, day 12, year 1 hundred 18, Antiochus and Antiochus his son, the kings. Whoever fears Anu and Antu shall not deliberately take it away.	AO 6476 + Ist U 104 (Thureau-Dangin 1922, no. 28; Neugebauer 1955, no. 600 L)
17	SE 118	Ephemeris for Jupiter, System A, for SE [115]–181	[Tablet of Šamaš-ēṭir], son of Ina-qibīt-Anu, son of Šipqat-Anu descendant of Ekur-zākir, incantation priest of Anu and Antu, Urukean. [Hand of Anu-aba-utēr, son of] Anu-bēlšunu, son of Nidinti-Anu, descendant of Sîn-lēqi-unninni, lamentation priest of Anu and Antu, Urukean. [Uruk, Month . . . , day . . .], year 1 hundred 18, Antiochus and Antiochus his son, the kings.	A 3434 (Neugebauer 1955, no. 601 M)
18	—	Auxiliary table of lunar syzygies for SE 115–124	[Tablet of Šamaš-ēṭir, son of Ina]-qibīt-Anu, son of Šipqat-Anu descendant of Ekur-zākir, incantation priest of Anu and Antu, Urukean. Hand of Anu-aba-utēr, son of [Anu-bēlšunu, son of] Nidinti-Anu, descendant of Sîn-lēqi-unninni, scribe of *Enūma Anu Ellil*, Urukean. Whoever fears Anu and Antu shall not take it away.	Ist U 109 + (Neugebauer 1955, no. 171 F)

TABLE 8.5 (*Continued*)

No.	Date	Contents	Colophon	Publication
19	—	Lunar auxiliary table for at least SE 117	[Tablet of] Šamaš-ēṭir, son of Ina-qibīt-Anu, son [of Šipqat-Anu] descendant of Ekur-zākir, incantation priest [of Anu and Antu], chief [priest] of the Rēš temple, scribe of *Enūma Anu Ellil*, [Urukean]. Hand of Anu-aba-utēr, son of Anu-bēlšunu, descendant of [Sîn-lēqi]-unninni, Urukean.	Ist U 135 (Neugebauer 1955, no. 163 H)

Anu-aba-utēr, son of Anu-bēlšunu I, and Anu-uballiṭ of the Ekur-zākir family

No.	Date	Contents	Colophon	Publication
20	SE 124	Ephemeris for Mars, System A, for SE 123–202	[Tablet of] Anu-aba-utēr, son of Anu-bēlšunu, descendant of Sîn-lēqi-unninni. Hand of Anu-uballiṭ son of Ina-qibīt-Anu, descendant of Ekur-zākir. Uruk, Month IX, day 4, year 1 hundred 24, Antiochus and Seleucus his son, the kings.	AO 6481 (Thureau-Dangin 1922, no. 27; Neugebauer 1955, no. 501 Y)
21	SE 124	Ephemeris for Saturn, for at least SE 123–182	[Tablet of Anu-aba-utēr, son of] Anu-bēlšunu, [descendant of Sîn-lēqi-unninni], lamentation priest of [Anu] and Antu, Urukean. [Hand of Anu-uballiṭ son of] Ina-qibīt-Anu, descendant of Ekur-zākir, incantation priest of Anu and Antu, Urukean. [Uruk, Month . . . , day . . .], year 1 hundred 24, Antiochus and Seleucus his son, the kings.	VAT 7819 (Neugebauer 1955, no. 702 Z)
22	—	Ephemeris for Mercury, System A1, for SE 118–143	Tablet of Anu-aba-[utēr, Hand of Anu-uballiṭ son of] Ina-qibīt-[Anu, descendant of Ekur-zākir . . .].	A 3424 + (Neugebauer 1955, no. 300 P)

Table 8.5 (*Continued*)

No.	Date	Contents	Colophon	Publication
Anu-aba-utēr, son of Anu-bēlšunu I, and Anu-balāssu-iqbi, son of Nidinti-Anu II				
23	se 130	Auxiliary table for daily lunar motion, System B, for se 130	Tablet of Anu-aba-utēr, son of Anu-bēlšunu, son of Nidinti-Anu, descendant of Sîn-lēqi-unninni, scribe of *Enūma Anu Ellil,* lamentation priest of Anu and Antu, Urukean. Hand of Anu-balāssu-iqbi son of Nidinti-Anu his brother. He wrote for the life of his breath, for the prolonging of his days, for the health of his seed, for the strengthening of his foundations, for him getting no illness. Whoever fears Anu and Antu shall not deliberately take it away. May Adad and Šala take away whoever does take it away. [Uruk], Month VI, day 28, year 1 hundred 30 [Seleucus] the king.	AO 6492 (Thureau-Dangin 1922, no. 25; Neugebauer 1955, no. 194 Zc)
24	se 136	Bilingual *šu-ila* prayer from the standard lamentation series *Kalûtu,* on repairing a *lilissu* drum	Tablet of Anu-aba-utēr, son of Anu-bēlšunu, descendant of Sîn-lēqi-unninni. Hand of Anu-balāssu-iqbi son of Nidinti-Anu, son of Anu-bēlšunu, descendant of Sîn-lēqi-unninni, Urukean. Uruk, Month VI, day 21, year 1 hundred 36, Seleucus the king.	W 20030/1 (Mayer 1978)
Anu-aba-utēr alone				
25	—	Lunar auxiliary table for at least se 137–156	Tablet of [. . .]. Hand of Anu-aba-utēr, son of Anu-bēlšunu, son of Nidinti-Anu, descendant of Sîn-lēqi-[unninni, . . .].	Ist U 153 + VAT 7828 (Neugebauer 1955, no. 165 Ze)
26	—	Compilation of mathematical problems	Tablet of Anu-aba-utēr, scribe of *Enūma Anu Ellil,* son [of Anu-bēlšunu . . .].	AO 6484 (see table B.20)

TABLE 8.5 (*Continued*)

No.	Date	Contents	Colophon	Publication
27	—	Procedure for Saturn ephemerides, Systems A and B	Tablet of Anu-aba-utēr, son of Anu-bēlšunu, son of [Nidinti-Anu], descendant of Šin-lēqi-unninni, lamentation priest of Anu and Antu [. . .].	A 3418 (Neugebauer 1955, no. 802 Zd)
28	—	Fragment of astronomical table	[Tablet of Anu-aba-utēr] son of Anu-bēlšunu, son of Nidinti-Anu, Urukean.	W 20030/111 (Van Dijk and Mayer 1980, no. 86)
Anu-bēlšunu II, son of Nidinti-Anu II				
29	SE 147	List of scholars and sages	[Tablet of] Anu-bēlšunu, son of Nidinti-Anu, descendant of Sîn-lēqi-unninni, [lamentation priest of] Anu and Antu, [Uruk], Month II, day 10, year 1 hundred 47, Antiochus the king. Whoever fears Anu shall not take it away.	W 20030/7 (Van Dijk and Mayer 1980, no. 89)
30	SE 150	Ritual for repairing a *lilissu* drum, from the standard series *Kalûtu*	[Tablet of] Anu-bēlšunu, son of Nidinti-[Anu . . .] [Uruk, Month . . . , day . . . ,] year 1 hundred 50 [. . . Whoever fears . . .] shall not take it away.	W 20030/4, (Van Dijk and Mayer 1980, no. 5; Linssen 2004, 270–4)

ᵃ Neugebauer 1955, 18, restores the text in the break as '[descendant of Sîn-lēqi-unninni, scribe of]', omitting mention of Ana-aba-utēr. But the break is long enough to accommodate more text (Neugebauer 1955, pl. 248) and the date and professional designation 'scribe of *Enūma Anu Ellil*' (table 8.6) both point to Ana-aba-utēr as writer of the tablet.

problems of a Late Babylonian mathematical compilation from Babylon, BM 34568 (§8.2).

Anu-aba-utēr's father, Anu-bēlšunu, copied the 'Esagila tablet' (table 8.5: 3; and see §8.1) for a member of the Aḫi'ūtu scribal family and both men wrote and owned a large number of mathematical astronomical tablets (table 8.5: 5–7, 9, 12–3, 15–25, 27–8). Two others of theirs use a complicated system of twelve 'micro-signs' within each zodiacal sign to identify days of the year with magical and medical ingredients (table 8.4: 11, 14). Unlike the other scribal families, the Sîn-lēqi-unninnis put their name to very few copies of the standard series of omens (table 8.5: 4) or medical compilations, but concentrated instead on *Kalûtu*, the standard series of incantations and rituals associated with their profession as lamentation priests (table 8.5: 2, 8, 10, 24, 29–30).[57]

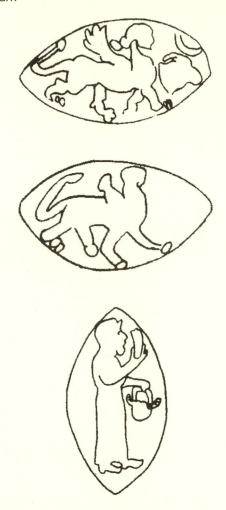

Figure 8.8 Impressions of the stamp seals belonging to Nidinti-Anu of the Ḥunzû family and his sons Ina-qibīt-Anu andAnu-uballiṭ. (Wallenfels 1994, nos. 330, 208, 125, courtesy of Ron Wallenfels and Gebr. Mann Verlag, Berlin.)

Anu-bēlšunu's birth horoscope also survives. Given that only twenty-eight Mesopotamian horoscopes are known, all from the later first millennium, and that all but three of them simply refer to "the baby" without giving a name, this is an extraordinary piece of luck. I assume he drew it up for himself, or had one of his sons do it for him as a practice piece:

Year (SE) 63, Month X, evening of day 2, Anu-bēlšunu was born. That day, the sun was in 9;30° Capricorn, the moon was in 12° Aquarius: his

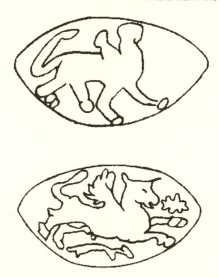

Figure 8.9 Impressions of the stamp seals belonging to Nidinti-Anu I of the Sîn-lēqi-unninni family and his son Anu-bēlšunu I. (Wallenfels 1994, nos. 306, 563, courtesy of Ron Wallenfels and Gebr. Mann Verlag, Berlin.)

days will be long. [Jupiter] was in the beginning of Scorpio: someone will help the prince. [The child] was born in Aquarius in the region of Venus: he will have sons. Mercury was in Capricorn; Saturn was in Capricorn; [Mars] in Cancer.[58]

Given his birth date—29 December 249 BCE—the pattern of writing and owning tablets now becomes clear. In his late teens and early twenties Anu-bēlšunu was writing tablets for other people: his father (table 8.5: 2, 4) and a non-family-member (table 8.5: 3). During this apprenticehood and later he was also writing legal documents for other people (table 8.6: 2, 4). Other people started to write tablets for him from at least SE 108, when he was forty-five: first his eldest son Nidinti-Anu, named for his paternal grandfather (table 8.5: 7). If Nidinti-Anu followed the same career trajectory as his father, he would have been between eighteen and twenty-two at this point, and Anu-bēlšunu would have had him between the ages of twenty-three and twenty-seven. Did he marry and start a family once his apprenticeship had ended? It was at this time that he first inherited family land (table 8.6: 3).

Anu-bēlšunu's younger son Anu-aba-utēr started to write astronomical tablets for his father in at least SE 119, when Anu-bēlšunu was fifty-six (table 8.5: 9),[59] and continued to do so until at least SE 121. Before working for his father he had spent at least a year compiling ephemerides and auxiliary tables for Šamaš-ēṭir of the Ekur-zākir family in SE 118 (table 8.5: 16–9): exactly

TABLE 8.6
Legal Tablets Mentioning Anu-bēlšunu I

No.	Date	Description	Publication
1	—	Receives rights to temple offerings with his brother Anu-uballiṭ	A 3681 (Weisberg 1991, no. 9)
2	SE 83	Scribe of sale of shares in Temple Enterer's prebend	VAT 9180 (Schroeder 1916, no. 11; Corò 2005, 162–4)
3	SE 88	Divides inherited land in the Adad Temple district of Uruk with his brother Anu-uballiṭ and cousin, also called Anu-uballiṭ	MLC 2170 (Weisberg 1991, no. 53)
4	SE 93	Scribe of sale of land in Uruk	VAT 8553 (Schroeder 1916, no. 34)
5	SE 98	Divides inherited land in the Adad Temple district of Uruk with his brothers Anu-uballiṭ and Anu-aḫḫē-iddin	NCBT 1958 (Doty 1977, 40)
6	SE 99	Witnesses sale of shares in Temple Enterer's prebend	AO 8556 (Contenau 1929, no. 242; Corò 2005, 365–7)
7	SE 110–9	Witnesses sale of shares in Temple Enterer's prebend—with Anu-uballiṭ of the Ḫunzû family	Ash 1923.78 (McEwan 1982, no. 57; Corò 2005, 350–2)
8	SE 116	Buys shares in Temple Attendant's prebend	MLC 2201 (Doty forthcoming, no. 55)
9	SE 119	Buys shares in Temple Attendant's prebend	HSM 913.2.181 (7498) (Wallenfels 1998, no. VI) = VAT 7534 (Schroeder 1916, no. 32); Corò 2005, 195–7
10	SE 126	Witnesses sale of warehouse in the Irigal temple	Ash 1923.75 (McEwan 1982, no. 53)

the same man who had been apprenticed to Anu-uballiṭ of the Ḫunzû family (table 8.3: 6). Anu-aba-utēr took an apprentice himself in SE 124, just three years after writing the latest tablet for his father, teaching Anu-uballiṭ of the Ekur-zākir family (perhaps Šamaš-ēṭir's nephew) how to write ephemerides (table 8.5: 20–2).[60] Later he took on his own nephew, Nidinti-Anu's son Anu-balāssu-iqbi (table 8.5: 23–4). There was, in short, a complex of professional relationships between the Aḫi'ūtu, Ḫunzû, Ekur-zākir, and Sîn-lēqi-unninni families (figure 8.10). It was a very tight-knit community indeed.

Anu-bēlšunu was economically active until at least his mid-sixties: the latest known attestation of him is as witness to an urban land sale in SE 126 = 185 BCE (table 8.6: 10). Legal records to which he is party also show his socio-economic status. He inherited land in SE 88 at the age of twenty-five (table 8.6: 3) and again in SE 98 (table 8.6: 5) on the death of his father. In his mid-fifties he twice purchased shares in the Temple Enterers' prebend (table 8.5: 38–9) but this was not his first involvement in these rights to temple income: earlier in his career he had written or witnessed at least three other such documents (table 8.6: 2, 6–7). Prebend sales were much more complex than in Iqīšâ's day two centuries earlier (§8.3): the income rights were now split into unit fractions of each day of the month and were paid for in coinage (see §7.3) rather than in the goods needed to carry out the prebendary activities. In this instance Anu-bēlšunu purchases $\frac{1}{6} + \frac{1}{9} = \frac{5}{18}$ day for three days of the month and $\frac{1}{3}$ on another, in total $1\frac{1}{6}$ days a month, or 14 days a year (table 8.6: 9):

> Anu-uballissu, son of Illut-Anu, son of Anu-uballiṭ, descendant of Kurî, of his own free will has sold one-sixth plus one-ninth of a day [on the 1st] day, 24th day, and 30th day, a total of one-sixth plus one-ninth of one day on those days, and one-third of a day on the 27th day of the Temple Attendant's prebend before the gods Anu, Antu, Enlil, Ea, Papsukkal, Ištar, Bēlet-ṣēri, Nanaya, Bēlet-ša-Rēš, Šarraḫītu, and all the gods of their temple, monthly, [through all the year], the *guqqû* and *eššēšu* offerings and everything else that pertains to the Temple Attendant's prebend which is with his brothers and all of his partners, for $\frac{1}{3}$ mina of refined silver, staters of Antiochus in good condition, as the [full] sale price, to Anu-bēlšunu, son of Nidinti-Anu, son of Anu-uballiṭ, descendant of Sîn-lēqi-unninni, in perpetuity. . . .

(10 witnesses)

Šamaš-ēṭir, the scribe, son of Ina-qibīt-Anu, son of Šipqāt-Anu, descendant of Ekur-zākir. Uruk, Month X day 21, year 119, Antiochus and Antiochus his son, kings.

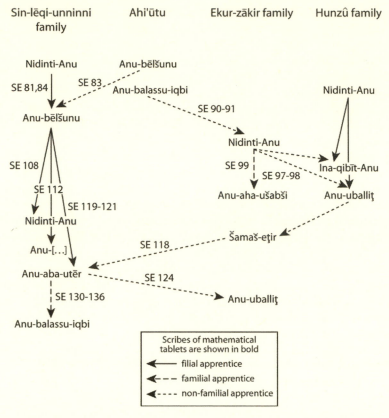

Figure 8.10 Professional relationships among the scholars of Seleucid Uruk.

The witnesses are almost all drawn from Anu-bēlšunu's close circle of scribal colleagues such as the Ḥunzûs and Aḥi'ūtus—and the scribe is none other than Šamaš-ēṭir, for whom Anu-bēlšunu's son Anu-aba-utēr had been writing scholarly tablets just the previous year (table 8.5: 16–9). The seller and remaining witnesses are from only two further families: the descendants of Kurî and of Luštammar-Adad. No marriage contracts survive from Hellenistic Uruk, but occasionally female buyers or sellers of temple prebends stated the name of their husbands. Thus several inter-marriages are known: Maqartu and Belēssunu Ḥunzû both wed Ekur-zākirs (before 217 and 207 BCE respectively), while Antu-banât Luštammar-Adad married into the Aḥi'ūtu family (before 157 BCE).[61] The fact that the scholars of the Sîn-lēqi-unninni family added injunctions to the colophons of their tablets, such as 'Whoever fears Anu and Antu shall not deliberately take it away' (table 8.5: 16, 23, 30), or 'The knowing may show the knowing; the unknowing shall not see' (table 8.5: 3), suggests that the restricted circle

THE LATER FIRST MILLENNIUM

of scholars attested in the colophons was strictly enforced in practice: a tiny number of men from a tiny social group were allowed access to this material.

The scholarly families all had long Uruk pedigrees. Descendants of Aḫi'ūtu, Ḫunzû, Ekur-zākir, and Sîn-lēqi-unninni are attested as scribes in the city from the sixth century onwards; members of the latter two families were also intellectually active from those times.[62] But the scholars themselves viewed their families as more than simply deep-rooted: a tablet copied by Anu-aba-utēr's son Anu-bēlšunu (who was named for his grandfather) in 165 BCE lists his ancestor Sîn-lēqi-unninni as the first of the post-diluvian sages, supposedly during the reign of Gilgameš himself (table 8.5: 29). Outside the cosy family network, however—and, increasingly, within it—Urukean society was ever more Hellenised. Antu-banât's Aḫi'ūtu bridegroom, for instance, went by the name of Antiochus, though his father and son used the familiar Babylonian names Ina-qibīt-Anu and Anu-balāssu-iqbi; their family's use of Greek names increased generation by generation.[63] It is likely (though not provable) that non-cuneiform users were even more prone to adopting Greek nomenclature. That names and language were a matter of choice can be seen from two building inscriptions of Antiochus Aḫi'ūtu's paternal uncle, who oversaw repairs to Anu's temple Rēš in 202 BCE. A cuneiform inscription, found on bricks in the ruins, states in part, 'Anu-uballiṭ, whose other name is Kephalon, son of Anu-balāssu-iqbi' rebuilt this temple 'for the sake of the life of Antiochus, king of the lands, my lord.' Another brick inscription, from the goddess Ištar's temple Ešgal, simply reads 'Anu-uballiṭ, whose other name is Kephalon'—neither in cuneiform nor Greek but in alphabetic Aramaic.[64]

The Rēš temple was uncovered during the course of German archaeological excavations at Uruk (figure 8.11). The Rēš may date back to the seventh century, but the excavated building is all attributable to Anu-uballiṭ/Kephalon's work in 202. It was a truly monumental construction of baked and mud brick, towering over the city. The entire complex measured some 210 × 170 × 7 metres, organised around at least nine principal courtyards with single or double ranges of rooms on each side. Its façade was elaborately recessed and buttressed, with white and gold-glazed brick friezes of stars, plants, and animals. The shrines of Anu and his divine spouse Antu, which backed onto an enormous mud-brick ziggurat some 100 metres square, were the focus of the complex, but the Rēš also housed at least twenty further shrines and instalments for divine statues.[65]

From one storeroom the excavators recovered the clay seals of long-perished papyrus rolls, bearing brief descriptions in Greek. In another, which had already been subject to clandestine excavations, were nearly 140 cuneiform tablets dated to 322–162 BCE: hymns and rituals, as one might expect, but also horoscopes, collections of omens, and astronomical

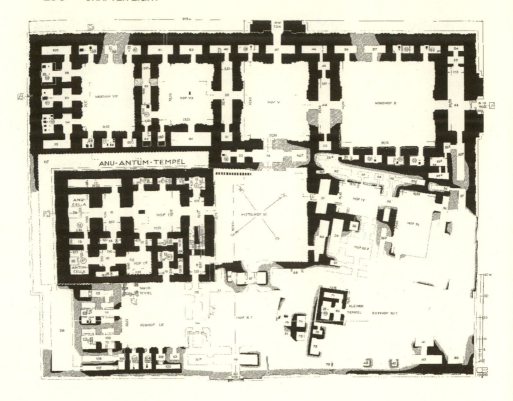

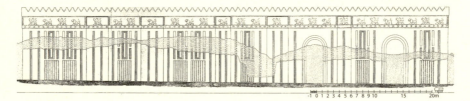

Figure 8.11 Reconstructions of the interior and façade of the Rēš temple in Seleucid Uruk. (Jordan 1928, pls. 12, 32.)

works, as well as a significant number of legal documents.[66] Several of the scholarly tablets belonged to Anu-aba-utēr of the Sîn-lēqi-unninni family (table 8.5: 24, 28–30), and it is widely agreed that many of the other tablets discussed here were illicitly dug from this area too, early in the twentieth century.

The colophons of these tablets reveal that the scholars were priests of one kind or another (table 8.7): after their apprenticeships, the Ḫunzûs and Ekur-zākirs all carried the title 'incantation priest of Anu and Antu',

TABLE 8.7
Professional Designations of Scholars in Seleucid Uruk

Family	Name	Title	Tablet(s)	Role	Date
Ekur-zākir	Nidinti-Anu	Incantation priest of Anu and Antu	8.3: 2, 4–5	Owner	SE 97–8
	Šamaš-ēṭir	Incantation priest of Anu and Antu	8.5: 16–8	Owner	SE 118
		Incantation priest of Anu and Antu, chief priest of Rēš, scribe of *Enūma Anu Ellil*	8.5: 19	Owner	—
	Anu-uballiṭ	Incantation priest of Anu and Antu	8.5: 21	Scribe	SE 124
Ḥunzû	Nidinti-Anu	Incantation priest of Anu and Antu	8.3: 1, 3	Owner	—
	Anu-uballiṭ	Incantation priest of Anu and Antu	8.3: 4–5	Scribe	SE 97–8
		Incantation priest of Anu and Antu	8.3: 6	Owner	—
Sîn-lēqi-unninni	Anu-bēlšunu I	Junior lamentation priest	8.5: 2	Scribe	SE 81
		Lamentation priest of Anu and Antu	8.5: 7–8	Owner	SE 108, 112
		Lamentation priest of Anu	8.5: 9–15	Owner	SE 119–21
	Abu-aba-utēr	Lamentation priest of Anu and Antu	8.5: 16–7	Scribe	SE 118
		Scribe of *Enūma Anu Ellil*	8.5: 10–4, 18	Scribe	SE 120–1
		Lamentation priest of Anu and Antu	8.5: 21	Owner	SE 124
		Scribe of *Enūma Anu Ellil*	8.5: 26	Owner	—
		Scribe of *Enūma Anu Ellil*, lamentation priest of Anu and Antu	8.5: 27, 30	Owner	SE 130
	Anu-bēlšunu II	Lamentation priest of Anu and Antu	8.5: 29	Owner	SE 147

while the Sîn-lēqi-unninnis were 'lamentation priests of Anu and Antu'. Šamaš-ēṭir rose to be chief priest of the Rēš and in addition bore the title 'Scribe of *Enūma Anu Ellil*', while Anu-aba-utēr earned this latter epithet through his astronomical apprenticeship with Šamaš-ēṭir in 192 BCE.

Finally, there is the conundrum of the scholars' astronomical activity to solve. It is clear to see why, as priests, they would need to possess and understand complex temple rituals (table 8.3: 6; table 8.5: 2, 7–8, 24, 30). But the primary object of their worship was Anu, the sky god, and thus many of those rituals had to be performed at celestially significant times. Some of those were a regular part of the calendar, such as the equinoctial rituals written by Šamaṣ-ēṭir for Anu-uballiṭ Ḫunzû (table 8.3: 6). Others, such as the propitiatory rituals surrounding the repair of the temples, so as not to disturb their divine inhabitants, had to be performed 'on a propitious day, in a propitious month', when the sun, moon, and planets were in auspicious configurations in the sky.[67] And finally, on the occasion of particularly inauspicious celestial events, such as lunar eclipses, the priests were charged with soothing the angered gods through ritual public lamentation.[68] All these rituals were elaborate and costly, requiring much preparation and expenditure, such as in the ritual manufacture of kettledrums.[69] On one infamous occasion in 531 the lamentation priests of Uruk, all from the Sîn-lēqi-unninni family, had been the subject of an official temple enquiry after a lunar eclipse they had lamented over had not taken place.[70] That put an onus on succeeding generations to dramatically improve their predictive ability—and to jealously guard their knowledge as inherited professional secrets. By the time of the tablets examined here, their mathematical astronomy was so accurate that they could be confident of fulfilling their priestly duties on time, every time, without embarrassment to their professional and familial heritage or their temple.[71]

8.5 CONCLUSIONS

Whereas for the Neo-Babylonian period, the preponderance of evidence is for professional numeracy and elementary pedagogical mathematics (chapter 7), this chapter has focused almost exclusively on mathematics as an intellectual activity. That is not to say that the professionally numerate disappeared in the Late Babylonian period; rather, the relative paucity of evidence indicates a shift towards the use of perishable recording media, especially for institutional purposes: the few extant cuneiform tablets are arguably all copies of parchment originals.[72] Houses could still be measured prior to sale,[73] and prebends divided into complex sums of unit fractions of a day.[74]

It is perhaps dangerous to generalise on so little evidence, but the currently known archaeologically contextualised sources suggest that Babylonian mathematics underwent a major conceptual shift between the fifth century and the mature Hellenistic period. The mathematical writings of the Šangû-Ninurta family are predominantly concerned with metrologies: not only did they own several substantial metrological tables, but the relationship between different systems of area measure, and between lengths and areas, is the overarching theme of their word-problem series *Seed and reeds* (§8.3). In this respect their mathematics sits comfortably within the mathematical tradition first attested in the late fourth millennium (§2.1). By contrast, the word problems of the Seleucid period, whether those of Ana-aba-utēr Sîn-lēqi-unninni or the anonymous compilations from Babylon, are thoroughly demetrologised. Cubits and minas are given rather trivial passing mention, but essentially numbers have become separated from the objects and sets they quantify for the first time in cuneiform culture.[75] Neither are any metrological tables currently dateable to the last three centuries BCE. How can this dramatic rethinking of the status of number as an entity in its own right be accounted for? The two obvious candidates are the concomitant development of mathematical astronomy and a more general influence of Hellenistic intellectual culture.

Mathematics and mathematical astronomy were central components of the last flowering of cuneiform culture. But although the two were written by the same scholars in the Seleucid period, the fifth- and fourth-century evidence from Uruk suggests that that there was no simple relationship between them: in pre-Hellenistic times at least, scholars could study one but not the other (§8.3). Indeed, there is very little of astronomical relevance in Late Babylonian mathematics. Although multiplication tables for a new standard set of multiplicands were copied in both Uruk and Babylon, none of those numbers has any astronomical significance.[76] Reciprocals play no role in the calculation methods laid down by astronomical precepts, and neither do squares or fourth powers.[77] Yet those are by far the commonest types of Late Babylonian arithmetical lists and tables known. It is only in the shared structure, format, and terminology of the word problems and precepts that similarities between mathematics and astronomy are clear—and many of the structural similarities are also identifiable in the ritual instructions written by the same scholars[78] and in Mesopotamian instructional texts more generally. Mathematics, then, does not seem to have been co-opted by the needs of Seleucid mathematical astronomy.

Neither it is likely that mathematics changed as a result of outside influence,[79] given that its Seleucid practitioners were the very last men to cling stubbornly to the last vestiges of the old ways of thinking and doing while society around them was increasingly in thrall to Hellenistic culture. The astronomers and priests of Esagila and Rēš comprised a tiny number of individuals from a restricted social circle, intermarrying, working closely together to train each successive generation, and highly valuing privacy and secrecy. (Ironically, if indeed one of them ever did drop a scholarly tablet in the street, it is highly unlikely that any passer-by would be sufficiently cuneiform-literate to decipher the colophon's injunction to return it on pain of divine wrath, never mind to read its top-secret contents.) Their sole aim was to uphold the belief systems and religious practices of ancient times: indeed Bēl-rē'ûšunu (known to posterity by the Greek name Berossus), chief administrator of Esaǧila in the 250s BCE, wrote a compendious history of Babylonian culture aimed at a Greek-speaking audience.[80] The few late cuneiform tablets bearing Greek script are exclusively transliterations or translations of Babylonian scholarship; there are no known representations of Greek intellectual writings in cuneiform.[81] In this context a tiny cadre of highly intellectual priests—probably no more than half a dozen at any one time in Babylon, two or three in Uruk—developed increasingly mathematically sophisticated means to ensure the calendrical accuracy of their rituals, whose improving accuracy can be tracked generation by generation.[82] No-one has suggested that those improvements were anything but indigenous. In this light, there is no reason to suppose that the Mušēzibs, Ḫunzûs, and Sîn-lēqi-unninnis were not perfectly capable of mathematical creativity too.

CHAPTER NINE

Epilogue

In the light of the preceding chapters, the cuneiform mathematics of ancient Iraq and its neighbours can be divided into three phases: numerate apprenticeship, metrological justice, and divine quantification (§9.1). Such socio-historical interpretations have been slow to develop because of the chronology of modern decipherment and interpretation. The translation and then popularisation of Babylonian mathematics lagged a generation behind scholarship on the classics of ancient Greek mathematics and thus got shoehorned into the paradigm of the Greek 'miracle' (§9.2). But a comparison of the internal characteristics of Old Babylonian, Late Babylonian, and Euclidean mathematics shows few points of commonality (§9.3). Neither are there convincing historical arguments for transmission from one mathematical culture to another (§9.4). But that does not lessen the importance of cuneiform mathematics. Rather, the radical reappraisal of the three millennia of ideas, practices, individuals, and communities presented in this book aims instead at understanding and appreciating the mathematics of ancient Iraq within its own rich cultural contexts and on its own terms (§9.5).[1]

9.1 THE BIG PICTURE: THREE MILLENNIA OF MATHEMATICS IN ANCIENT IRAQ

This book does not pretend to be a definitive history of mathematics and numeracy in and around ancient Iraq. It has simply scratched the surface of a rich and complex three-thousand-year tradition in the hope of encouraging both Assyriologists and historians of mathematics to consider cuneiform mathematics not as a single closed, abstract system but as a complex of ideas and practices that are best understood when contextualised within the social, intellectual, and political history of the ancient Middle East. The chronological and geographical range presented here is both vast and artificially limited, defined as it is by the spread of a particular writing technology rather than any intrinsically mathematical or historical concern. But that is not a major handicap given that cuneiform-on-clay was developed in the late fourth millennium specifically to record numbers of things and that right until its dying days at the end of the first millennium the cuneiform literati were always numerati too. As the previous eight chapters have shown, the tradition was far from monolithic. Roughly speaking it can be divided into three phases.

The idea of *numerate apprenticeship* characterises the first millennium or so of surviving evidence (chapter 2). A small cadre of bureaucrats managed the domestic economies of big institutions of the late fourth and early

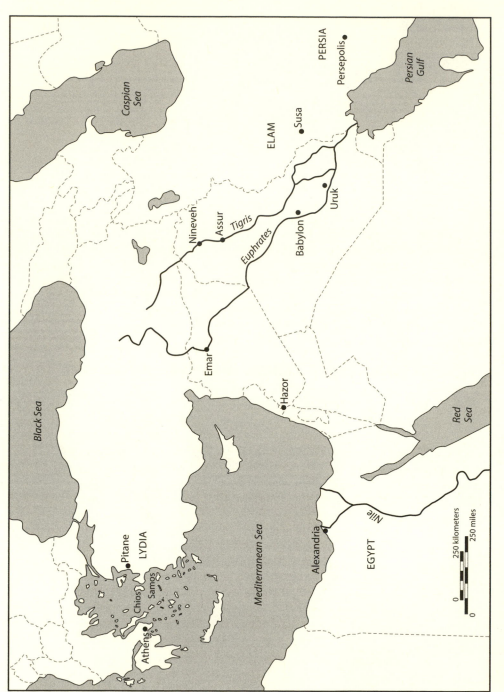

Figure 9.1 The locations mentioned in this chapter, including findspots of mathematical tablets.

third millennia, first the temples and later also royal palaces. They recorded aspects of the quantitative management of land, livestock, and labour, not only to account for what had already been acquired or produced but also to make short-term predictions of costs and yields. New members of the group were trained and enculturated primarily through practical experience but also through simple exercises that to some extent mimicked the content, format, and structure of the accounts that the competent professionals produced. Paradoxically, such exercises can be identified only by virtue of their differences from working records: they focus on the interfaces between metrologies (grain and labour; length and area) as well as on two types of number: the key round numbers in each metrological system and, increasingly, numbers that were co-prime to those key numbers and therefore difficult to divide by. There is very little direct information about the teachers and learners of these early mathematical exercises, which are almost always anonymous and without useful archaeological context.

In the later third millennium the management needs of large territorial kingdoms drove major metrological reforms and cognitive developments towards the concept of *metrological justice* (chapters 3–4). New metrologies were invented; existing ones were extended upwards and downwards, often using sexagesimally based units, and adapted for better integration between systems. In the twenty-first century the idea of sexagesimal fractions and multiples was generalised into the sexagesimal place value system—an all-purpose number system which enabled scribes to add and multiply without regard for the metrology to which the numbers pertained. Generalisation of the unit fraction, from $\frac{1}{2}$–$\frac{1}{3}$ to the reciprocal of any sexagesimally regular number of any length, enabled easy bottom-up division for the first time. Further, tabulation allowed not only the visual separation of quantitative and qualitative data but also encouraged the categorisation of different quantitative data types, further divorcing numbers from the sets or objects to which they pertained and the units in which those objects were counted and measured (§6.2). Together these three innovations (tabulation, reciprocals, the SPVS) were the invisible support system for a fearsomely involved system of annual balanced accounts for various aspects of the Ur III economy—invisible because the scribes were trained to destroy their intermediate work records and submit only the finished accounts in which quantities were recorded in traditional metrological units within sentence-like structures.

It is probably at this time that pedagogical curricula began to be formalised (chapter 4). Increased numbers of bureaucrats were needed, who had to be trained in consistent trans-territorial accounting methods and in writing Sumerian, the language of bureaucracy and state. Again, the contemporary evidence is meagre, but as late as the 1740s scribal training still used royal hymns of the twenty-first and twentieth centuries, as well as Ur III work rates that were no longer in practical use. These curricula no longer aimed at simply producing numerate and literate competence; it was equally important that individual scribes identified both with the scribal community as a whole and with the needs and ideals of the state and its ruler.

Early Mesopotamian kingship was constructed on the intertwined ideals of piety, military might, and justice. And justice—'making straight'—was conceptualised through fairness and equality. Equality could be measured, and measurement was the key to ensuring justice. It was the gods who bestowed the quality of justice on kings, symbolised through the presentation of land-surveying equipment. In curricular Sumerian literature the great gods, such as Enki and Enlil, presented goddesses, especially Nisaba, Ninlil, and Inana, with the reed and rope and the accoutrements of writing. These goddesses both used their equipment to ensure the smooth running of divine households and presented those tools to kings to endow them with the means of administering justice. In monumental imagery, by contrast, it was often male deities, associated with the dynastic line, who were shown bestowing the symbols of justice on kings. In the human realm, kings oversaw metrological standards, promulgated laws about fair measurement, and punished the fraudulent use of metrological equipment—a royal duty still valued by the Assyrian king Aššurbanipal, over a thousand years later.

Beyond the rote memorisation of standard elementary series of metrological and arithmetical relationships, scribal students learned a style of mathematics that encapsulated the principles of metrological justice in a manner that complemented the messages embedded in law and literature. On the one hand word problems about land and labour portrayed images of the ideal scribe in action that were just as effective as those in pithy proverbs or vivid debate poems. On the other hand word problems about the manipulation of lines and areas to find unknown quantities encapsulated the very essence of metrological justice: in solving abstruse puzzles about measured space, the true scribe demonstrated his or her technical capability and moral fitness for upholding justice and maintaining social and political stability on behalf of king and god.

The schools themselves were not large institutional buildings, however, but small domestic spaces. Some students, such as the trainee priest Bēlānum/Ur-Utu, were home-schooled by peripatetic tutors; in other instances literary letters suggest that the students lodged with the teacher. For Inana-amaĝu and her siblings, home and school, father and teacher, were probably one and the same. Parental-teacherly relationships could also be created formally, as attested by Ba'al-mālik and his adoptee apprentices in thirteenth-century Emar.

Although the collapse of the Old Babylonian kingdom in the late seventeenth century resulted in a break in the Babylonian evidentiary record for the following century or so, later second-millennium sites on the peripheries

of the cuneiform world—such Assur, Elamite Kabnak, Canaanite Hazor—
show that Old Babylonian–style arithmetic and metrology had a long-lived
and widespread, if patchy, diffusion. However, that mathematics never really caught on in the wider Middle East, presumably because it had little
relevance to local counting systems, metrologies, or political ideologies.
In fourteenth- and thirteenth-century Babylonia itself there are meagre
traces of continuity with Old Babylonian pedagogy but strong continuities
in the culture of professional numeracy. Land surveyors and accountants
worshipped the (newly) numerate goddess Nin-sumun and employed familiar techniques of tabulation and land survey. Most prominent amongst
them were the descendants of Arad-Ea, the first family line to self-identity
with a named ancestor, himself a numerate professional. However, much
mathematics (along with other intellectual traditions) had been lost or
abandoned as no longer relevant. Rod-and-reed imagery continued to be
deployed on royal monuments, but it is doubtful whether it retained its
original significance as a marker of metrological justice.

Early Mesopotamian intellectuals used writing primarily as an aid to
memorising a tradition that was passed on through direct person-to-person
communication. By the later second millennium collection, edition, and compilation of scholarly works into long series led to a mechanism of tablet-to-
tablet transmission, in which students would create libraries of increasingly standardised works through copying and extracting. Scholars such as
Sîn-lēqi-unninni and Nabû-zuqup-kēna were also highly creative: comment-
ing, interpreting, and improving on the tradition rather than passively receiving it uncritically. From Tukulti-Ninurta to Aššurbanipal, several Assyrian kings sought to acquire Babylonian cultural capital by force, deporting
both tablets and scholars to Assyrian libraries. However, neither mathematics nor the quantified observations of nascent Babylonian astronomy
are attested amongst the scholarly loot. Were the Assyrians not interested,
or were these genres too important to the Babylonians to relinquish?

The Babylonian elementary educational curriculum that had developed
by the sixth century appears equally indifferent to mathematics; only basic
metrological lists and squares of numbers were ever written on the tablets
covered in writing exercises that were dedicated to the god Nabû of Accounts. Late Babylonian word problems were often based on scenarios similar to Old Babylonian ones, but used different metrologies and new methods
of solution that relied less on the manipulation of lines and areas than before, even when those problems specifically expressed metrological concerns.
By Seleucid times all sense of number as measure had been abandoned;
metrological tables were no longer copied, and number was purely arithmetical. The ancient tradition of metrological justice had long disappeared,
as had any meaningful role for numerate goddesses; there was no longer
any ideology of metrological justice for mathematics to support.

Rather, from the mid-seventh century onwards, Assyrian scholars like Nabû-zuqup-kēna and the anonymous Babylonian compilers of eclipse records and astronomical diaries had begun to think in terms of *divine quantification*. They understood that the gods' management of time and space was deeply mathematised. Apparently random events of great ominous significance were observed, quantified, and recorded in the hope that numerical patterns could be detected amongst them. The ultimate aim was to understand the will of the gods, to ensure that they were propitiated and would act benignly to the king and humanity. Thus in later Babylonia mathematics became a priestly concern, and even when it was not immediately practical it was an expression of the deeply mathematised nature of the universe as the gods had created it; even the gods themselves were associated with particular numbers (chapter 8). The secret knowledge of mathematics, astronomy, and ritual was communicated through apprenticeship within a tightly restricted social circle, either down the bloodline or between male members of a tiny number of families, all of whom had close associations with the Marduk's temple in Babylon or Anu's in Uruk. These men resisted the allure of new-fangled Persian and Hellenistic culture, clinging to the old ways of belief while constantly renewing and improving their mathematical methodologies.

By the Seleucid period intellectual mathematics had divorced itself from the concerns of professional numeracy. Accountants and land surveyors continued to earn their living by managing temple economies and recording transfers of property. But as for the most part they wrote their records on perishable media, we have little access to their thoughts and actions. Nevertheless, those lost records are a useful reminder that what survives is not the sum total of what was written: tablets were habitually recycled, while wooden writing boards, parchment, and papyri have not withstood the environmental conditions under which they were stored, lost, or abandoned. Beyond the written there was doubtless a strong oral component to mathematical and numerate communication, whether for teaching apprentices, reporting to superiors or patrons, collaborating with colleagues, or communing with the divine. Further, there are vast areas of embodied and instrumental practice that are currently mysterious: how were calculations carried out, measurements made, equipment constructed? Doubtless there are hints in the written record, but such questions are very hard to answer given the present state of knowledge.

9.2 ANCIENT MATHEMATICS IN THE MODERN WORLD

The image of cuneiform mathematics presented in this book is radically different from the received picture, especially that portrayed in the secondary literature on the history of mathematics. That is because although the events and ideas presented here all happened long ago, the recovery of that history is unusually recent, and the academic field startlingly underpopulated. A brief survey was given in §1.1; here a more critical analysis is attempted. Iraq and its neighbours have been mathematically literate for over five millennia, but the pertinent primary sources have been known to modern scholarship for just over a century. It is on the one hand 'the first half of history' and yet at the same time a 'new antiquity'.[2] Scholarly narratives of the early history of mathematics have thus had to be rewritten dramatically over the last hundred years to accommodate the ancient Middle East. This rewriting, particularly in more populist, generalist accounts, has tended either to write it off as not proper mathematics, or to characterise it as proto-Greek. Morris Kline's influential *Mathematical thought from ancient to modern times* (1972) opens: 'Mathematics as an organized, independent, and reasoned discipline did not exist before the classical Greeks of the period from 600 to 300 BC entered upon the scene'.[3] Three millennia of 'Babylonian' mathematics are dismissed as essentially trivial and unchanging within the first 12 pages of the 1350-page, three-volume work. An explicitly stated aim of B. L. van der Waerden's *Science awakening* (1954) is 'to explain clearly how Thales and Pythagoras took their start from Babylonian mathematics but gave it a very different, a specifically Greek character'.[4] Even Neugeubauer's *The exact sciences in antiquity* (1952) presents 'Babylonian' mathematics as the precursor to Greek achievements, if not always explicitly—for Neugebauer acknowledged that the necessary evidence was not yet available—then implicitly through juxtaposition: 'All that we can safely say is that a continuous tradition must have existed, connecting Mesopotamian mathematics of the Hellenistic period with contemporary Semitic (Aramaic) and Greek writers and finally with Hindu and Islamic mathematicians'.[5] The Klinean attitude need hardly be dignified with a rebuttal in the light of the foregoing eight chapters, but the characterisation of the cuneiform tradition as proto-Greek mathematics remains ubiquitous and powerful. In order to problematise it, we first need to understand how it came about.

Between the gradual abandonment of cuneiform script some two millennia ago and its re-decipherment in the mid-nineteenth century, stories and fantasies about Mesopotamia retained an enduring allure. Greek writers began to take a serious interest in the Middle East in the 540s BCE, when the conquest of Lydia (now western Turkey) brought the Persians to the edges of the Classical world. They sought at once to explain how such a mighty and influential polity had come into being and at the same time to belittle and ridicule it.[6] Persia and its satrapies Assyria and Babylonia became confused in the Classical imagination, so that mathematics and astronomy were attributed to the 'Magi' and the 'Chaldaeans', apparently interchangeably.[7] In reality the Magi (Old Persian *maguš*) were a class of Zoroastrian priests closely connected to the Persian court while the Chaldaeans (Babylonian *kaldû*) were a group of powerful nomadic tribes based in the marshes of southern Iraq, often in conflict with central authority.[8] Neither group were themselves mathematicians or astronomers, according to the cuneiform record itself. Rather, in the later first millennium mathematical activity was the domain of tiny groups of specialists based in the ancient temples of Babylonia (chapter 8). The native professional terms *āšipu* 'incantation priest', *kalû* 'lamentation priest', and *ṭupšar enūma anu ellil* 'scribe of *Enūma Anu Ellil*' were unknown to the Greek tradition; instead the terms *magos* 'priest' and *chaldaios* 'southern Babylonian' approximated their status. Such mathematical and astronomical knowledge that did reach the Graecophone world in the last centuries BCE became absorbed into the mainstream of intellectual culture and its Babylonian origins forgotten. Soon only fables and rumours remained of ancient Middle Eastern mathematical traditions. The topos of the scholarly Chaldaean lasted through Late Antiquity, in the writings of commentators such as Iamblichus of Chalcis, and into the writings of medieval and early modern mathematicians such as Robert Recorde and John Wallis.[9]

The received image changed only in the aftermath of the rediscovery of ancient Middle Eastern culture. French and English scholars had deciphered cuneiform script in the 1840s and '50s, spurring on a headlong rush to collect and publish vast quantities of ancient tablets dug from the ruin mounds at the eastern edge of the Ottoman empire, around Mosul and south of Baghdad.[10] But for mathematics the breakthrough did not come until early in the twentieth century as the sexagesimal place value system and its arithmetic were interpreted and then the first mathematical procedures translated. Roughly speaking, there were two schools of mathematical cuneiformists in the first decades of the twentieth century. French Assyriologists such as Vincent Scheil (1850–1940) and François Thureau-Dangin (1872–1944) were deciphering on internal evidence the tablets they had excavated. Because they published their careful findings about metrology, numeration, and arithmetic in new, small-circulation specialist journals such as *Revue d'Assyriologie* their results had little impact on the wider academic community.[11] Much more influential—for history of mathematics, if not for Assyriology—was a monograph called *Mathematical, metrological and chronological tablets from the Temple Library of Nippur* (1906), written by Herman Hilprecht (1859–1925), the German curator of the University of Pennsylvania's cuneiform collection.[12] According to his survey of the previous literature, just twenty-four Old Babylonian mathematical and metrological tablets had been published hitherto. Fully half of them were metrological lists and tables, while the rest would now be described as multiplication tables; tables of squares, inverse squares, and inverse cubes; compilations of word problems; and a calculation.[13] The new volume almost tripled the known corpus, adding forty-four tablets to the group, including the hitherto unknown standard OB arithmetical series of reciprocals and multiplications, as well as calculations and compilations of problems.

David Eugene Smith (1860–1944), undoubtedly America's foremost historian of mathematics at that time, corresponded with Hilprecht about the book, drawing comparisons with both Egyptian and Roman mathematics.[14] He reviewed it for the *Bulletin of the American Mathematical Society* and incorporated six pages on Babylonian mathematics in his two-volume *History of mathematics* textbook, first published in 1923 and still in print today.[15] It represents the first popular diffusion of Babylonian mathematics in the English-speaking mathematical community. By contrast, ancient Egyptian mathematics had been revealed to the modern world in the 1870s through the decipherment of the Rhind Papyrus,[16] while Euclid's *Elements* had been circulating in the West for centuries, in many rival formats.[17] Johan Heiberg's scholarly edition appeared over the course of 1883–1916 and Thomas Heath's English translation of Heiberg came out in 1908; the 1926 edition is still in widespread use.[18] Heath's seminal *History of Greek mathematics* was published in 1921. It was thus natural for Smith's account—which necessarily was essentially of Old Babylonian arithmetic—to draw comparisons with both Egyptian and ancient Greek mathematics, as they were then understood.

However, Smith's work came too early to take into account the great wave of decipherments undertaken in the 1920s to '40s by Thureau-Dangin and Otto Neugebauer (1899–1990), in which many hundreds of new tablets, especially those containing word problems, were translated and analysed mathematically.[19] Babylonian mathematics reached a broader Anglophone public only in the 1950s, in the wake of Neugebauer's *The exact sciences in antiquity* (1952, hereafter *ESA*), his phenomenally popular and influential synthesis based on his massive programme of edition and translation over the previous twenty years. (He had written *Vorgriechische Mathematik* for a popular Germanophone audience already in 1934.)[20] Indeed, I would contend that a direct line of descent can be traced from *ESA* to the contents and form of what soon became the obligatory Babylonian chapter in every history of mathematics text book for the following forty years. What Neugebauer chose to highlight in *ESA*, selected from the several hundred manuscript sources that he had deciphered and published since the mid 1930s, was not in any sense representative of the cuneiform corpus as a whole, or even of Old Babylonian mathematics. Instead, three particular Old Babylonian texts became canonised in the secondary literature during the course of the 1950s and '60s, further reinforced by Asger Aaboe's influential *Episodes from the early history of mathematics* (1964):[21]

- Plimpton 322, the famous table that includes fifteen pairs of numbers belonging to 'Pythagorean' triples (§4.3), and which has been subjected to all manner of imaginative interpretations;
- YBC 7289, a numbered diagram on a Type IV tablet which displays a highly accurate approximation to the 'square root' of 2 (§4.3);[22]
- IM 55357, an illustrated but unfinished problem which shows an understanding of the properties of similar triangles—and interestingly the first mathematical cuneiform tablet to be published by an Iraqi, in an article tellingly called 'An important mathematical problem text . . . (on a Euclidean theorem)'.[23]

The narrow focus on these three texts led Old Babylonian mathematics to be viewed through the lens of early Greek mathematics, whose received image was at that point no less partial: Heath and his contemporaries had created a narrative in which the so-called crisis of incommensurability had led to the rejection of arithmetic in favour of formal Euclidean-style axioms and deductive proofs (see §9.4). Just as one strand of ancient Greek mathematics had become privileged over the rest,[24] a generation later Old Babylonian mathematics was cherry-picked for resemblances to that tradition. The first propositions of Book II of Euclid's *Elements* were found to have Old Babylonian precursors (see §9.3), as did 'Pythagoras' theorem'. Amongst hundreds of extant sources, those relating to the supposed preoccupations of the early Greeks—the square root of 2, the Pythagorean theorem, *Elements* II.1–10—and which were primarily visual or tabular, rather than prosaic, received priority in publication and interpretation, while other, more typical features were downplayed or simply ignored.

The chronology of decipherment was not the only factor in the modern construction of (Old) Babylonian mathematics. Throughout the twentieth century historians and popularisers of the subject were primarily historically minded mathematicians, with little or no professional expertise in the languages, history, or archaeology of ancient Iraq. Van der Waerden for instance was a highly regarded algebraist, while Kline worked on electromagnetism and a critique of the New Math educational programme of the 1950s.[25] Neugebauer too had started his career as a mathematical physicist before turning his considerable talents to decipherment, from the 1940s in collaboration with the Assyriologist Abe Sachs.[26] Throughout the twentieth century mathematical realism, or Platonism, was powerfully argued by such influential philosophers and mathematicians as Frege, Gödel, and Quine. On this view, abstract mathematical objects such as numbers and sets in some sense exist, independent of our beliefs about them.[27] Whatever the powerful philosophical arguments in its favour, as a historiographical stance it is problematic because it implies that mathematical ideas and techniques are *found* or *discovered* by individuals or groups. Thus at is simplest the realist historical enterprise consists of identifying Platonic mathematical objects in the historical record and equating the terminology used to describe and manipulate them with their modern-day technical counterparts. The emphasis is on tracing mathematical *sameness* across time and space. For instance Aaboe opens his *Episodes* by inviting the reader to imagine a modern 'schoolboy' transported back in time:

> Mathematics alone would look familiar to our schoolboy: he could solve quadratic equations with his Babylonian fellows and perform geometrical constructions with the Greeks. This is not to say he would see no differences, but they would be in form only and not in content; the Babylonian number system is not the same as ours, but the Babylonian formula for solving quadratic equations is still in use. The unique permanence and universality of mathematics, its independence of time and cultural setting, are direct consequences of its very nature.[28]

Further, real people are entirely absent from this style of history, whether as individuals, professional groups, or societies.[29] For nearly two millennia mathematical achievements in the ancient Middle East had been attributed to either the Magi or the Chaldaeans, both terms that approximately acknowledged the reality of southern Babylonian priestly practitioners. Postdecipherment, paradoxically, agency was lost, and cuneiform mathematics became attributed to anonymous, context-less Babylonians, rarely dated or located, and assumed to be unproblematic (male) 'mathematicians'.

If in the twentieth century mathematics was eternally unchanging then so was the ancient Orient. The often unwitting European intellectual denigration and appropriation of the Middle Eastern past has long been acknowledged and analysed by historians and anthropologists of the region, especially in the wake of Edward Said's seminal book *Orientalism* (1978). From the Greek writers onwards, westerners have read both infantilism and decadence into the East:

> The Orient existed for the West, or so it seemed to countless Orientalists, whose attitude to what they worked on was either paternalistic or candidly condescending—unless of course they were antiquarians, in which case the 'classical' Orient was a credit to *them* and not to the lamentable modern Orient.[30]

Many mid-twentieth-century Assyriologists belonged to this mindset too, treating tablets as deracinated primary sources with little or no regard to their archaeological context, and/or using the modern West as their interpretative model. Witness the misidentification of the Mari storeroom as a 'schoolroom', for instance (§5.2). As Zainab Bahrani notes, the use of the term 'Mesopotamia', which has no meaning in the geography of the modern Middle East, encouraged the treatment of ancient Iraq as a never-never land which could be appropriated as the 'cradle of civilisation' (Western of course, for in this view there is no other):

> [Mesopotamia] is at once the earliest phase of a universal history of mankind in which man makes the giant step from savagery to civilization, and it is an example of the unchanging nature of Oriental cultures. [In the Orientalist view] the Mesopotamian past is the place of world culture's first infantile steps: first writing, laws, architecture and all the other firsts that are quoted in every student handbook and in all the popular accounts of Mesopotamia.[31]

Innocence and corruption, saming and othering: the Orientalist reading of the East became inscribed even into the mathematics. At the same time as (some) Old Babylonian mathematics was written into the Western tradition as laying the foundation for the supposed 'Greek miracle' of ancient Athens, its counterpart from the later first millennium was roundly ignored. It was irrelevant to the standard teleological narrative by virtue of being contemporaneous with, or even later than, Classical Greek mathematics from Aristotle to Euclid. In any case, the consensus went, it was by and large a degenerate misunderstanding, or at best a static transmission, of the earlier material, which had lost its conceptual autonomy and therefore the right to be treated as real mathematics.[32]

9.3 INSIDE ANCIENT MATHEMATICS: TRANSLATION, REPRESENTATION, INTERPRETATION

So what of Babylonian mathematics' status as the origins of the Western tradition? Whether or not much of twentieth-century historiography was naïve or misguided, are there any current reasons for rejecting or accepting such a claim? There are two potential lines of approach: to compare the content, structure, and concepts of (Old and Late) Babylonian and Greek mathematics; and to consider the opportunities and mechanisms for transmission. Putting aside historical concerns for the moment, let us begin with internal features. Historians can be confident of direct transmission of a particular mathematical idea, technique, or problem through identification of 'index fossils'—features that remain stubbornly unchanged over time and place.[33] In the case of Old Babylonian mathematics, for instance, methods for finding the volume of a grain pile were refined and generalised, while the parameters of the grain pile stayed the same.[34] On a larger geographical and chronological scale, pre-modern mathematical riddles about repeated doubling are always about grain, and the doubling takes place either on every day of the month or, later, on every square of a chess board (§5.2).

Most recently it has been claimed that Euclid's *Elements* II.1–10, VI.28–29, *Data* 84–86, and Diophantos' *Arithmetic* I.27–30 were all influenced by the Old Babylonian mathematical tradition.[35] But although to modern mathematical eyes some small parts of Greek deductive mathematics may appear to treat the same mathematical objects as do Old Babylonian word problems, there are no index fossils, whether numerical, conceptual, or structural, to demonstrate transmission rather than independent invention. That has become increasingly clear since the 1990s when translators began to treat the words of ancient mathematics as seriously as its numbers (or letters). For most of the twentieth century, the primary aim of historical research on ancient mathematics was to discover *what* the ancients knew and to analyse it as mathematics pure and simple. As Karin Tybjerg puts it:

> Treatments such as those of Thomas Heath, Otto Neugebauer and Hieronymus Zeuthen [. . .] were primarily concerned with mathematical content and because content was thought to be independent of presentation, the Greek geometrical texts were not just translated to a modern language, but also rewritten in the modern mathematical idiom of geometrical algebra.[36]

More recently historians have been concerned to recover *how* the ancients thought about mathematics. In order to recover ancient mathematical *concepts*, historians have gone back to the original sources and retranslated them in a way which tries to stay as faithful as possible to the texture of the original vocabulary and syntax. To use translators' jargon, the mid-twentieth-century translations are *domesticating*, in that they aim to make ancient mathematics familiar and comfortable. The newer translations, on the other hand, are *alienating*, in that they try to maintain the intellectual distance between the sources and us. Reviel Netz has recently summarised his rationale for translating deductive Greek mathematics:

> There are many possible barriers to the reading of a text in a foreign language, and the purpose of a scholarly translation is as I understand it to remove all barriers having to do with the foreign language itself, leaving all other barriers intact.[37]

Jens Høyrup expresses similar concerns in his advocacy of 'conformal translation' of Babylonian word problems, to capture the original meaning as closely as possible; both advocate structural analysis through close reading.[38] It may be thought that Høyrup and Netz succeed all too well in their alienating aims: there is no doubt that their translations are harder to read than their predecessors'. But although the reader has to work at understanding them, the rewards are satisfying and often surprising.

To illustrate this, compare two translations of an Old Babylonian word problem on the difference between the length and width of a rectangle (table 9.1). In van der Waerden's translation of 1954 the problem is entirely numerical, as he explains:

> The first lines formulate the problem: two equations with two unknowns, each represented by a symbol, *ush* and *sag*, length and width. The Sumerian symbols are dealt with as our algebraic symbols x and y; they possess the same advantages of remaining unchanged in declension. We can therefore safely put the problem in the form of 2 algebraic equations:

$$(1) \qquad xy + x - y = 183$$
$$x + y = 2$$

> The last 4 lines of the text merely verify that the resulting numbers $x = 15$ and $y = 12$ indeed satisfy (1). [. . .] What the Babylonians do, step by step, in numbers, amounts indeed to application of the formulas. [. . .] But they do not give these formulas; they merely give one example after another, each of which illustrates the same method of calculation.[39]

In his reading 'the Babylonians' are just like modern mathematicians: they use 'symbols' and 'equations', which means that the problem can 'safely' be expressed as modern algebra. Høyrup, by contrast, opens his comments on the same problem with an interpretative diagram that does not appear on the cuneiform tablet (figure 9.2) and continues:

> The text starts by stating that a rectangular surface or field (obliquely hatched in the diagram [on the left]) is built, that is, marked out: after pacing off its dimensions, the speaker 'appends' the excess of the length over the width (regarded as a broad line—vertically hatched) to it; the outcome is 3 03. Even this is done quite concretely in the terrain. Then he 'turns back' and reports the accumulation of the length and the width to be 27. The procedure starts by 'appending' these latter 'things accumulated' (dotted) to the hatched surface, from which we get a new rectangle with the same length and width that has been augmented by 2. The surface is 3 03 + 27 = 3 30, and the sum of the length and the new width is obviously 27 + 2 = 29. This standard problem is solved by means of the usual procedure, as shown in the [diagram on the right]. The length turns out to be 15, and the width 14; the original or 'true' width is therefore 12.[40]

TABLE 9.1
Old Babylonian Word Problem AO 8862 (1) in Two Translations

van der Waerden's Translation (1954, 63)	Høyrup's Translation (2002b, 164–5)[a]
Length, width. I have multiplied length and width, thus obtaining the area. Then I added to the area, the excess of the length over the width: 3 03 (i.e., 183 was the result). Moreover, I have added the length and width: 27. Required length, width, and area.	Length, width. I have made length and width hold each other. I have built a surface. I turned around (it). As much as length went beyond width, I have appended to inside the surface: 3 03. I turned back. I have accumulated length and width: 27. What are the length, width, and surface?

<table>
<tr><td>(given:) 27 and 3 03, the sums</td><td>27 3 03 the things accumulated</td></tr>
<tr><td>(result:) 15 length   3 00, area</td><td>15   the length</td></tr>
<tr><td>12 width</td><td>12   the width   3 00 the surface</td></tr>
</table>

One follows this method:	You, by your proceeding, append 27, the things accumulated, length and width, to inside 3 03: 3 30. Append 2 to 2: 29. You break its moiety, that of 29: 14;30 steps of 14;30 is 3 30;15. From inside 3 30;15 you tear out 3 30: 0;15, the remainder. The equal-side of 0;15 is 0;30. You append 0;30 to one 14;30: 15, the length. You tear out 0;30 from the second 14;30: 12, the true width. I have made 15, the length, and 12, the width, hold each other: 15 steps of 12 is 3 00, the surface. By what does 15, the length, go beyond 12, the width? It goes beyond by 3. Append 3 to inside 3 00, the surface: 3 03, the surface.
27 + 3 03 = 3 30 2 + 27 = 29	
Take one half of 29 (this gives 14;30).	
14;30 × 14;30 = 3 30;15 3 30;15 − 3 30 = 0;15	
The square root of 0;15 is 0;30.	
14;30 + 0;30 = 15 length 14;30 − 0;30 = 14 width.	
Subtract 2, which has been added to 27, from 14, the width. 12 is the actual width.	
I have multiplied 15 length by 12 width.	
15 × 12 = 3 00 area 15 − 12 = 3 3 00 + 3 = 3 03.	

[a] The syntax of Høyrup's translation has been changed to reflect natural English word order.

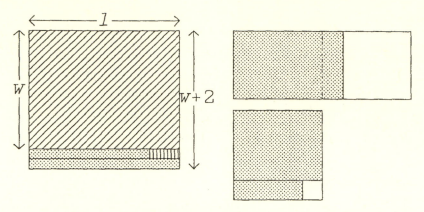

Figure 9.2 Høyrup's geometrical analysis of AO 8862 (1). (Høyrup 2002b, 169, courtesy of Jens Høyrup and Springer Verlag.)

All of van der Waerden's apparently arithmetical numbers turn out to have dimension: they are particular lengths and areas that are manipulated very physically. In particular, there are two forms of addition. Similar objects can be 'accumulated'—the literal meaning of the verb is 'to heap up'—and small objects can be 'appended' to larger ones. In this case, a line is added to an area by implicitly giving it a width of 1; in other contexts that conversion is explicitly made. Similarly, objects can be torn apart, or made to hold each other to form an area. And although there are no such cut-and-paste diagrams attached to Old Babylonian mathematical problems, it is clear that they were implicit in the text.

Late Babylonian mathematics, which rarely made it into the twentieth-century mainstream of mathematics history, undergoes a similarly de-arithmetising transformation under a structuralist close reading. Like problem (1) of AO 8862, problems (9) and (15) of the Seleucid compilation BM 34568 explore the relationships between lengths, widths, and areas of rectangles (table 9.2). Neugebauer's commentary (translated from his German) reads in its entirety:

(9) From $l + w$ and $A = lw$, w and l are accordingly calculated:

$$w = \frac{1}{2}\left((l + w) - \sqrt{(l + w)^2 - 4A}\right)$$
$$l = w + \sqrt{(l + w)^2 - 4A}$$

TABLE 9.2
Late Babylonian Word Problems BM 34568 (9) and (15) in Two Translations

Neugebauer's Translation (1935–7, III 18–9)[a]	Høyrup's Translation (2002b, 393, 395–6)[b]
(9) Length and breadth added is 14 and 48 (is) the area. The sizes are not known. 14 times 14 (is) 3 16. 48 times 4 (is) 3 12. You subtract 3 12 from 3 16 and it leaves 4. What times what should I take in order (to get) 4? 2 times 2 (is) 4. You subtract 2 from 14 and it leaves 12. 12 times 0;30 (is) 6. 6 (is) the width. To 2 you shall add 6, it is 8. 8 (is) the length.	(9) I have accumulated the length and the width: 14, and 48 the surface. Since you do not know, 14 steps of 14, 3 16. 48 steps of 4, 3 12. You lift 3 12 from 3 16: 4 is remaining. What steps of what may I go so that 4 (results)? 2 steps of 2, 4. You lift 2 from 14: 12 is remaining. 12 steps of 0;30, 6. The width is 6. You join 2 to 6: 8. The length is 8.
(15) The length goes beyond the width by 7 cubits. 2 00 (is) the area. What are length and width? 2 00 times 4 (is) 8 00. 7 times 7 (is) 49. 8 00 and 49 added together (is) 8 49. What times what should I take in order (to get) 8 49? 23 times 23 (is) 8 49. You subtract 7 from 23 and it leaves 16. Take 16 times 0;30. (It is) 8. 8 (is) the width. 7 and 8 added is 15. 15 (is) the length.	(15) The length goes beyond the width by 7 cubits. The surface is 2 00. What are the length and width? 2 00 steps of 4, 8 00. 7 steps of 7, 49. You join 8 00 and 49 with each other: 8 49. What steps of what may I go so that 8 49 (results)? 23 steps of 23, 8 49. You lift 7 from 23: 16 is remaining. You go 16 steps of 0;30, 8. The width is 8. Join 7 and 8: 15. The length is 15.

[a] This is an English translation of Neugebauer's original German.
[b] The syntax of Høyrup's translation has been changed to reflect natural English word order.

(15) From $l - w$ and A

$$w = \frac{1}{2}\left(\sqrt{4A + (l-w)^2} - (l-w)\right)$$
$$l = (l - b) + b$$

are calculated, which is correct, as $4A - (l-w)^2 = (l+w)^2$.[41]

According to these brief notes, there is little to distinguish Neugebauer's Late Babylonian problems from van der Waerden's Old Babylonian ones. Contrast Høyrup's analysis, which, as before, is prefaced with an explanatory diagram that encapsulates the method of solution (figure 9.3).[42]

#9 and #15 are not new as problems, but their method is innovative. The procedure is based on the principle shown in [figure 9.3]. In #9 the complete square □($l + w$) is found to be 3 16, from which 4 times

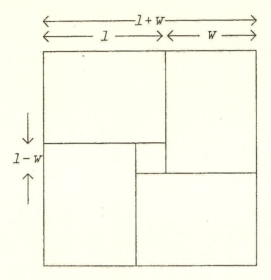

Figure 9.3 Høyrup's geometrical analysis of BM 34568 (9) and (15). (Høyrup 2002b, 395, courtesy of Jens Høyrup and Springer Verlag.)

the area $[](l,w)$ is removed, that is, 3 12; this leaves $\square(l - w) = 4$, whence $l - w = 2$. Subtracting this difference from $l + w = 14$ leaves twice the width. The width is therefore 6, and addition of the difference gives the length. Not the faintest trace of the method of average and deviation is left, nor of any cut-and-paste procedure.

#15 is strictly analogous [. . .], using the same configuration but with inserted diagonals [see figure 9.3], as demonstrated by the apparently roundabout calculation of $23^2 - 2\bullet(23^2 - 17^2)$ instead of $2\bullet17^2 - 23^2$. At first the square on the diagonal $\square(d) = 4\ 49$ is lifted from $\square(l + w) = 8\ 49$, which leaves 4 00 as the area of 4 half-rectangles. This remainder is doubled, which tells us that 4 times the rectangular area is 8 00—and lifting this again from 8 49 leaves 49 for $\square(l - w)$. Thereby we are brought to the situation of #9, and the rest follows exactly the same pattern.

Thus in Høyrup's reading, cut-and-paste procedures have disappeared completely from the repertoire of available techniques. Rather, the areas of two figures are compared: nothing is accumulated, nothing torn out or held together. The mathematical objects remain geometrical figures with specific lengths and areas, but mathematically the reader simply observes and compares them. Their Old Babylonian predecessors, by contrast, were actively interfered with. But even their geometrical status is unstable: Old

Babylonian mathematics distinguishes quite clearly between making four identical copies of a single object (here the rectangle) and multiplying lengths and widths to make areas. Here, both operations are treated simply as arithmetical multiplication with the phrase 'steps of'.

Finally, Netz offers a new reading of Greek deductive mathematics. Proposition 5 of Euclid's *Elements* Book II also explores the same relationship between the length, width, and area of a rectangle (table 9.3). But there the similarities with either of the Babylonian examples end. Heath states that 'perhaps the most important fact about [this proposition] is [its] bearing on the geometrical solution of a quadratic equation'.[43] That is, it is simply modern algebra in disguise. But since Sabetai Unguru's groundbreaking analysis of 1975, that interpretation is no more convincing than pre-Høyrupian readings of Babylonian 'algebra'.[44] (Again, decipherment lags a generation behind.) As Netz's new analysis shows, the diagram is not a mere illustration but a necessary feature of the whole—the text refers to it constantly, and the lettered points act as hyperlinks between the two components. The lines are no abstract objects but exist in real geometrical space: they have dimension but no quantitative properties. While the Babylonian word problems directly exhort the reader to observe or manipulate the lines and areas for themselves, the Euclidean proposition imagines a construction that is assembled and manipulated without human agency.[45]

In sum, then, language-sensitive translation reveals three very different mathematical cultures, even when matters of language, script, media, and numeration are dismissed as mere surface presentation. David Fowler coined the aphorism that 'Greek mathematics is to draw a figure and tell a story about it';[46] Høyrup shows that Babylonian mathematics is at one level to *not* draw a figure but tell a story about it anyway. And the stories told are very different in character, even if the subject matter is superficially identical. Old Babylonian mathematics is inherently metric: all parameters have both quantity and measure, explicit or implicit, as well as dimension. Late Babylonian mathematics is increasingly demetrologised, as geometrical operations are replaced by arithmetical ones. Both depend on word problems in which the question is formulated explicitly with numbers, and numbers are used throughout the solution. Conversely, there is no overt reference to diagrams, points, or lines in the text, unlike the Euclidean tradition. Both Old and Late Babylonian mathematics are entirely inductive: solutions to specific problems serve as generic examples from which generalisations are inferred (not always correctly); and starting assumptions (axioms or postulates) are not stated explicitly. In contrast, the classical Greek tradition is inherently geometric: parameters have dimension but no quantity or measure. Euclid's is a geometry without numbers: the points, lines, and areas are described by letters but have no particular sizes

TABLE 9.3
Euclid's *Elements*, Book II Proposition 5, in two translations

Heath's Translation (1926, I 382–3)	*Netz's Translation (1999, 9–10)*
If a straight line be cut into equal and unequal segments, the rectangle contained by the unequal segments of the whole together with the square on the straight line between the points of section is equal to the square on the half.	If a straight line is cut into equal and unequal (segments), the rectangle contained by the unequal segments of the whole, with the square on the (line) between the cuts, is equal to the square on the half.
For let a straight line AB be cut into equal segments at C and into unequal segments at D; I say that the rectangle contained by AD, DB together with the square on CD is equal to the square on CB (see figure 9.4)	For let some line, (namely) the (line) AB, be cut into equal (segments) at the (point) Γ and into unequal (segments) at the (point) Δ;
	I say that the rectangle contained by the (lines) AΔ ΔB with the square on the (line) ΓΔ, is equal to the square on the (line) ΓB (see figure 9.5).

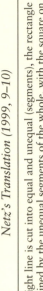

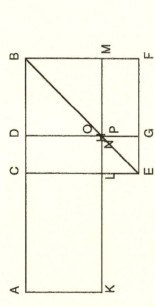

Figure 9.4 Heath's diagram of Euclid's *Elements* II.5. (Heath 1926, 382.)

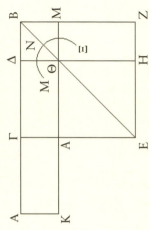

Figure 9.5 Netz's diagram of Euclid's *Elements* II.5. (Netz 1999, 9, courtesy of Cambridge University Press.)

For let the square CEFB be described on CB, [I. 46]ᵃ and let BE be joined; through D let DG be drawn parallel to either CE or BF; through H again let KM be drawn parallel to either AB or EF, and again through A let AK be drawn parallel to either CL or BM. [I. 31]

Then, since the complement CH is equal to the complement HF, [I. 43] let DM be added to each; therefore the whole CM is equal to the whole DF.

But CM is equal to AL, since AC is also equal to CB; [I. 36] therefore AL is also equal to DF.

Let CH be added to each; therefore the whole AH is equal to the gnomon NOP.

But AH is the rectangle AD, DB, for DH is equal to DB, therefore the gnomon NOP is also equal to the rectangle AD, DB.

Let LG, which is equal to the square on CD, be added to each; therefore the gnomon NOP and LG are equal to the rectangle contained by AD, DB and the square on CD.

But the gnomon NOP and LG are the whole square CEFB, which is described on CB; therefore the rectangle contained by AD, DB together with the square on CD is equal to the square on CB.

Therefore etc. Q. E. D.

For, on the (line) ΓB, let a square be set up, (namely) the (square) ΓEZB and let the (line) BE be joined, and, through the (point) Δ, let the (line) ΔH be drawn parallel to either of the (lines) ΓE BZ and, through the (point) Θ, again let the (line) KM be drawn parallel to either of the (lines) AB EZ, and again, through the (point) Λ, let the (line) AK be drawn parallel to either of the (lines) ΓΛ BM.

And since the complement ΓΘ is equal to the complement ΘZ, let the (square) ΔM be added (as) common; therefore the whole ΓM is equal to the whole ΔZ. But the (area) ΓM is equal to the (area) AΛ, since the (line) AΓ, too, is equal to the (line) ΓB; therefore the (area) AΛ, too, is equal to the (area) ΔZ. Let the (area) ΓΘ be added (as) common; therefore the whole AΘ is equal to the gnomon MNΞ. But the (area) AΘ is the (rectangle contained) by the (lines) AΔ ΔB; for the (line) ΔΘ is equal to the (line) ΔB; therefore the gnomon MNΞ too, is equal to the (rectangle contained) by the (lines) AΔ ΔB. Let the (area) AH be added (as) common (which is equal to the (square) on the (line) ΓΔ. Therefore the gnomon MNΞ and the (area) AH are equal to the rectangle contained by the (lines) AΔ ΔB and the square on the (line) ΓΔ. But the gnomon MNΞ and the (area) AH, (as a) whole, is the square ΓEZB, which is (the square) on the (line) ΓB; therefore the rectangle contained by the (lines) AΔ ΔB with the square on the (line) ΓΔ, is equal to the square on the (line) ΓB.

Therefore if a straight line is cut into equal and unequal (segments), the rectangle contained by the unequal segments of the whole, with the square on the (line) between the cuts, is equal to the square on the half; which it was required to prove.

ᵃ Notations in square brackets are Heath's cross-references to other propositions of the *Elements*, on which these statements depend.

attached to them. It is also heavily deductive and axiomatic: the emphasis is on deriving general proofs from explicitly stated theorems and axioms. There are precious few instances of index fossils here.[47]

9.4 THE WORLDS OF ANCIENT MATHEMATICS: HISTORY, SOCIETY, COMMUNITY

Even if the shared arithmetical character of Euclidean and Babylonian mathematics has now given way to three very different geometries, for twentieth-century historians there were also historical grounds for postulating a connection between them. As David Fowler cogently summarises:

> What, then, is the principal interpretation today of [. . .] the complete absence of anything like the rational or real numbers from [deductive, Euclidean mathematics]? To describe it briefly and brutally, it is that the early Greek mathematicians did start with an arithmetised geometry, but then reformulated it in this non-arithmetical way. Moreover, though this is not often said, they did this so thoroughly that no trace of this earlier arithmetic survives in this corpus of deductive mathematics. In other words, we have here an invention of its supposed presence, followed by an invention of its complete disappearance![48]

In the standard twentieth-century account, this reformulation grew out of the so-called crisis of incommensurability, when Pythagoras and his followers were grappling with the realisation that not all lines could be measured in terms of any other line. The diagonal of a square, for instance, was not arithmetically expressible in terms of the length of that square. Thus the Pythagoreans gave up in disgust and retreated to the safety of exploring the properties and relationships of points, lines, and areas without recourse to badly behaved numbers.[49] As Fowler acknowledges:

> Of course there are reasons for these proposals. First there is the existence of the earlier sophisticated Babylonian mathematics, which is thoroughly and visibly arithmetised, though more subtly and less obviously geometrical. Here the next invention of modern scholarship is of a Babylonian influence on the origins of Greek deductive mathematics, and hence its arithmetical basis at this earliest stage, and again I say 'invention' because the evidence is indirect, mainly by comparing some Babylonian procedures with those of some Euclidean propositions. But, just to point to one missing feature, there is no evidence known so far of any sexagesimal numbers, the formulation of Babylonian mathematics, to be found in Greek scientific thought before the time of Hypsicles in the second century BC.[50]

Further, as the previous section has shown, there was little arithmetical about Babylonian mathematics until the third century BCE, about three centuries after Pythagoras is supposed to have lived. Admittedly, various stories associate Pythagoras with Babylonian learning, such as those from Iamblichus' *On the Pythagorean life*:[51]

> He [Pythagoras] was captured by the expedition of Kambyses [r. 530–523] and taken to Babylon. There he spent time with the Magi, to their mutual rejoicing, learning what was holy among them, acquiring perfected knowledge of the worship of the gods and reaching the heights of their mathematics and music and other disciplines.[52]

Crucially, though, such myths were the product of the fourth century CE and later, nearly a millennium after Pythagoras's time. Over forty years ago Walter Burkert found that historians can confidently attribute to the early Pythagoreans some impressive developments in musical theory and acoustics, but nothing at all in mathematics.[53] More recently D. J. O'Meara's study of the Neo-Pythagoreans of Late Antiquity has found no trace of concern with commensurability and incommensurability in their writings either. Such tales should be understood as part of the Late Antique construction of Pythagoras as a legendary figure who encompasses all wisdom in order to teach and benefit all humanity, not as a true story of cultural transmission.[54]

If Pythagoras and the 'crisis of incommensurability' are to be abandoned as the catalysts of Babylonian influence on Greek mathematics, and there is little trace of direct textual transmission, perhaps at least some similarities in mathematical culture more broadly can be detected.

Mainland Greece supported a literate culture from perhaps as early as the eighth century BCE, but literate mathematics as an intellectual discipline developed only in the period 440–360, in and around Athens. The direct evidence for this early mathematics is patchy and circumstantial: material evidence of professional numeracy from accountants, land surveyors, and architects; the use of mathematical imagery and ideas by historians such as Herodotus, playwrights such as Aristophanes, and philosophers such as Plato and Aristotle.[55] The earliest Greek mathematician whose work survives directly is Autolycus of Pitane, on the Turkish coast of the Aegean, who probably lived between about 360 and 290 BCE and whose extant writing is a geometrical treatise on astronomy.[56] Earlier work, such as that by Hippocrates of Chios (c. 425 BCE?), is known only through descriptions by Simplicius and Eutocius, commentators of the sixth century CE. Their sources may have included a history of mathematics by Eudemus (c. 325 BCE), now lost.[57] During Autolycus' lifetime, between 336 and 323 BCE, Alexander the Great's conquests overturned the political configuration of the Mediterranean and Middle East. After Alexander's death, his vast empire was divided into three separate kingdoms, centred on Egypt, Babylonia, and Asia Minor (modern Turkey). The following century was the golden age of Greek mathematics. Archimedes was active in Sicily, Euclid and Apollonius in Egypt, at the library and museum founded for Alexander himself.

Yet across the eastern Mediterranean as a whole, society at large was not caught up in the excitement of the new, deductive-axiomatic mathematics. Reviel Netz, after a careful and imaginative survey of ancient Greek mathematicians as a social group, concludes that 'very few bothered at all in antiquity with mathematics, let alone became creative mathematicians'.[58] A detailed prosopographical study leads him to suggest that mathematicians—good, bad, and indifferent—were born at a rate of at most three a year across the antique Mediterranean world.[59] They tended to be from leisured, wealthy backgrounds with private incomes that allowed them to pursue their very unfashionable interests:

> [Mathematics] was an enterprise pursued by *ad hoc* networks of amateurish autodidacts—networks for which the written form is essential; constantly emerging and disappearing, hardly ever obtaining any institutional foothold. . . . Our expectations of a 'scientific discipline' should be forgotten. An 'intellectual game' will be a closer approximation.[60]

By contrast Old Babylonian mathematics and its practitioners present a completely different social picture (chapter 4; §9.1). There is no archaeological or textual evidence of mathematics as a leisure pursuit or prestige occupation in the early second millennium BCE: it was taught directly to some scribal students, but not all of them, as part of their professional formation and closely tied to the ideology of just rule. Comparisons of Euclidean or Archimedean treatises with Old Babylonian word problems thus do not compare like with like, for the earlier tradition was written by and for pre-pubescents, not research-active adults. The collapse of the Old Babylonian state in 1600 BCE entailed a massive rupture of all sorts of scribal culture. Much of Sumerian literature was lost from the stream of tradition, for instance, and most of Old Babylonian mathematics too. When cuneiform culture spread west to Egypt and the eastern Mediterranean coast in the second millennium, mathematics did not travel with it (§6.1). In this light it becomes clear that Old Babylonian mathematics cannot have influenced early Greek developments: it was part of a scribal culture that all but died out nearly a millennium before the earliest Greek literate culture, over a thousand miles away.

The much-neglected later Babylonian mathematics, however, is circumstantially a much better candidate to be seen as proto-Greek: the evidence dates from the mid-seventh century onwards, comfortably predating the earliest Greek sources and stories, and belongs to a world in which Greece is much closer to Mesopotamia than it was in the early second millennium. Recent commentators have posited transmission in the fifth century BCE, via (unattested) Aramaic mathematics in Phoenicia and Syria.[61] But given the intellectual climate of the time such a scenario is unlikely. Since the late sixth century mainland Greece had been continually threatened by the enormous Persian empire, which encompassed all of the Middle East including Egypt, Turkey, and of course Babylonia. Not surprisingly, Greek writers had very little good to say about Persia and Babylonia at this period. They were fascinated and repelled in equal measure, and most accounts came in garbled form from travellers to the Persian court, or with the Persian army. Amongst the propaganda about effete Eastern potentates and barbaric social practices there were stories of astronomers and magicians, but the Greek intelligentsia found very little in these narratives worth engaging with.[62] As summarised above, there is precious little reliable evidence of Greek mathematics before about 440 BCE, certainly not enough on which to build an argument of transmission or influence via an entirely unattested intermediary.

The situation changed to some extent after the conquests of Alexander the Great, by which time Mediterranean mathematics had matured into the recognizably axiomatic-deductive 'Greek' style. Much of the Middle East was settled by Greek colonies, and ethnically Greek rulers and elites spread Greek culture and social practices, especially in the cities. But it remained a thin Greek veneer over a deeply rooted indigenous civilisation. In Babylon, for instance, Alexander's successors built a Greek theatre and hippodrome, but the heart of the city's intellectual traditions remained the temple of the god Marduk, who had been worshipped there for over one and a half thousand years. This temple and the temple of the sky god Anu in the southern city of Uruk continued to be bastions of Babylonian religion, literacy, and civilisation for generations after the rest of the populace had become more or less Persianised and later Hellenised.[63]

A great deal of Babylonian mathematical astronomy, and presumably mathematics too, reached the Greek astronomical community—or at least Hypsicles and the astronomer Hipparchus—in the second century BCE, probably via Alexandria.[64] But that was a century too late to have influenced Euclid and his contemporaries. Earlier than that the Greeks remained largely indifferent to Babylonian culture, convinced of their own intellectual supremacy. Most famously, the *Babyloniaca* of Bel-Rē'ûšunu/Berossos, written in the third century BCE, seems to have sunk like a lead balloon (§8.4); at least, none of the later historians who cite it are ethnically Greek. Fragments of it survive in Armenian and Jewish writings, and that is all.[65]

Assyriologists are fond of Leo Oppenheim's phrase 'the stream of tradition' to describe the continuity of Babylonian intellectual culture. But, in the mathematical case at least, those streams were more like trickles, liable to dry up at any moment. Ancient mathematics, whether Babylonian or Greek, was maintained and transmitted across the centuries by tiny communities of experts, each clinging to life in a niche socio-intellectual ecology. It is hardly surprising that the few mathematicians of Classical antiquity should have been entirely ignorant of the mathematical achievements of Old Babylonian scribes (as indeed were their Late Babylonian successors); they were more distant from Euclid in time and space than the oldest surviving manuscript of Euclid's *Elements* is from us.[66] Equally, the dwindling, isolated, conservative community of numerate astronomer-priests in Hellenistic Babylonia had little in common with the patrician, leisured mathematicians of the contemporary Mediterranean and neither was particularly interested in learning from the other—until well after the floruit of deductive Greek mathematics. But the mechanism of that particular intellectual contact and transfer, which took place in the second century BCE, is a problem for historians of astronomy to tackle.

9.5 CONCLUSIONS

The status of cuneiform mathematics, of any period, as the precursor to the axiomatic-deductive Greek tradition is now unsustainable on present evidence. It was an artefact of mid-twentieth-century historiography, constructed through accepting some texts into the popular canon and rejecting others. The favoured few—Plimpton 322, YBC 7289, and company—share more than the fact that they can be read as proto-Greek and thereby proto-modern. Most are non-narrative, all have a pictorial or tabular component, and in particular all can be read as 'pure', abstract mathematics, enabling them to be represented as artefacts familiar to mid-twentieth-century mathematicians. The rejected majority members of the corpus, on the other hand, are deeply culturally embedded, imbued with local knowledge, verbose compilations fully comprehensible only within their linguistic, social, and intellectual context. It may be more productive, if Babylonian mathematics must be located within any external tradition, to explore its place within the sciences in the Middle East, and Asia more generally. As a problem-based, algorithmic, rhetorical, concrete discourse, it seems to share many basic formal features with the mathematics of other pre-modern societies, including Egypt, India, and China.

However, perhaps more pressing is the need to break down the monolithic twentieth-century construction of Babylonian (or Mesopotamian or cuneiform) mathematics into more historically manageable pieces. Strictly speaking the only common features of the corpus as a whole are the script and medium in which it is written. Of course, the basic archaeo-linguistic periodisations enable useful talk about, say, Old Babylonian mathematics, even if lack of provenance means that historians cannot date or localise a particular source more closely than southern Iraq in the early second millennium BCE. Yet the linguistic and archaeological approaches pioneered in recent years have demonstrated that it is already possible to write about short-term diachronic change and regional variation, as well as the long-range shifts from geometrical to arithmetical conceptualisation outlined here. A grounding in and sensitivity to language, history, and archaeology are all prerequisites for would-be historians of the subject. Purportedly 'historical' analyses of Babylonian mathematics, whether of single word problems or larger bodies of material, have no claim to be considered history if they are solely rational reconstructions based on twentieth-century English translations as their primary sources—§9.2 has shown how dangerous that can be. History of mathematics must be historical as well as mathematical, and aim to be explanatory as well as descriptive.

It is not straightforward even to define the limits of the cuneiform mathematical corpus. It should come as no surprise that there is no unproblematic Sumerian or Akkadian equivalent to the word 'mathematics', though several technical terms were used by particular individuals or within specific contexts:

- *imšukkum*, lit. 'hand tablet' for rough work (Type IV or R: see table 4.5) => 'word problem', found on some unprovenanced Old Babylonian compilations;[67]
- *kibsātum* 'steps, footprints' => 'process' for solving a word problem, found on compilations from late Old Babylonian Sippar, probably all written by Iškur-mansum, son of Sîn-iqīšam (table 4.4);[68]
- *arû* 'steps, of times' => mathematical table, calculation; seventh century onwards (§5.4).[69]

Finally, a recently published word problem begins with the Akkadian word *nikkassum*, from Sumerian *niĝ₂-kas₇*, which is usually translated 'account', 'calculation'.[70] This new attestation shows that, at least for one Old Babylonian scribe, the semantic range of *nikkassum* went beyond the activities of professional numeracy and into the realm of pedagogy and intellectual creativity.

A few years ago I formulated what I then thought was an uncontroversial working definition of mathematical texts, designed to distinguish intellectual and pedagogical activity from the routine applications of numeracy in bureaucracy and law:

Clearly not all cuneiform texts with numbers in them are mathematical, and mathematical texts do not necessarily contain many numbers. A distinction must be made between texts which are of mathematical interest and texts which are primarily *about* mathematics, i.e. which have been written for the purpose of communicating or recording a numerical technique or aiding a quantitative procedure to be carried out.[71]

But while that formulation has mostly gained acceptance, or at least has never been rejected, I have been slow to realise the consequence that the cuneiform mathematical corpus constructed from it is much richer, more varied, more alien than has generally been acknowledged. The procedures, calculations, and tables of the Old Babylonian and Hellenistic periods are of course retained. But into the corpus now comes—for instance—numerical manipulation of omens (§5.4), instructions for intercalating months, cultic topography, not to mention astronomical precepts (chapter 8). There are several fascinating consequences of this reorientation. Perhaps most importantly, it is no longer necessary to account for the strange phenomenon of the sudden appearance of 'mathematics' in the early second millennium, its absence for some twelve hundred years, followed by its resurrection in the Hellenistic period. Instead, as this book has demonstrated, there was a wide variety of mathematical activity in every major period of cuneiform culture, closely bound up to the socio-intellectual concerns of the time.

It seems highly unlikely that a proto-Euclidean *Elements* will turn up amongst the hundreds of Neo-Babylonian tablets still to be deciphered. But that hardly matters. While it is no longer accurate or appropriate to view Babylonian scholarship through the lens of Euclidean-style mathematics, and even if modern mathematical aesthetics dismisses it as simplistic and alien compared to the Greek tradition, it still deserves a prominent place in the history of mathematics. Compared to the difficulties of grappling with fragmentary and meagre *n*th-generation sources from other ancient cultures the cuneiform evidence is concrete, immediate, and richly contextualised. We can often name and date individuals precisely; we have their autograph manuscripts, their libraries and household objects. This opens a unique window onto the material, social, and intellectual world of the mathematics of ancient Iraq that historians of other ancient cultures can only dream of.

Metrological Systems

TABLE A.1
Some Uruk III Metrologies

System	Unit	Later Name and Approximate Modern Value
Discrete objects and lengths	▷	1 object or 1 rod (c. 6 m)
	●	10 objects or 10 rods (c. 60 m)
	▶	60 objects or 60 rods (c. 360 m)
	▶•	600 objects or 600 rods (c. 3.6 km)
	⬤	3600 objects or 3600 rods (c. 21.6 km)
	◉	36,000 objects or 36,000 rods (c. 216 km)
Rationed goods, including jugs of beer	▷	1 object
	•	10 objects
	▶	60 objects
	⬟	120 objects
	⬟	1200 objects
	◉	7200 objects
Areas of land	▷	*iku*, c. 3600 m²
	▶•	*eše* = 6 *iku*, c. 2.2 ha
	•	*bur* = 3 *eše*, c. 6.5 ha
	◉	*buru* = 10 *bur*, c. 65 ha
	⬤	*šar* = 6 *buru*, c. 390 ha
Measures of grain[a]	⌣	Unknown
	▷	= 5 ⌣
	•	= 6 ▷

TABLE A.1 (*Continued*)

System	Unit	Later Name and Approximate Modern Value
	●	= 10 •
	▷	= 3 ●
	▷•	= 10 ▷
Unit fractions in the grain and ration systems	⧖	a 2nd part
	⧖	a 3rd part
	⧖	a 4th part
	⧖	a 5th part
	⧖	a 6th part
	⧖	a 10th part

Note: For more detail, see Nissen, Damerow, and Englund 1993, 28–9.

[a] Units of malted barley are marked with a diagonal slash across the bottom right of each sign; units of barley groats are marked with small flecks across each sign.

Table A.2
Some Third-Millennium Metrologies

System	Unit	Approximate Modern Value
Capacity measures in Early Dynastic Ebla	1 *anzam*	
	1 *sila* = 6 *anzam*	
	1 shekel = 4 *sila*	
	1 *barizu* = 2½ shekels	
	1 *gubar* = 2 *barizu*	
Early Dynastic and Sargonic lengths	1 finger	17 mm
	1 double-hand = 10 fingers	17 cm
	1 cubit = 3 double-hands	50 cm
	1 seed-cubit = 2 cubits	1 m
	1 reed = 3 seed-cubits	3 m
	1 rod = 2 reeds	6 m
Early Dynastic to Old Babylonian discrete counting	1	
	10	
	ǧeš	Sixty
	ǧešu	Six hundred
	šar	Three thousand, six hundred
	šaru	Thirty-six thousand

Note: Area, volume, and weight metrologies are identical to those of the Ur III and Old Babylonian periods (table A.3). Compound metrological signs, which denote both number and unit, are represented in translation by the numerical value followed by the sign in parentheses, e.g., 2(*ǧešu*) 3(*ǧeš*) for 1380. For more detail, see Powell 1987–90.

TABLE A.3
Classic Ur III and Old Babylonian Metrologies

System	Unit	Approximate Modern Value
Length[a]	1 finger	17 mm
	1 cubit = 30 fingers	0.5 m
	1 **rod** = 12 cubits	6 m
	1 chain = 1 00 cubits or 5 rods	30 m
	1 cable = 1 00 rods	360 m
	1 league = 30 00 rods or 30 cables	10.8 km
Area and volume	1 area *sar* = 1 rod square[b]	36 m^2
	1 volume *sar* = 1 area *sar* × 1 cubit[b]	18 m^3
	1 *ubu* = 50 *sar*[c]	1800 m^2 or 900 m^3
	1 *iku* = 2 *ubu* = 1 40 *sar*	3600 m^2 or 1800 m^3
	1 *eše* = 6 *iku*	2.16 ha or 108,000 m^3
	1 *bur* = 3 *eše*	6.48 ha or 324,000 m^3
Capacity[d]	1 **sila**[e]	1 litre
	1 *ban* = 10 *sila*[f]	10 litres
	1 *bariga* = 6 *ban*	60 litres
	1 *gur* = 5 *bariga*[g]	300 litres
Weight	1 grain	0.05 g
	1 shekel = 3 00 grains	8.3 g
	1 **mina** = 1 00 shekels	0.5 kg[h]
	1 talent = 1 00 minas[i]	30 kg
Bricks	1 brick **sar** = 12 00 bricks[j]	720 bricks
	Small, unbaked brick = 15 × 10 × 5 fingers	25 × 17 × 8 cm
	Number of small bricks in 1 volume *sar* = 1 26 24 = 7;12 brick *sar*	5184 bricks

TABLE A.3 (*Continued*)

System	Unit	Approximate Modern Value
	Square, baked brick = 20 × 20 × 5 fingers	33 × 33 × 8 cm
	Number of square bricks in 1 volume *sar* = 32 24 = 2;42 brick *sar*	1944 bricks

Note: Standard units of calculation (which are often implicit within Old Babylonian problems) are shown in **bold**. For more detail see Powell 1987–90.

[a] 1 **cubit** was the standard unit of height. Assyrian length metrology also had a span of ½ cubit.

[b] The *sar* could be divided into 60 shekels and the shekel into 180 grains.

[c] Units from the *ubu* upwards were not written explicitly, but used special unit-specific notation. These writings are indicated in the translations by putting the units in parentheses, thus: '2(*bur*) 1(*eše*) 3(*iku*) area' or, where space is short, as, e.g., '2.1.3, 25 *sar*'. Special, absolute value signs were used to write multiples of the *bur*: 10, 60, 600, 3600, and occasionally the 'big 3600', or 3600 × 3600.

[d] Kassite land surveyors additionally used capacity units as area measures, on the basis that 1 *ban* of grain could seed a field of 1 *iku*, or 10 *gur* per 10 *iku*.

[e] The *sila* could be divided into 60 shekels and the shekel into 180 grains.

[f] *Ban* and usually *bariga* units were not written explicitly, but used special unit-specific notation. These writings are indicated in the translations by putting the units in parentheses, thus: '1 *gur* 2(*bariga*) 3(*ban*) 4 *sila*' or, where space is short, as '1.2.3, 4 *sila*'.

[g] Multiples of the *gur* were written with the sexagesimal place value system.

[h] Early Dynastic weight standards were much more variable, and heavier (in the range 0.55–0.68 kg), than later (Powell 1987–90, 508).

[i] Multiples of the talent were written with the sexagesimal place value system.

[j] Multiples of the brick *sar* were written in the same way as large area and volume measures.

TABLE A.4
First-Millennium Metrologies

System	Unit	Approximate Modern Value
Arû measure for lengths and areas	1 big finger	3.1 cm
	1 *arû*-cubit = 24 big fingers	75 cm
	1 rod = 12 cubits	9 m
	1 rod × 1 rod = 1 *sar*	81 m^2
	1 *ubu* = 50 *sar*	0.4 ha
	1 *iku* = 2 *ubu*	0.81 ha
	1 *eše* = 6 *iku*	4.86 ha
	1 *bur* = 3 *eše* = 18 *iku*	14.6 ha
	1 *šar* = 60 *bur*	875 ha
Arû 'seed measure' for areas	1 *qû*	270 m^2
	1 *sūtu* = 10 *qû*	0.27 ha
	1 *pānu* = 6 *sūtu* = 2 *iku*	1.62 ha
	1 *kurru* = 5 *pānu* = 10 *iku*	8.1 ha
Cable 'reed measure' for lengths and areas	1 finger	2 cm
	1 cable-cubit = 24 fingers	0.5 m
	1 rod = 12 cubits	6 m
	1 chain = 5 rods = 50 cubits	30 m
	1 cable = 2 chains = 10 rods	60 m
	1 rod × 1 rod = 1 *sar*	36 m^2
	1 cable × 1 cable = 25 *sar*	900 m^2
	1 *iku* = 1 40 *sar*	0.36 ha
Cable 'seed measure' for area	1 grain	70 cm^2
	1 *akalu* = 18 00 grains	7.5 m^2
	1 *qû* = 10 *akalu*	75 m^2
	1 *sūtu* = 6 *qû*	450 m^2
	1 *pānu* = 6 *sūtu*	0.27 ha
	1 *kurru* = 5 *bariga*	1.35 ha

TABLE A.4 (*Continued*)

System	Unit	Approximate Modern Value
'Reed measure' for lengths and areas	1 finger	2 cm
	1 cubit = 24 fingers	0.5 m
	1 reed = 7 cubits	3.5 m
	1 rod = 2 reeds or 14 cubits	7 m
	1 finger × 1 finger = 1 small finger	4 cm^2
	1 cubit × 1 finger = 1 grain	100 cm^2
	1 reed × 1 finger = 1 area finger	730 cm^2
	1 cubit × 1 cubit = 1 small cubit	0.25 m^2
	1 reed × 1 cubit = 1 area cubit	1.75 m^2
	1 reed × 1 reed = 1 area reed	12.25 m^2

Note: For more detail, see Powell 1987–90. In seed measure, *ban* and usually *bariga* units were not written explicitly but used special unit-specific notation. These writings are indicated in the translations by putting the units in parentheses, thus: '1 *gur* 2(*bariga*) 3(*ban*) 4 *sila*' or, where space is short, as '1.2.3, 4 *sila*'. In reed measure, units from the *ubu* upwards were not written explicitly, but used special unit-specific notation. These writings are indicated in the translations by putting the units in parentheses, thus: '2(*bur*) 1(*eše*) 3(*iku*) area' or, where space is short, as e.g. '2.1.3, 25 *sar*'.

APPENDIX B

Published Mathematical Tablets

TABLE B.1
Mathematical Problems from Late Fourth-Millennium Uruk (Uruk III)

Tablet Number	Findspot[a]	Description[b]	Publication
W 19408,76	Nb XVI/3	Two problems set	Englund and Nissen 2001, pl. 14; Nissen, Damerow, and Englund 1993, 58 fig. 50; Friberg 1997–8, text FS U9
W 20044,9	Nd XVI/3	One problem set; annotation	Friberg 1997–8, text FS U7
W 20044,11	Nd XVI/3	One problem set	Friberg 1997–8, text FS U2
W 20044,20	Nd XVI/3	One problem set; annotation (figure 2.2)	Englund and Nissen 2001, pl. 32
W 20044,27	Nc XVI/4	One problem set; annotation	Friberg 1997–8, text FS U6
W 20044,28	Nc XVI/4	One problem set; annotation	Friberg 1997–8, text FS U3
W 20044,29	Nc XVI/4	One problem set	Friberg 1997–8, text FS U8
W 20044,33	Nc XVI/4	One problem set; annotation	Friberg 1997–8, text FS U5; Englund and Nissen 2001, pl. 34
W 20044,35	Nc XVI/4	One problem set; annotation	Friberg 1997–8, text FS U4

Note: All problems are to find the area of an irregular quadrilateral field.

[a] Findspot information is taken from the CDLI database <http://cdli.ucla.edu>. See most conveniently Nissen, Damerow, and Englund 1993, 6 fig. 5, or Englund 1998, 36 fig. 6, for an excavation plan of the centre of Uruk.

[b] 'Annotation' refers to 1–3 proto-cuneiform word signs, typically written at the bottom of the tablet, that may represent the name of the student who has been set the work (cf. the Sargonic problems, table B.5).

TABLE B.2
Other Mathematical Exercises from the Late Fourth Millennium (Uruk IV–III)

Tablet Number	Findspot[a]	Description	Publication
Berlin 5 (MSVO 3, 2)	Unknown	Metrological table? Script period Uruk III (figure 2.7).	Nissen, Damerow, and Englund 1993, 42–3 fig. 38; Englund 1998, 192, 196 fig. 77
IM 23426 (MSVO 4, 66)	Larsa?	Metrological table? Script period Uruk III.	Friberg 1987–90, 539; 1999, 111–5; Englund 1996, no. 66; 1998, 188– 3 fig. 75; 2001, 13–17
W 9393,d	Uruk Qa XVI/3	Accounting exercise? Large round numbers. Script period Uruk IV.	Englund 1994, pl. 57; 1998, 110 fig. 36
W 9393,e	Uruk Qa XVI/3	Accounting exercise? Grain, oil, textiles. Script period Uruk IV.	Englund 1994, pl. 57; 1998, 110 fig. 36
W 20517,1	Uruk Nc XVII/1	Accounting exercise? Vessels. Script period Uruk III.	Englund 1998, 107 n237; Englund and Nissen 2001, pl. 75

[a] Findspot information is taken from the CDLI database <http://cdli.ucla.edu>. See most conveniently Nissen, Damerow, and Englund 1993, 6 fig. 5, or Englund 1998, 36 fig. 6, for an excavation plan of the centre of Uruk.

Table B.3
Mathematical Tablets from Šuruppag/Fara (Early Dynastic IIIa)

Tablet Number	Description	Publication
VAT 12593 (SF 82)	Metrological table of lengths and areas	Deimel 1923, no. 82; Neugebauer 1935–7, I 91; Nissen, Damerow, and Englund 1993, 136–8 fig. 119; Robson 2003, 27–30
TSŠ 50	One problem stated and answered; grain rations	Jestin 1937, no. 50; Powell 1976a; Høyrup 1982; Melville 2002; Friberg 2005b, §4.10
TSŠ 81	On problem stated and answered; flour rations	Jestin 1937, no. 81; Melville 2002
TSŠ 188	One problem stated and solved; square field	Jestin 1937, no. 188; Friberg 1986, 10 n18
TSŠ 671	One problem stated and answered; grain rations	Jestin 1937, no. 671; Powell 1976a; Høyrup 1982; Melville 2002; Friberg 2005b, §4.10

Source: Based on the CDLI database <http://cdli.ucla.edu> and Krebernik 1998, 313.

Note: Powell 1976a, 436 n19, also lists as possibly mathematical the tablets TSŠ 81, 91, 242, 245, 260, 554, 613, 619, 648, 649, 725, 748, 758, 775, 780, 828, and 969 (Jestin 1937), as well as VAT 12590 and 12450 (Deimel 1923, nos. 93, 125). But Pomponio and Visicato 1994, 3–4, have since reclassified them variously as registers of grain-based products, wool, vegetable fibres, reeds, copper, boats, sheep, and personnel; and lists of parcels of land. The mathematical status of TSŠ 51, 190, 251, 632, 926, and 930 remains uncertain; they are very difficult to read. The findspots of the TSŠ tablets are unknown (Martin 1988, 98). TSŠ 77, long thought to be from Early Dynastic Šuruppag, now turns out to be from Old Babylonian Kisurra (table B.10; Krebernik 2006).

Table B.4
Mathematical Tablets from Adab and Ebla (Early Dynastic IIIb)

Tablet Number	Findspot	Description	Publication
A 681 (OIP 14, 70)	Adab	Metrological list; colophon 'Nammaḫ wrote the calculation'	Luckenbill 1930, no. 70; Edzard 1969; Friberg 2005b, §4.9
TM 75.G.1392	Ebla, archive L.2769	One calculation	Friberg 1986; Archi 1989
TM.75.G.1572	Ebla, archive L.2769	One calculation (?)	Edzard 1981, no. 11; Foster 1983, 305
TM 75.G.1693	Ebla, archive L.2769	Metrological list; colophon 'Established by the scribes of Kiš; Išmeya'	Archi 1980; Friberg 1986
TM 75.G.2346	Ebla, archive L.2769	One calculation	Archi 1989

Note: Foster 1983, 304–5, also identifies TM.75.G.1394 (Edzard 1981, no. 33), another tablet from the Ebla archive L.2769, as possibly mathematical. He points out that this very difficult text moves from the third person to the first to the second, as do some OB mathematical word problems (§4.1), which 'at least raises the possibility that [it] is a tricky hypothetical exercise in Ebla accounting'. He provides a provisional translation of the first several lines, which involve purchasing grain for rations.

TABLE B.5
Mathematical Problems from Sargonic Adab, Ǧirsu, Nippur, and Elsewhere

Tablet Number	Provenance	Description[a]	Publication
A 786	Adab	One problem solved: type (c)	Luckenbill 1930, no. 116; Friberg 2005b, §1.1
A 5443	Ǧirsu	One problem solved: type (b)	Whiting 1984
A 5446	Ǧirsu	Two problems set: type (b); mentions Meluḫḫa and Ur-Ištaran	Whiting 1984
AO 11404	Ǧirsu	One problem solved: type (c)	Gelb 1970a, no. 163; Foster and Robson 2004, 11 fig. 4
AO 11405	Ǧirsu	One problem solved: type (c)	Gelb 1970a, no. 164; Foster and Robson 2004, 12 fig. 5
AO 11409	Ǧirsu	One problem solved: type (c); mentions Kaya the felter	Gelb 1970a, no. 166; Foster and Robson 2004, 10 fig. 3
Ash 1924.689	Ǧirsu	Two problems set: type (b); mentions Ur-Ištaran	Gelb 1970b, no. 112
CMAA 016-C0005	Unknown	One problem solved: type (a)	Friberg 2005b, §2.1 fig. 1
HS 815	Nippur	One problem solved: type (a)	Pohl 1935, no. 65; Friberg 2005b, §3.3
Ist L 2924	Girsu	Two problems set: type (c); mentions Barran, Uruna, and Dudu, administrators (figure 3.3)	Genouillac 1910, no. 2924; Donbaz and Foster 1982, no. 80; Foster 1982a, 28
Ist L 4505	Girsu	Two problems set: type (e)	Genouillac 1910, no. 4505; Foster 1982c, 239
NBC 6893	Unknown	One problem solved: type (d); mentions Ur-Iškur	Hackman 1958, no. 24

TABLE B.5 (*Continued*)

Tablet Number	Provenance	Description[a]	Publication
NBC 7017	Ğirsu	Two problems solved: type (d)	Foster 1982c, no. 1; Foster and Robson 2004, 9 fig. 2
ZA 94, 3	Ğirsu region	One problem solved, with partial working: type (a)	Foster and Robson 2004, 3 fig. 1; Friberg 2005b, §2.2
PUL 26	Ğirsu region	One problem solved: type (c)	Limet 1973, no. 34; Friberg 2005b, §1.1
PUL 27	Ğirsu region	One problem solved: type (b)	Limet 1973, no. 36
PUL 28	Ğirsu region	One problem solved: type (b)	Limet 1973, no. 37
PUL 29	Ğirsu region	One problem solved: type (a)	Limet 1973, no. 38; Friberg 2005b, §3.2
PUL 31	Ğirsu region	One problem set: type (a) (figure 3.2)	Limet 1973, no. 39; Friberg 2005b, §3.3

Source: Based on Foster 1982c, 239–40; Foster and Robson 2004, 2; and Friberg 2005b, §1.1; see further bibliography there.

Note: Ist L 9395 and Ist L 9404, catalogued as 'mesures de longeur' and 'mesures à trois dimensions' (Genouillac 1921, 37) but not yet published, may also bear problems of types (c)/(d) and (e) respectively (Foster 1982c, 239).

[a] Problem types: (a) to find the short side of a rectangle, given the long side and area; (b) to find the area of a square, given its side; (c) to find the area of an irregular quadrilateral; (d) to find the area of a rectangular field and associated agricultural data; (e) to find the volume of a rectangular prism.

TABLE B.6
Other Types of Mathematical Tablet from the Sargonic Period

Tablet Number	Provenance	Description	Publication
IM 58045 (2N-T 600)	Nippur, temple of Enlil[a]	Geometrical diagram (figure 3.6)	Friberg 1987–90, 541; Damerow 2001, 263
Ist L 4598	Girsu	Accounting exercise with large round numbers	Genouillac 1910, no. 4598; Foster 1982c, 240
MAD 1, 188	Ešnuna, J 21: 23 level IVb	Metrological list of silver weights, with names attached, written twice	Westenholz 1974–7, 100 no. 11

Note: Foster 1982c, 239, lists 17 more Sargonic tablets that may bear mathematical exercises. As they are either fragmentary or show no evidence of calculation, they have been omitted from this list. IM 50682, from Gasur, is a learner's writing exercise with four repetitions of the simple multiplicative phrase *n a-ra$_2$ m-kam*, literally 'it is *m* steps of *n*', or 'it is *n*, *m* times', where *n* = 40 or 50 and *m* = 1 and possibly 2 and 3 (Meek 1935, no. 214).

[a] Findspot information from the Nippur excavation notebooks held by the Babylonian Section of the University of Pennsylvania Museum of Anthropology and Archaeology.

TABLE B.7
Mathematical Tablets from the Ur III Period

Tablet Number	Findspot	Description	Publication
AOT 304	Ğirsu	Exercise in brick accounting; calculation	Thureau-Dangin 1903, no. 413; Friberg 1987–90, 541; 2001, 92–3; Robson 1999, 66
BM 106425	Umma	Reciprocal table (figure 3.10)	Robson 2003–4
BM 106444	Umma	Reciprocal table	Robson 2003–4
HS 201	Nippur	Reciprocal table	Oelsner 2001; Robson 2003–4
Ist Ni 374	Nippur	Reciprocal table	Friberg 2005b, §4.1
Ist T 7375	Ğirsu	Reciprocal table	Delaporte 1911; 1912, no. 7375; Robson 2003–4
Walker 47a	Unknown	Exercise in labour accounting	Jones and Snyder 1961, no. 318; Civil 1994, 119, 123, 137 n21; Robson 1999, 105–6
YBC 9819	Umma	Exercise in brick accounting	Legrain 1935; Dunham 1982; Robson 1996; 1999, 68; Friberg 2001, 133–6

TABLE B.8
Old Babylonian Metrological Lists and Tables

Tablet Number	Findspot or Provenance	Description[a]	Publication
(Ist Si)	Sippar	Composite text of lists of capacities, weights, and areas	Scheil 1902b, 49–54 nos. 269, 511
2N-T 384 = A 29979	Nippur, TB House D, locus 63	Type II tablet: obverse Sumerian Proverbs 2; reverse table of capacities	Gordon 1959, pl. 74 (2 MMMM)
2N-T 530 = UM 55-21-78	Nippur, TA locus unknown	Type I tablet: fragmentary table of weights, areas, and lengths	Neugebauer and Sachs 1984
3N-T 316 = A 30211	Nippur, TA House F, locus 205	Type II tablet: obverse lexical list (OB Diri); reverse table of capacities and lengths	Robson 2002b, fig. 5 (detail of reverse)
3N-T 594 = IM 58573	Nippur, TA House F, locus 205	Type II tablet: obverse verbose list of reciprocals; reverse list of capacities (figure 4.5)	Robson 2002b, fig. 7
3N-T 729 = IM 58658	Nippur, TA House F, locus 184	Type II tablet: obverse Sumerian Proverbs 1; reverse list of capacities	Gordon 1959, pl. 16 (1 II)
3N-T 906, 246	Nippur, TA House F, locus 205	Type II tablet: obverse (?) list of capacities; reverse (?) Sumerian Proverbs 2	Gordon 1959, pl. 73 (2 LLLL); Heimerdinger 1979, pl. 69
3N-T 917, 378	Nippur, TA House F, locus 191	Type II tablet: list of capacities	Heimerdinger 1979, pl. 76
A 21948 = Ish 34-T 149	Nērebtum (Ishchali), city gate	Type I tablet: table of areas	Greengus 1979, no.92

TABLE B.8 (*Continued*)

Tablet Number	Findspot or Provenance	Description[a]	Publication
AO 8865	Larsa?	Six-sided prism: tables of lengths [and heights], tables of inverse squares and inverse cubes; colophon	Neugebauer 1935–7, I 69 no. 24, 71 no. 23, 73 no. 52, 88 no. 1; Proust 2005
Ash 1923.366	Larsa?	Six-sided prism: tables of lengths and heights, tables of inverse squares and inverse cubes	van der Meer 1938, no. 156; Robson 2004c, no. 9
Ash 1923.410	Larsa?	Type III tablet: table of weights	Robson 2004c, no. 10
Ash 1923.414	Larsa?	Type III tablet: table of areas	Robson 2004c, no. 11
Ash 1924.564	Unknown	Fragment of Type I or II tablet: list of capacities	Dalley and Yoffee 1991, no. 35; Robson 2004c, no. 23
Ash 1924.1341	Kiš, Ingharra	Type III tablet: table of weights	Robson 2004c, no. 26
Ash 1931.137	Kiš?	Type I tablet: list of capacities, weights, and areas	Robson 2004c, no. 19
Ash 1932.526n	Unknown	Type II tablet: obverse lexical list; reverse list of capacities	Gurney 1986, no. 95; Robson 2004c, no. 12
Ash 1933.180	Unknown	Fragment of Type I or II tablet: list of capacities	Dalley and Yoffee 1991, no. 221; Robson 2004c, no. 13

TABLE B.8 (*Continued*)

Tablet Number	Findspot or Provenance	Description[a]	Publication
BE 20/1 31	Nippur	Type III tablet: table of capacities or weights (shekels)	Hilprecht 1906, no. 31
BE 20/1 32	Nippur	Type III tablet: table of capacities or weights (shekels)	Hilprecht 1906, no. 32
BE 20/1 36	Nippur	Type III tablet: table of capacities	Hilprecht 1906, no. 36
BE 20/1 37	Nippur	Type II tablet: obverse lexical list; reverse list of capacities	Hilprecht 1906, no. 37
BE 20/1 39	Nippur	Type III tablet: table of areas	Hilprecht 1906, no. 39
BE 20/1 40	Nippur	Type III tablet: table of areas	Hilprecht 1906, no. 40
BE 20/1 41	Nippur	Type III tablet: table of lengths	Hilprecht 1906, no. 41
BE 20/1 42	Nippur	Type III tablet: table of lengths	Hilprecht 1906, no. 42
BE 20/1 43	Nippur	Type III tablet: table of lengths	Hilprecht 1906, no. 43
Berlin 69	Unknown	Type I tablet: list of areas; colophon	Nissen, Damerow, and Englund 1993, fig. 127
BM 17403	Sippar?[b]	Type III tablet: table of lengths	Nissen, Damerow, and Englund 1993, fig. 125
BM 17567	Sippar?[b]	Type III tablet: table of weights	Nissen, Damerow, and Englund 1993, fig. 126

TABLE B.8 (*Continued*)

Tablet Number	Findspot or Provenance	Description[a]	Publication
BM 92698	Larsa	Type I tablet: tables of lengths, heights; verbose tables of squares, inverse squares, and inverse cubes	Rawlinson 1891, pl. 37 [40]; Neugebauer 1935–7, I 69 no. 3
BM 96949	Sippar?	Type I tablet: table of capacities; colophon	Robson 2004c, 35–7; see table 4.1
CBS 3937	Nippur	Type II tablet: obverse Sumerian Proverbs 1; reverse list of capacities	Gordon 1959, pl. 18 (1 NN)
CBS 4505	Nippur	Type III tablet: table of capacities or weights (shekels)	Hilprecht 1906, no. 33
CBS 6841	Nippur	Type II tablet: obverse Sumerian Proverbs 2; reverse list of capacities	Gordon 1959, pl. 55 (2 QQ)
CBS 10801+	Nippur	Type II tablet: obverse lexical list; reverse list of capacities; same tablet as Ist Ni 10135 (to be published by C. Proust)	Hilprecht 1906, no. 38 (CBS 10207 only)
CBS 10990+	Nippur	Type I tablet: list of capacities, weights, areas, and lengths	Hilprecht 1906, no. 29
CBS 11080+	Nippur	Type II tablet: obverse Sumerian Proverbs 2; reverse table of lengths	Gordon 1959, pl. 48 (2 II)

TABLE B.8 (*Continued*)

Tablet Number	Findspot or Provenance	Description[a]	Publication
CBS 19814	Nippur	Type II tablet: obverse blank; reverse table of capacities	Hilprecht 1906, no. 35
CBS 19820	Nippur	Type II tablet: obverse blank; reverse table of capacities	Hilprecht 1906, no. 34
Columbia (Plimpton) 317	Purchased from E. J. Banks, n. d.	Type III tablet: table of grain; catch-line and colophon	Robson 2002a, fig. 9
Columbia (Plimpton) 319	Purchased from E. J. Banks, 1923	Type III tablet: table of capacities	Robson 2002a, 278 (no copy)
CUA 36	Unknown	Type I tablet: list or table of areas	Robson 2005a, 392
Di 95	Sippar Amnānum, recycling bin in courtyard of *galamahs'* house, phase IIId	Type I tablet: table of capacities and weights	Tanret 2002, no. 45
Di 98 (+) Di 99	Sippar Amnānum, recycling bin in courtyard of *gala-mahs'* house, phase IIId	Type I tablet: table of capacities	Tanret 2002, nos. 42–3
Di 101 (+) Di 108	Sippar Amnānum, recycling bin in courtyard of *gala-mahs'* house, phase IIId	Type I tablet: table of weights	Tanret 2002, nos. 46–7

TABLE B.8 (*Continued*)

Tablet Number	Findspot or Provenance	Description[a]	Publication
Di 102	Sippar Amnānum, recycling bin in courtyard of *gala-mahs*' house, phase IIId	Type I tablet: table of capacities	Tanret 2002, no. 44
Di 746	Sippar Amnānum, courtyard of *gala-mahs*' house, phase IIIe1	Type IV tablet: three-line extract of list of capacities, duplicated on reverse	Tanret 2002, no. 57
IB 1505	Isin, NO III, Room 3	Type I tablet: list of lengths	Wilcke 1987, 111–2
IB 1625+	Isin, NO IV, Room 5	Type I tablet: table or list of lengths, weights, and areas	Wilcke 1987, 111 (description only)
Ist O 4108	Kiš (Uhaimir)	Type I tablet: list or table of lengths, table of inverse squares	Neugebauer 1935–7, I 69 no. 25, 70 no. 19, 73 no. 55, II pl. 68
MLC 1531	Unknown	Type III tablet: table of capacities; colophon	Clay 1923, no. 40; see table 4.1
MLC 1854	Unknown	Type I tablet: list of weights	Clay 1923, no. 41; Neugebauer 1935–7, I 92–4
N 4919	Nippur	Type II tablet: obverse Sumerian Proverbs 2; reverse list of capacities	Alster 1997, pl. 7

TABLE B.8 (*Continued*)

Tablet Number	Findspot or Provenance	Description[a]	Publication
NBC 2513	—	Type I tablet: table of weights; colophon	Nies and Keiser 1920, no. 36
UET 7, 114	Ur, findspot unknown	Type I tablet: table of lengths; colophon	Gurney 1974, no. 114; Friberg 2000, 155
UET 7, 115	Ur, findspot unknown	Type I tablet: table of heights	Gurney 1974, no. 115; Friberg 2000, 156
VAT 1155	—	Type I tablet: table of weights	Meissner 1893, pls. 56–7
VAT 2596	—	Type I tablet: list of capacities	Meissner 1893, pl. 58
VAT 6220	—	Type I tablet: terse reciprocal table and multiplication tables ($\times 50$ to $\times 6$), table of squares, list or table of lengths	Neugebauer 1935–7, I 90
W 16743cq = VAT 21586	Uruk, Pe XVI/4–5	Type I tablet: table of areas	Cavigneaux 1996, no. 282
W 16743fy	Uruk, Pe XVI/4–5	Fragment: table of weights	Cavigneaux 1996, no. 286
W 20248,12	Uruk, Sîn-kāšid's palace, De XIV/2	Fragment: list of areas	Cavigneaux 1982, 29–30
W 20248,15	Uruk, Sîn-kāšid's palace, De XIV/2	Fragment: list of areas	Cavigneaux 1982, 30 pl. 6b

TABLE B.8 (*Continued*)

Tablet Number	Findspot or Provenance	Description[a]	Publication
YBC 4700	Larsa?	Type III tablet: table of heights; colophon (see table 4.3, figure 4.6)	Previously unpublished
YBC 4701	Larsa?	Type III tablet: table of lengths; colophon (see table 4.3, Figure 4.6)	Previously unpublished

Note: Some published metrological lists and tables may have inadvertently been omitted from this table. Many more, especially from Nippur and Kiš, will soon be published by Ignacio Márquez Rowe and the author, and by Christine Proust.

[a] For tablet typology, see tables 4.5, 4.8.

[b] See Sigrist, Figulla, and Walker 1996, x, for the probable origin of BM tablets from the 1894-1-15 collection.

TABLE B.9
Old Babylonian Arithmetical Lists and Tables Published or Identified Since 1945

Tablet Number	Provenance	Description[a]	Publication
2N-T 585 = A 30016	Nippur, TB locus 190	Type III tablet: verbose table of squares; colophon	Neugebauer and Sachs 1984
3N-T 261 = UM 55-21-289	Nippur TA House F, locus 205	Type III tablet: verbose multiplication table ($\times$ 1;40)	Robson 2002b, fig. 8 (obverse only)
3N-T 494 = IM 58519	Nippur, TA House F, locus 205	Type II tablet: obverse Sumerian Proverbs 2; reverse terse multiplications	Gordon 1959, pl. 69 (2 RRR)
3N-T 594 = IM 58573	Nippur, TA House F, locus 205	Type II tablet: obverse verbose list of reciprocals; reverse list of capacities (figure 4.5)	(See table B.8)
3N-T 608 = UM 55-21-360	Nippur, TA House F, locus 205	Type III tablet: terse multiplication table ($\times$ 3)	Robson 2002b, fig. 8 (obverse only)
3N-T 918, 432	Nippur, TA House F, locus 191	Type III tablet: verbose table of reciprocals	Heimerdinger 1979, pl. 77
3N-T 919, 474	Nippur, TA House F, locus 191	Type I or II tablet: terse multiplication tables	Heimerdinger 1979, pl. 77
AO 8865	Larsa?	Six-sided prism: metrological tables (lengths, heights); tables of inverse squares and inverse cubes	(See table B.8)
Ash 1922.178	Larsa?	Type III tablet: verbose multiplication table ($\times$ 25); colophon (figure 4.2)	Robson 2004c, no. 1
Ash 1923.318	Larsa?	Type III tablet: verbose multiplication table ($\times$ 3)	Robson 2004c, no. 6

TABLE B.9 (*Continued*)

Tablet Number	Provenance	Description[a]	Publication
Ash 1923.366	Larsa?	Six-sided prism: metrological tables (lengths, heights); tables of inverse squares and inverse cubes	(See table B.8)
Ash 1923.447	Larsa?	Type III tablet: verbose multiplication table ($\times$ 24); colophon (table 4.3)	Robson 2004c, no. 2
Ash 1923.450	Larsa?	Type III tablet: verbose multiplication table ($\times$ 12)	Robson 2004c, no. 4
Ash 1923.451	Larsa?	Type III tablet: verbose multiplication table ($\times$ 24); colophon (table 4.3)	Robson 2004c, no. 3
Ash 1924.447	Marad	Type III table: verbose multiplication table ($\times$ 10); colophon	Dalley and Yoffee 1991, no. 5; Robson 2004c, no. 7
Ash 1924.472	Unknown	Type III tablet: verbose multiplication table ($\times$ 10); colophon	Robson 2004c, no. 5
Ash 1924.573	Kiš, Uhaimir	Type III tablet: terse multiplication table ($\times$ 4;30)	Robson 2004c, no. 20
Ash 1924.590	Kiš	Type III tablet: verbose reciprocal table; colophon	Robson 2004c, no. 15
Ash 1924. 1214	Kiš	Type I tablet? verbose table of squares	Robson 2004c, no. 17
Ash 1929. 833	Kiš	Type III tablet: terse multiplication tables ($\times$ 2;24 and 2)	Robson 2004c, no. 16
Ash 1932. 180	Kiš, Ingharra	Type III tablet: verbose multiplication table ($\times$ 12;30)	Robson 2004c, no. 25

TABLE B.9 (*Continued*)

Tablet Number	Provenance	Description[a]	Publication
BM 22706	Unknown	Type L tablet: table of powers	Nissen, Damerow, and Englund 1993, fig. 128
BM 80150	Unknown	Type I tablet: terse reciprocal table and multiplication tables	Neugebauer 1935–7, I 11 no. 12, 23 no. 3, 49 no. 105; Nissen, Damerow, and Englund 1993, fig. 124
CBS 3882+	Nippur	Type II tablet: obverse Sumerian Proverbs 8; reverse terse multiplication tables	Alster 1997, pl. 53
CBS 4805a	Nippur	Type II tablet: obverse Sumerian Proverbs 2; reverse terse multiplication tables	Alster 1997, pls. 4–5
CBS 6832	Nippur	Type II tablet: obverse 'Minor' Sumerian Proverbs; reverse terse multiplication tables	Alster 1997, pl. 97
CBS 13852+	Nippur	Type II tablet: obverse verbose multiplication table ($\times$ 18); reverse Sumerian Proverbs 1	Gordon 1959, pl. 13 (1 Y)
CBS 14158+	Nippur	Type II tablet: obverse multiplication table ($\times$ 9) and Sumerian Proverbs 2; reverse lexical list	Gordon 1959, pl. 53 (2 PP)
Columbia (Plimpton) 318	Purchased from R. D. Messayah, 1915?	Type III tablet: terse table of inverse squares; traces of colophon	Robson 2002a, fig. 10

Table B.9 (*Continued*)

Tablet Number	Provenance	Description[a]	Publication
CUA 63	Unknown	Type IV tablet: lexical extract and terse multiplication table ($\times$ 30)	Robson 2005a, 390–1
IB 1558	Isin, NO, room 4	Type I tablet: terse multiplication tables (including 2;13 20, 1;40, and 1;20) and table of squares	Wilcke 1987, 111 (description only)
IM 52001 = HL$_1$-46	Me-Turan	Triangular prism: table of inverse squares	Isma'el 1999a
IM 73355 = L.33.98	Larsa, square E.XII	Type L tablet: table of powers; colophon	Arnaud 1994, no. 55
IM 73365 = L.33.121	Larsa, square E.F. XIII	Type III tablet: verbose multiplication table ($\times$ 40)	Arnaud 1994, no. 66
IM 73381 = L.33.191	Larsa, 'scribe's house'	Type I tablet: verbose multiplication tables ($\times$ 25, 22;30, and 20)	Arnaud 1994, no. 81
IM 92092 = S$_2$-698	Tell es-Seeb	Type I tablet: multiplication tables (including 50, 45, 44;26 40, 1;40, and 1;15)	Isma'el 1999b
IMJ 80.20.201	Unknown	Type III tablet: verbose multiplication table ($\times$ 40)	Horowitz and Tammuz 1998
Kelsey 89406	Unknown	Type III tablet: terse multiplication table ($\times$ 5); colophon	Moore 1939, no. 94
MM 367	Unknown	Type III tablet: verbose multiplication table ($\times$ 8); colophon	Márquez Rowe 1998, 286
MM 726	Unknown	Type III tablet: verbose multiplication tables ($\times$ 25, $\times$ 22 30)	Márquez Rowe 1998, 288

TABLE B.9 (*Continued*)

Tablet Number	Provenance	Description[a]	Publication
MMA 57. 16.2	Unknown	Type III tablet: terse multiplication table (× 5)	Spar 1988, no. 64
N 3928	Nippur	Type III tablet: terse list of reciprocal pairs from 2 05 ~ 28 48	Sachs 1947a, 229
N 3977	Nippur	Type II tablet: obverse Sumerian Proverbs 1; reverse terse multiplication tables	Gordon 1959, pl. 17 (1 JJ)
N 4959	Nippur	Type II tablet: obverse Sumerian Proverbs 2; reverse terse multiplication tables	Gordon 1959, pl. 45 (2 V)
N 6025	Nippur	Type II tablet: obverse Sumerian Proverbs 2; reverse terse multiplications	Gordon 1959, pl. 60 (2 CCC)
SÉ 95	Unknown	Type III tablet: terse multiplication table (× 25)	Jursa and Radner 1995–6, text B3
VAT 21620	Uruk, Pe XVI/4–5	Type III tablet: verbose multiplication table (× 9)	Cavigneaux 1996, no. 280
*W 16603af	Uruk, Pe XVI/4–5	Type III tablet: terse multiplication table (× 8;20)	Cavigneaux 1996, no. 277
*W 16603an = VAT 21636	Uruk, Pe XVI/4–5	Type I tablet: terse multiplication tables (including 44;26 40)	Cavigneaux 1996, no. 289
*W 16603b	Uruk, Pe XVI/4–5	Type III tablet: verbose multiplication table (× 45)	Cavigneaux 1996, no. 281
*W 16603s = VAT 21638	Uruk, Pe XVI/4–5	Fragment: verbose multiplication tables (including 3)	Cavigneaux 1996, no. 278

Table B.9 (*Continued*)

Tablet Number	Provenance	Description[a]	Publication
*W 16743ck = VAT 21565	Uruk, Pe XVI/4–5	Type III tablet: terse multiplication table (× 22;30)	Cavigneaux 1996, no. 279
W 20248,21	Uruk, Sîn-kāšid's palace, De XIV/2	Type III tablet: terse multiplication table (× 22;30)	Cavigneaux 1982, 30, pl. 6
W 20248,22	Uruk, Sîn-kāšid's palace, De XIV/2	Type I tablet: terse multiplication tables (including 45); does not join W 20248,23	Cavigneaux 1982, 30, pl. 6
W 20248,23	Uruk, Sîn-kāšid's palace, De XIV/2	Type I tablet: terse multiplication tables (including 45); does not join W 20248,22	Cavigneaux 1982, 30, pl. 6
W 20248,24	Uruk, Sîn-kāšid's palace, De XIV/2	Type I tablet: terse multiplication tables (including 15)	Cavigneaux 1982, 30, pl. 6
W 20248,25	Uruk, Sîn-kāšid's palace, De XIV/2	(No photo or description published)	Cavigneaux 1982, 30
YBC 4702	Sippar	Type III tablet: verbose multiplication table (× 50); colophon; sealed	Neugebauer 1935–7, II 36; Nemet-Nejat and Wallenfels 1994
YBC 11924	Larsa?	Type III tablet: verbose multiplication table (× 4); colophon (table 4.3)	Neugebauer and Sachs 1945, 23 no. 99, 13b; Robson 2004c, 13–6

Note: Neugebauer 1935–7, I 32–67, II 36–7, III 49–51, and Neugebauer and Sachs 1945, 19–33, list a further 270-odd arithmetical tablets. Some tablets published since then may inadvertently have been omitted from this table.

[a] For tablet typology, see tables 4.5, 4.8.

TABLE B.10
Old Babylonian Calculations and Diagrams Published or Identified Since 1999

Tablet Number	Findspot or Provenance	Description[a]	Publication
2N-T 500 = A 29985	Nippur, TB House B, locus 10	Type III tablet: obverse calculation of regular reciprocal; reverse Sumerian Proverbs 2 and further calculation	(See table 1.5)
3N-T 362+ = IM 58446+	Nippur, TA House F, locus 205	Type S tablet: obverse Sumerian literature, ETCSL 5.1.3; reverse calculation of regular reciprocal pair	Robson 2000a, fig. 2; 2002b, fig. 16
3N-T 605 = UM 55-21-357	Nippur, TA House F, locus 205	Type S tablet: calculation of regular reciprocal pair	Robson 2000a, no. 1; 2002b, fig. 19
3N-T 611 = A 30279	Nippur, TA House F, locus 205	Type R tablet: calculation of square	Robson 2002b, fig. 18
Ash 1924.586	Kiš, Uhaimir	Type IV tablet: fragmentary calculation	Robson 2004c, no. 21
Ash 1930.365	Kiš	Recycled tablet: tabular calculations	Robson 2004c, no. 18
Ash 1931.91	Kiš, Ingharra	Type IV tablet: numbered diagram	Robson 2004c, no. 24
CBS 3551	Nippur	Type R tablet: calculation of square	Robson 2000a, no. 4
CBS 7265	Nippur	Type R tablet: calculation of square	Robson 2000a, no. 5
CBS 11601	Nippur	Type R tablet: calculations	Robson 2000a, no. 9
Columbia (Plimpton) 320	Purchased from R. D. Messayah, 1915?	Type R tablet: numbered diagram and calculation	Robson 2002a, fig. 11

Table B.10 (*Continued*)

Tablet Number	Findspot or Provenance	Description[a]	Publication
Di 2207	Sippar Amnānum, *gala-mahs'* house, locus 23, phase IIIe2	Type IV tablet: multiplications	Tanret 2002, no. 59
Di 2232	Sippar Amnānum, *gala-mahs'* house, locus 11, phase IIId	Type IV tablet: fragmentary calculation?	Tanret 2002, no. 55
Di 2234	Sippar Amnānum, *gala-mahs'* house, locus 11, phase IIId	Type IV tablet: model contract and calculation	Tanret 2002, no. 56
IM 85943	Sippar Yaḥrūrum	Type IV tablet: calculations	Al-Rawi and Dalley 2000, no. 84
IM Si 337	Sippar Yaḥrūrum	Type IV tablet: calculations	Al-Rawi and Dalley 2000, no. 79
MLC 1731	Unknown	Type III tablet: calculations	Sachs 1946
N 837	Nippur	Type R tablet: calculation	Robson 2000a, no. 8
N 3914	Nippur	Type R tablet: tabular calculations	Robson 2000a, no. 10
N 4942	Nippur	Type R tablet: numbered diagram, rectangular	Robson 2000a, no. 11
TSŠ 77	Kisurra	Type IV tablet: unnumbered diagram	Jestin 1937, no. 77; Krebernik 2006

TABLE B.10 (*Continued*)

Tablet Number	Findspot or Provenance	Description[a]	Publication
U 10144 = UET 6/3 877	Ur, no findspot	Type IV tablet: Sumerian proverb and calculation	Shaffer 2006, no. 877
UM 29-13-173	Nippur	Type R tablet: calculations	Robson 2000a, no. 14
UM 29-15-709	Nippur	Type R tablet: numbered diagram (two copies)	Robson 2000a, no. 12
UM 29-16-401	Nippur	Type R tablet: calculation of a square	Robson 2000a, no. 7
W 16743av	Uruk, Pe XVI/4–5	Fragment: calculation of regular reciprocal pair	Cavigneaux 1996, no. 283
W 16743k	Uruk, Pe XVI/4–5	Type IV tablet: fragmentary calculation	Cavigneaux 1996, no. 265
YBC 7345	Unknown	Type IV tablet: obverse Sumerian proverb; reverse multiplications	Alster 1997, pl. 130
YBC 7359	Unknown	Type R tablet: numbered diagram (two copies)	Nemet-Nejat 2002, no. 16

Note: Over 60 earlier publications are listed in Robson 1999, 11 Table 1.1; some 50 more are also edited there in Appendix 5. Nemet-Nejat 2002 provides new photographs of 17 Type R tablets from the Yale Babylonian Collection that were first edited by Neugebauer and Sachs 1945. Robson 2000a, nos. 3, 4, 7, gives new hand-copies and editions of three calculations previously published in transliteration by Neugebauer and Sachs 1945; 1984. Robson 2004c, no. 22, presents a new hand-copy and edition of a tabular calculation first published in Dalley and Yoffee 1991, no. 64.

[a] For tablet typology, see tables 4.5, 4.8.

TABLE B.11
Old Babylonian Word Problems and Catalogues Published or Identified Since 1999

Tablet number	Provenance	Description[a]	Publication
Bodleian AB 216	Unknown	Type S tablet: problem about fields, with worked solution and diagram	Robson 2004c, no. 14
CBS 43	Purchased from J. Shemtob, 1888	Type S tablet: 6 statements of problems about squares	Robson 2000a, no. 17
CBS 154+	Purchased from J. Shemtob, 1888	Type S tablet: 5 statements of problems about squares	Robson 2000a, no. 18
CBS 165	Purchased from J. Shemtob, 1888	Fragment: list of problems about rectangles	Robson 2000a, no. 19
CBS 11681	Nippur	Type S tablet: 2 problems about cubes	Robson 2000a, no. 15
CBS 12648	Nippur	Type M2 tablet: problems about volumes	Obverse, Hilprecht 1906, no. 25a; reverse, Robson 2000a, no. 14
CBS 19761	Nippur	Type M2 tablet: problems about squares; joins Ist Ni 5175 (to be published by C. Proust)	Robson 2000a, no. 16
Columbia (Plimpton) 322	Larsa?	Type L tablet: tabular catalogue of 15 sets of parameters for problems about diagonals	Neugebuaer and Sachs 1945, text A; Friberg 1981b; Robson 2001a; 2002c
UET 6/3 685	Ur, no findspot	Fragment: word problem; contents unclear	Shaffer 2006, no. 685
UET 6/3 686	Ur, no findspot	Fragment: word problem, possibly about a circle (mentions a constant of value 5)	Shaffer 2006, no. 686
YBC 4692	Unknown	Type S tablet: catalogue of 24 sets of parameters for word problems about rectangles	Neugebauer and Sachs 1945, text Sa

Note: For the 150 or so Old Babylonian problem tablets published in and before 1999, see Nemet-Nejat 1993, 108–12, with additions and corrections by Robson 1999, 7 n30. Nemet-Nejat 2002, nos. 20–22, also provides new photographs of three tablets edited in hand-copy by Neugebauer and Sachs 1945, texts Ub, M, and Q. Previously published and identified catalogues are listed in Robson 1999, 9–10.

[a] For tablet typology, see tables 4.5, 4.8.

TABLE B.12
Mathematical Tablets from the Old Babylonian Kingdom of Mari

Tablet Number	Findspot	Description[a]	Publication
ARM 9, 299	Mari palace, room 5	Type III tablet: terse table of reciprocals; colophon (see §5.2)	Birot 1960, no. 299
A 3514	Mari palace, room 108	Type III tablet: terse multiplication table ($\times$ 44;26 40)	Soubeyran 1984, no. 1
Deir ez-Zor 2961	Terqa	Type IV tablet: calculation	Rouault and Masetti-Rouault 1993, no. 298
M 288	Mari palace	Type IV tablet: numbered diagram	Charpin 1993
M 6199b	Mari palace, room 108	Type I tablet: terse multiplication tables ($\times$ 22;30, $\times$ 20)	Soubeyran 1984, no. 4
M 7857	Mari palace	Type R tablet: calculation of powers (figure 5.3)	Guichard 1997; Proust 2002
M 8613	Mari palace, room 115	Type R tablet: list of geometrical progression	Soubeyran 1984, no. 6
M 10681	Mari palace, room 156	Type III tablet: verbose multiplication table ($\times$ 22;30)	Soubeyran 1984, no. 2
M 12306	Mari palace, room 108	Type I tablet: multiplication tables ($\times$ 10, 9, 8;20, 8)	Soubeyran 1984, no. 3
M 12598	Mari palace, room 108	Type S tablet: verbose table of inverse squares	Soubeyran 1984, no. 5
TH 84.008	Mari, Asqudum the diviner's house	Type I tablet: list of weights, from ½ grain to several talents	Chambon 2002
TH 84.009	Mari, Asqudum the diviner's house	Type IV tablet: extract from list of capacities	Chambon 2002
TH 84.046	Mari, Asqudum the diviner's house	Type IV tablet: extract from list of capacities	Chambon 2002

[a] For tablet typology, see tables 4.5, 4.8.

TABLE B.13
Old Assyrian Mathematical Tablets from Kaneš and Assur

Tablet Number	Provenance or Findspot	Description	Publication
Kt a/k 178	Kaneš, 1948 excavations	Calculation on round tablet	Hecker 1996
Kt t/k 76+	Kaneš, 1968 excavations	List of weights from 1 shekel to perhaps 100 talents; vocabulary of metals and stone	Hecker 1993, 286–90; Michel 1998, 253; 2003, 139
Kt 84/k 3	Kaneš, 1984 excavations	Calculation on round tablet	Donbaz 1985, 7; Ulshöfer 1995, no. 510
Ass 13058e	Assur, library M7	Calculation on round tablet	Donbaz 1985, 6; Pedersén 1985–6, I 77 no. 13
Ass 13058f	Assur, library M7	Calculation on round tablet	Donbaz 1985, 5; Pedersén 1985–6, I 77 no. 14
Ass 13058i	Assur, library M7	Calculation on round tablet	Donbaz 1985, 5; Pedersén 1985–6, I 77 no. 17
Ass 13058k	Assur, library M7	Calculation on round tablet (figure 5.4)	Donbaz 1985, 6; Pedersén 1985–6, I 77 no. 19
Ass 13058q	Assur, library M7	Calculation on round tablet	Donbaz 1985, 6; Pedersén 1985–6, I 77 no. 27; Michel 1998, 251 n1
Ass 14479 = A 1001	Assur, findspot unpublished	Calculation on fragment of round tablet	Donbaz 1985, 6

Source: Based on Michel 1998; 2003, 139–40.

TABLE B.14
Mathematical Tablets from the Middle and Neo-Assyrian periods

Tablet Number	Provenance[a]	Description	Publication
Ist A 20 + VAT 9734	Middle Assyrian Assur, library M2/N1	Complete series of multiplication tables, in classic OB style, with a partially legible colophon dating it to c. 1178–1076 BCE	Neugebauer 1935–7, II pl. 65; Pedersén 1985–6, II 22 no. 46; Freydank 1991, 115 (colophon)
VAT 9840	Neo-Assyrian Assur	List of Assyrian capacities and weights, from 1 *sila* to 1,000,000 *imeru*; from ⅓? shekel to 600? talents, with a badly damaged colophon	Schroeder 1920, no. 184; Friberg 1993, no. 1
Ass 13956dr[b]	Neo-Assyrian Assur, library N4	Table of ratios between length units, from 6 grains = 1 finger to 30 cables = 1 league	Thureau-Dangin 1926; Pedersén 1985–6, II 68 no. 365; Friberg 1993, no. 4
SU 52/5	Neo-Assyrian Huzirīna	Standard reciprocal table written in syllabic Sumerian	Hulin 1963; Gurney and Hulin 1964, no. 399
K 2069	Neo-Assyrian Nineveh	Unique sexagesimal reciprocal table on a base of 1 10	Neugebauer 1935–7, I 30–2, II pl. 10; Curtis and Reade 1995, no. 223
Sm 162	Neo-Assyrian Nineveh	Fragmentary version of OB-style word problem (HS 245, table B.15), on the reverse of an astronomical tablet; colophon (see §5.4)	L. W. King 1912b, pl. 11; Koch 2001; colophon: Hunger 1968, no. 308
Sm 1113	Neo-Assyrian Nineveh	Fragmentary version of OB-style word problem (HS 245, table B.15)	Weidner 1957–8, 393; Koch 2001

[a] The findspots of these tablets are given and discussed in more detail in the course of chapter 5.
[b] Ass 13956ea, a tiny unpublished tablet from the same library, with two columns of single-digit numbers, may also be mathematical (Pedersén 1985–6, II 68 no. 374).

TABLE B.15
Mathematical Tablets from Kassite Babylonia

Tablet Number	Provenance	Description	Publication
AO 17264	Uruk?	Word problem, on dividing a field into 6 parts so that pairs of brothers get equal shares; erroneous method of solution	Thureau-Dangin 1934; Neugebauer 1935–7, I 126–34; Brack-Bernsen and Schmidt 1990
Bab 36621	Babylon, library M6 in the Merkes district	Extract from standard Ur III-OB reciprocal list (first 7 lines); followed by repeated sign of unknown meaning	Pedersén 2005, 87 fig. 40, 89 no. 56 (obverse only); Oelsner 2006
CBS 10996	Nippur	List of mathematical and musical constants; may date to first millennium	Kilmer 1960; Robson 1999, list B
HS 245	Nippur	Word problem, on division by sexagesimally irregular numbers, with unfinished solution (see also Sm 162 and Sm 1113, table B.14)	Rochberg-Halton 1983; Horowitz 1998, 177–82; Koch 2001

TABLE B.16
Mathematical Tablets from Late Bronze Age Amarna, Hazor, and Ugarit

Tablet Number	Findspot	Description	Publication
EA 368 = Ash 1921.1154	Amarna, 1920–1 season, house O.49.23	Fragment of Egyptian-Akkadian vocabulary, including list of weights	Izre'el 1997, 77–81 pl. CL
IAA 1997-3306	Hazor, LBA palace, locus 7439	Fragment of four-sided prism with OB-style terse reciprocal and multiplication tables	Horowitz 1997; Goren 2000 (dating); Horowitz, Oshima, and Sanders 2006, 78–80, 212
RS 6.308 = AO 18896	Ugarit, Acropolis	Fragment of two-column tablet: list of areas; colophon 'Hand of Šaduya, servant of Nabû and Nisaba, finished and checked'	Nougayrol 1968, no. 152/J
RS 20.14	Ugarit, Rap'ānu's House, room 5	Two-column tablet: list of weights with catch-line for areas; colophon 'Hand of Šaduya, servant of Nabû and Nisaba'	Nougayrol 1968, no. 149/G
RS 20.160N + 8 fragments	Ugarit, Rap'ānu's House, rooms 5 and 7	Three-column tablet: list of capacities, weights, and areas; colophon 'Hand of Yanāḫan, scribe, student of Nūr-mālik'	Nougayrol 1968, nos. 143, 145–7, 150–1/A, C–E, H–I
RS 21.10 (+) 20.222	Ugarit, Rap'ānu's House, rooms 5 and 7	Three-column tablet: list of capacities, weights, and areas	Nougayrol 1968, no. 144/B
RS 25.511B	Ugarit, Lamaštu archive	Fragment of list of weights	Nougayrol 1968, no. 173
Rs Pt 1844	Ugarit, Rap'ānu's House, room 5?	Fragment of list of weights from three-column tablet	Nougayrol 1968, no. 148/F

Table B.17
Mathematical Tablets from Old Babylonian Susa and Middle Elamite Kabnak/Haft Tepe

Tablet Number	Provenance	Description	Publication
HT 140	Kabnak	Fragment: table of capacities	Herréro and Glassner 1996, no. 277
HT 142	Kabnak	Fragment: table of areas	Herréro and Glassner 1996, no. 279
HT 215	Kabnak	Fragment: possibly catalogue of parameters for word problems	Herréro and Glassner 1991, no. 124
HT 222	Kabnak	Fragment: list or table of weights	Herréro and Glassner 1991, no. 126
HT 615	Kabnak	Fragment: terse table of squares of integers and half-integers	Herréro and Glassner 1996, no. 289
MDP 18, 14	Susa, no findspot details	Type I tablet: terse multiplication tables ($\times$ 16, [15], 12 30, 10, 9, 8 20, 8 extant)	Dossin 1927, no. 14
MDP 27, 59	Susa, VR[a]	Type I tablet: metrological list of weights	van der Meer 1935, no. 59
MDP 27, 60	Susa, VR	Type IV tablet: calculation	van der Meer 1935, no. 60
MDP 27, 61	Susa, VR	Type III tablet: verbose multiplication table ($\times$ 3); reverse extract from lexical list	van der Meer 1935, no. 61
MDP 27, 291	Susa, VR	Type IV tablet: calculation	van der Meer 1935, no. 291
MDP 27, 292	Susa, VR	Type III tablet: terse multiplication table ($\times$ 7 12)	van der Meer 1935, no. 292
MDP 27, 293	Susa, VR	Type I tablet: terse multiplication tables ($\times$ 7 30, 7 12, 7, 6 40, 6, 5, 4 30, 4)	van der Meer 1935, no. 293 (swap obv. and rev.)

Table B.17 (*Continued*)

Tablet Number	Provenance	Description	Publication
MDP 27, 294	Susa, VR	Type I tablet: terse multiplication tables (at least × 16 40 and 15)	van der Meer 1935, no. 294
MDP 27, 295	Susa, VR	Type II tablet? repeated terse multiplication table (× 22 30)	van der Meer 1935, no. 295
MDP 27, 296	Susa, VR	Type I tablet: terse multiplication tables (at least reciprocals, × 50, × 44 26 40)	van der Meer 1935, no. 296
MDP 27, 297	Susa, VR	Type III tablet: terse reciprocal table	van der Meer 1935, no. 297
RA 35, 1	Susa, find circumstances unknown	Type S tablet: catalogue of 33 sets of parameters for word problems about finding square areas, with answers	Scheil 1938, 92–6; Neugebauer and Sachs 1945, 6–7
RA 35, 2	Susa, find circumstances unknown	Type M3 tablet: catalogue of at least 52 sets of parameters for word problems about finding rectangular areas, with answers; colophon: '[Nisaba, my] lady, Haia, my lord! Ulû, servant of Nisaba-gina (?)'	Scheil 1938, 96–103; Neugebauer and Sachs 1945, 8–10
TS B V 115	Susa, Ville Royale, Chantier B Level 5	Type III tablet: verbose multiplication table (× 30)	Unpublished; described by Tanret 1986, 147

Note: Neither HT 63 nor HT 617 (Herréro and Glassner 1996, nos. 271, 290) is mathematical, although they appear so at first glance. No findspots are known for any of the Haft Tepe tablets. This list excludes the 26 OB mathematical tablets from Susa published in Bruins and Rutten 1961; for more recent bibliography see Robson 1999, 317; Høyrup 2002b. New photos of Bruins and Rutten 1961 nos. 2 and 4 are published in Harper, Aruz, and Tallon 1996, nos. 194–5.

[a]VR = Ville Royale, 1934–5 season.

TABLE B.18
Mathematical Tablets from the Neo-Babylonian Period

Tablet Number	Provenance	Description	Publication
79.B.1/59+	Babylon, temple of Nabû of Harû	Tablet type 1b: Syllabaries S^a, S^b; spellings; personal names; metrological list (capacities); colophon (see §7.2)	Cavigneaux 1981, 19, 49–50, 155
79.B.1/70	Babylon, temple of Nabû of Harû	Tablet type 1b: Syllabaries S^a, S^b; spellings; personal names; metrological list (capacities); fragmentary colophon	Cavigneaux 1981, 20, 52
79.B.1/119	Babylon, temple of Nabû of Harû	Tablet type 1a or 1b: Syllabaries S^a, S^b; metrological list (weights and capacities); fragmentary colophon	Cavigneaux 1981, 24, 63
79.B.1/186	Babylon, temple of Nabû of Harû	Tablet type 1b: Syllabary S^a; metrological list (capacities); spellings	Cavigneaux 1981, 29
79.B.1/194	Babylon, temple of Nabû of Harû	Tablet type 1b: Syllabaries S^a, S^b; spellings; personal names; metrological list (unclear); fragmentary colophon?	Cavigneaux 1981, 30
79.B.1/206	Babylon, temple of Nabû of Harû	Tablet type 1b: Syllabaries S^a, S^b; spellings; personal names; metrological list (capacities); fragmentary colophon	Cavigneaux 1981, 31, 74
Ash 1924.796+	Kiš	Table of squares on long library tablet, with colophon and firing holes	Robson 2004c, no. 28
Ash 1924.1048	Kiš	Tablet type 1: Vocabulary S^b A; metrological list (capacities)	van der Meer 1938, no. 127; Robson 2004c, no. 30
Ash 1924.1098+	Kiš	Tablet type 1b': Vocabulary S^b A; metrological list (capacities) and spellings	van der Meer 1938, no. 75; Robson 2004c, no. 34

TABLE B.18 (*Continued*)

Tablet Number	Provenance	Description	Publication
Ash 1924.1196	Kiš	Tablet type 1b: Vocabulary S[b] A and metrological list (weights)	Robson 2004c, no. 39
Ash 1924.1217	Kiš	Tablet type 1b: Vocabulary S[b] A; metrological list (capacities); spellings	van der Meer 1938, no. 34; Robson 2004c, no. 31
Ash 1924.1242	Kiš	Tablet type 1b; Syllabary A; spellings; metrological list (weights) (figure 7.5)	van der Meer 1938, no. 123 (obv.); Robson 2004c, no. 40
Ash 1924.1278	Kiš, Mound W	Small library tablet: metrological list; colophon	Robson 2004c, no. 29
Ash 1924.1450	Kiš	Tablet type 1b: Ur_5-*ra* I; contract metrological list (lengths); spellings	Robson 2004c, no. 35
Ash 1924.1464+	Kiš	Tablet type 1b: Ur_5-*ra* III; metrological list (lengths)	Gurney 1986, no. 9; 1989, no. 140 (obv.); Robson 2004c, no. 36
Ash 1924.1477	Kiš	Tablet type 1b: Syllabary A; personal names; metrological list (capacities)	Robson 2004c, no. 32
Ash 1924.1520+	Kiš	Tablet type 1b': Syllabary A; metrological list (capacities)	van der Meer 1938, no. 128; Robson 2004c, no. 33
Ash 1924.1760	Kiš	Tablet type 1b: [missing]; Ur_5-*ra* XVII; spellings; metrological list (lengths)	Gurney 1989, no. 139; Robson 2004c, no. 37
Ash 1924.1847	Kiš	Tablet type 1b': Ur_5-*ra* II; metrological list (lengths)	Gurney 1986, no. 62 (obv.); Robson 2004c, no. 38
Ash 1924.2214	Kiš	Tablet type 1b': Syllabary A; metrological list (weights)	Robson 2004c, no. 41

TABLE B.18 (*Continued*)

Tablet Number	Provenance	Description	Publication
Ash 1932.187h	Kiš, Ingharra Trench C-11	Tablet type 1a or 1b: Vocabulary S^b A; metrological list (capacities)	Robson 2004c, no. 27
BM 29440	Unknown	Excerpts of historical chronicle, astronomical data, and metrological table (weights) on small rectangular tablet	Leichty and Walker 2004, no. 2
BM 47431	Northern Babylonia	Word problem about *apsamikku* geometrical figure, with numbered diagram, on small rectangular tablet	Robson 2007d
BM 53174+	Northern Babylonia	Tablet type 1b: Syllabary S^a; metrological list (capacities); lexical list	Gesche 2000, 359
BM 53178	Northern Babylonia	Tablet type 1b: Vocabulary S^b A; metrological list (weights)	Gesche 2000, 360
BM 53563	Northern Babylonia	Fragment of metrological list	Gesche 2000, 366
BM 53939	Northern Babylonia	Tablet type 1b: lexical lists; repeated extract of table of squares	Gesche 2000, 378–80
BM 57537	Northern Babylonia	Tablet type 2b: literary extract; plant list; table of squares	Gesche 2000, 436–8
BM 59375	Northern Babylonia	Tablet type 1b: Vocabulary S^b A; metrological list (weights)	Gesche 2000, 438
BM 60185+	Northern Babylonia	Tablet type 1a: Syllabary S^a; metrological list (capacities); fragmentary colophon	Gesche 2000, 456–8
BM 65671	Northern Babylonia	Tablet type 1b: lexical lists; metrological list (lengths); model contract	Gesche 2000, 500–1

TABLE B.18 (*Continued*)

Tablet Number	Provenance	Description	Publication
BM 66894+	Northern Babylonia	Tablet type 1b: Vocabulary S^b A; lexical list; metrological list (weights)	Gesche 2000, 529–30
BM 68367	Northern Babylonia	Tablet type 1a: Syllabary S^a; Vocabulary S^b A; metrological list (weights)	Gesche 2000, 553–4
BM 71971+	Northern Babylonia	Tablet type 1b: Syllabary S^a; metrological list (capacities); personal names	Gesche 2000, 585
BM 72553	Northern Babylonia	Tablet type 1b: Syllabary S^a; proverbs; metrological list (lengths)	Gesche 2000, 595
BM 74049	Northern Babylonia	Tablet type 1b: Syllabary S^a; lexical list; metrological list (capacities)	Gesche 2000, 612–3
BM 78822	Northern Babylonia	Word problem about finding the side of a rectangular area, with calculations, on a small rectangular tablet (figure 7.2)	Jursa 1993–4; Friberg 1997, 295–6
CBS 11019	Nippur	Metrological table (fractions of the shekel); Persian according to Sachs	Sachs 1947b; Friberg 1993 no. 3
CBS 11032	Nippur	Metrological list (fractions of the shekel); Persian according to Sachs	Sachs 1947b; Friberg 1993 no. 2
IM 77122	Nippur, governor's archive	Letter; metrological list (capacities)	Cole 1997, no. 89
IM 77133	Nippur, governor's archive	Metrological list (capacities) with personal names	Cole 1997, no. 124
UET 7, 155	Ur, no excavation number	Tablet type 1b: Syllabary S^a, Vocabulary S^b A; lexical lists; proverbs; table of squares; literary extract	Gurney 1974, no. 155; Gesche 2000, 789

Note: See §7.2 for an explanation of the tablet typology.

Table B.19
Neo- and Late Babylonian Mathematical Tablets from Nippur, Sippar, and Elsewhere

Tablet Number	Provenance	Description	Publication
BM 51077 = 1882-3-23, 2073	Probably Sippar	Fragment of metrological table (capacities)	Friberg 1993, no. 6
BM 64696 = 1882-9-18, 4677	Sippar	Compilation of word problems; 6+ extant	Friberg, Hunger, and Al-Rawi 1990, 502–5
BM 67314 = 1882-9-18, 7310	Sippar	Compilation of word problems; 2+ extant	Friberg 1997, 296–7 (partial publication)
CBS 8539	Nippur	Metrological tables on multi-column tablet	Hilprecht 1906, no. 30; Friberg 1993, no. 10
N 2873	Nippur	Fragment of word problem	Robson 2000a, no. 20
Sippar 2175/12	Sippar, 'Sippar library'	Multi-place reciprocal table	Al-Jadir 1987 (photo)
VAT 3462	Unknown	Fragment of decimal reciprocal table	Neugebauer 1935–7, I 30, II pl. 13

TABLE B.20
Mathematical Tablets from Late Babylonian Uruk

Tablet Number	Findspot	Description	Publication
AO 6456	Uruk	Table of multi-place reciprocals; colophon (table 8.3: 1)	Thureau-Dangin 1922, no. 31
AO 6484	Uruk	Compilation of 17 word problems; colophon (table 8.5: 26)	Thureau-Dangin 1922, no. 33; Neugebauer 1935–7, I 96–107
AO 6555	Uruk	The 'Esaǧila tablet': compilation of 7 problems about Marduk's temple Esaǧila; metrological table; colophon (table 8.5: 3)	Thureau-Dangin 1922, no. 32; George 1992, no. 13
BM 141493	Probably Uruk	Multiplication table × 13	Nemet-Nejat 2001b
Ist U 91+	Uruk	Multi-column tablet: multiplication tables for 32, 28 48, 18 45, 11 15, 9 22 30, 6 45, 4 20, 3 30, 2 15, and 2 13 20	Aaboe 1998
VAT 7848	Uruk	Compilation of word problems; 7 + extant	Neugebauer and Sachs 1945, text Y
W 22260a	Uruk Ue XVIII/1	Obv. multi-place reciprocal table; rev. metrological tables	Hunger 1976b, no. 101; Friberg 1993, no. 7
W 22309a + b	Uruk Ue XVIII/1	Metrological tables	Hunger 1976b, no. 102; Friberg 1993, no. 5
W 22661/3a + b	Uruk, Ue XVIII/1, Level 3	Numerical fragments	von Weiher 1983–98, V no. 317
W 22715/2	Uruk Ue XVIII/1, Level 3 Room 3	Multiplication table	von Weiher 1983–98, IV no. 177
W 23016	Uruk Ue XVIII/1, Level 4 fill	Calculation of reciprocal pair (see §8.3)	von Weiher 1983–98, V no. 316

TABLE B.20 (*Continued*)

Tablet Number	Findspot	Description	Publication
W 23021	Uruk Ue XVIII/1, Level 4 fill	Calculations of reciprocal pairs (see §8.3)	Friberg 1999a
W 23273	Uruk Ue XVIII/1, Level 4 Room 4	Metrological tables; colophon (table 8.2: 10)	von Weiher 1983–98, IV no. 172; Friberg 1993, no. 11
W 23281	Uruk Ue XVIII/1, Level 4 Room 4	Obv. metrological tables, rev. multi-place reciprocal table (see §8.3)	von Weiher 1983–98, IV no. 173; Friberg 1993, no. 12
W 23283+	Uruk Ue XVIII/1, Level 4 Room 4	Multi-place reciprocal table; colophon (table 8.2: 11)	von Weiher 1983–98, IV no. 174
W 23291	Uruk Ue XVIII/1, Level 4 Room 4	Compilation of word problems; 26 + extant (see §8.3)	Friberg 1997
W 23291-x	Uruk (Ue XVIII/1)	Compilation of 23 word problems; colophon (table 8.2: 9)	Friberg, Hunger, and Al-Rawi 1990

TABLE B.21
Mathematical Tablets from Late Babylonian Babylon

Tablet Number	British Museum Collection Number	Description	Publication
BM 32178	1876-11-17, 1905 + 2228	Fragment of a list of multi-place squares	Aaboe 1965, text III
BM 33567	Rm 4, 123	Fragment of a list of multi-place squares	Aaboe 1965, text II
BM 34081+	Sp, 179 + Sp II, 102, + Sp II, 349 + 1881-7-6, 279 + 1881-7-6, 590	Word problem within collection of astronomical precepts for computation of the motion of Jupiter	Neugebauer 1955, no. 813 Section 5
BM 34517	Sp, 641	Unidentified fragment of numerical table	Pinches and Sachs 1955, no. 1646
BM 34568	Sp II, 40 + 85 + 990 + Sp III, 137 + 252 + 645	Collection of word problems (17 + extant) on two-column tablet	Neugebauer 1935–7, III 14–22; Høyrup 2002b, 391–9
BM 34577	Sp II, 49	Fragment of multi-place reciprocal table	Pinches and Sachs 1955, no. 1635
BM 34578	Sp II, 50	Fragment of list of multi-place squares	Pinches and Sachs 1955, no. 1641; Aaboe 1965, text VIII
BM 34592	Sp II, 65+	Standard OB-style verbose reciprocal table; table of three-place squares	Pinches and Sachs 1955, no. 1637; Aaboe 1965, text IV
BM 34596	Sp II, 70+	Fragment of multi-place reciprocal table	Pinches and Sachs 1955, no. 1633; Friberg 1983
BM 34601	Sp II, 75+	Fragmentary calculation of many-place fourth power	Pinches and Sachs 1955, no. 1644; Friberg 1983

Table B.21 (*Continued*)

Tablet Number	British Museum Collection Number	Description	Publication
BM 34612	Sp II, 91	Fragment of multi-place reciprocal table	Pinches and Sachs 1955, no. 1631
BM 34635	Sp II, 118	Fragment of multi-place reciprocal table	Pinches and Sachs 1955, no. 1634
BM 34714	Sp II, 203	Fragment of list of multi-place squares	Pinches and Sachs 1955, no. 1639; Aaboe 1965, text VII
BM 34724	Sp II, 214	Fragmentary collection of word problems (5+ extant)	Pinches and Sachs 1955, no. 1648
BM 34757	Sp II, 248	Word problem within collection of astronomical procedures for computation of the motion of Jupiter	Neugebauer 1955, no. 817
BM 34762	Sp II, 255	Fragment of multi-place reciprocal table	Pinches and Sachs 1955, no. 1632
BM 34764	Sp II, 257	Fragment of list of multi-place squares	Pinches and Sachs 1955, no. 1640; Friberg 1983
BM 34800	Sp II, 293	Fragmentary collection of word problems (2+ extant)	Pinches and Sachs 1955, no. 1647
BM 34875	Sp II, 382	Fragment of list of multi-place squares	Pinches and Sachs 1955, no. 1638; Aaboe 1965, text VI
BM 34901	Sp II, 413	Fragmentary calculation of many-place reciprocal pair	Pinches and Sachs 1955, no. 1645
BM 34907	Sp II, 421	Fragmentary calculation of many-place reciprocal pair	Pinches and Sachs 1955, no. 1643

TABLE B.21 (*Continued*)

Tablet Number	British Museum Collection Number	Description	Publication
BM 34958	Sp II, 479	Fragmentary calculation of many-place reciprocal pair	Pinches and Sachs 1955, no. 1642; Friberg 1983
BM 36654	1880-6-17, 386	Five-column table of arithmetical progressions	Neugebauer and Sachs 1967, text G
BM 36776	1880-6-17, 513	Fragmentary list of constants	Robson 1999, list K
BM 36849+	1880-6-17, 589	Multi-column tablet: OB-style table of squares; multiplication tables for 55, 54, 37 30, 35, 32, 28 48, and 18 45	Aaboe 1998
BM 37096	1880-6-17, 854	Fragmentary list of constants	Robson 1999, list L
BM 40107	—	Table of inverse squares	Neugebauer 1935–7, I 68 no. 4
BM 40813	1881-4-28, 359	A copy of the 'Esağila tablet', a set of mathematical problems about Marduk's temple Esağila	George 1992, no. 13
BM 41101	1881-4-28, 648	Fragment of a multi-place reciprocal table	Aaboe 1965, text I
BM 45668	1881-7-6, 63	Fragment of list of multi-place squares	Pinches and Sachs 1955, no. 1636; Aaboe 1965, text V
BM 55557	1882-7-4, 147 + 204	Fragment of a list or table of fourth powers	Britton 1991–3
BM 76984 (+) BM 77051	1883-1-18, 2356 + 2648	Fragments of a multi-place reciprocal table	Britton 1991–3, texts A (+) B

Table B.21 (*Continued*)

Tablet Number	British Museum Collection Number	Description	Publication
BM 78079	1886-5-12, 4	Fragment of multi-place reciprocal table	Britton 1991–3, text C
BM 78084	1886-5-21	Three short word problems on reciprocal pairs on small rectangular tablet	Nemet-Nejat 2001a
BM 99633	1884-2-1, 1995	Fragment of list of squares	Britton 1991–3, text D
MMA 86.11.404	—	Fragmentary collection of word problems (2+ extant)	Neugebauer and Sachs 1945, text X; Friberg 2005a, no. 77
MMA 86.11.406 (+) 410 (+) Liv 29.11.77.34	—	Fragment of multi-place reciprocal table	Neugebauer and Sachs 1945, 15; Friberg 2005a, nos. 75–6
MMA 86.11.407 (+) 408 (+) 409	—	Fragment of multi-place reciprocal table	Neugebauer and Sachs 1945, 15; Friberg 1983; 2005a, nos. 72–4

Table B.22
Overview of Published Cuneiform Mathematical Tablets, by Period and Genre

Period (Table)	Metrological Lists and Tables	Arithmetical Lists and Tables	Calculations and Diagrams	Word Problems	Reference Lists	Model Documents	And Non-Mathematical[a]	Total[b]
Uruk IV–III (B.1–2)	3	—	—	9	—	2	—	14
Early Dynastic IIIa–b (B.3–4)	3	—	3	4	—	—	—	10
Sargonic (B.5–6)	1	—	1	19	—	1	—	22
Ur III (B.7)	—	5	—	—	—	3	—	8
Old Babylonian (B.8–11)	76	364	173	127	13	5	99	712
OB Susa (B.17)	1	10	5	20	5	—	—	40
Old Assyrian and Mari (B.12–13)	4	7	11	—	—	—	1	22

TABLE B.22 (Continued)

Period (Table)	Metrological Lists and Tables	Arithmetical Lists and Tables	Calculations and Diagrams	Word Problems	Reference Lists	Model Documents	And Non-Mathematical[a]	Total[b]
Middle Babylonian, Middle Assyrian, and peripheral (B.14–17)	9	4	—	2	3	—	—	18
Neo-Assyrian (B.14)	2	2	—	2	—	—	1	6
Neo-Babylonian and Achaemenid (B.18–20)	39	7	5	5	1	—	35	58
Seleucid (B.20–21)	2	31	2	11	2	—	2	47
Total	140	430	200	199	24	11	138	957

Note: This table includes all mathematical cuneiform tablets published before 2007, including many not listed in tables B.1–21 because they are catalogued in standard works such as Neugebauer 1935–7; Neugebauer and Sachs 1945; Bruins and Rutten 1961; Robson 1999 (see individual tables for details). However, the entire corpus, with transliterations, English translations, and catalogue, is available online at <http://cdl.museum.upenn.edu/dccmt>.
[a] That is, tablets that contain both mathematical and non-mathematical content, usually other school exercises.
[b] Totals are often less than the sum of the individual entries in each row, as one tablet may contain more than one mathematical genre (e.g., a metrological table and an arithmetical list; a calculation and a word problem).

NOTES

PREFACE

1. Bos and Mehrtens 1977; Bloor 1976.
2. Neugebauer 1952; Høyrup 2002b; Robson 1999.
3. Robson 2007b.

CHAPTER ONE: SCOPE, METHODS, SOURCES

1. Tripp 2002, 8. The mathematics of Islamic Iraq cannot be covered in the book; see Berggren 1986 for a good introduction. For the more general history of modern Iraq see Tripp 2002.

2. An engaging description of 'the land and the life' of ancient Iraq is given by Postgate 1992, 3–21; a more comprehensive survey of its 'material foundations' is Potts 1998.

3. For a detailed history of decipherment and rediscovery, see Lloyd 1980; for the early history see also Kuklick 1996 and Larsen 1996.

4. See Robson 2002a. This overview is based on the more detailed historiography in Høyrup 1996.

5. Neugebauer 1935–7; Thureau-Dangin 1938.

6. Neugebauer and Sachs 1945; Bruins and Rutten 1961.

7. For instance, Baqir 1950a; 1950b; 1951; 1962; Bruins 1953; 1954; 1962; 1971.

8. For instance, Friberg 1986; Powell 1971; 1976a; their work is discussed in more detail in chapter 3.

9. Summarised in Schmandt-Besserat 1992 and discussed in chapter 2.

10. The seminal work is Høyrup 1990a; he has since collected and re-edited this and subsequent related work into a book, Høyrup 2002b.

11. Their work is beautifully summarised in Nissen, Damerow, and Englund 1993.

12. For instance, Friberg 1997; 1999a; Friberg et al. 1990. See chapter 8 for more details.

13. Photograph in F. R. Steele 1951, pl. 7; transliteration and translation by Neugebauer and Sachs 1984, 246.

14. As §4.4 demonstrates, some scribes were female; but the majority were men and boys.

15. The cuneiform writing system is described more fully by Walker 1990, 17–24.

16. For more on the Sumerian language, see Black 2007 or Michalowski 2003.

17. For more on the Akkadian language, see George 2007 or Huehnergard and Woods 2003.

18. For contrasting views, compare Edzard 2000 with Michalowski 2000.

19. See note 10.

20. The list of problems is BM 80209, probably from Sippar, with square problems in lines 1–2 (Friberg 1981a). The illustrated compilation of problems is BM 15285, perhaps from Larsa (see §2.4, figure 2.10).

21. RA 35, 1–2 (see table B.17).

22. Documented by Kuklick 1996.

23. See Stone 1987, 84–5, appendix II, for further details. The original publication is by McCown and Haines 1967.

24. Veldhuis 2000b has made a preliminary study of the school tablets in House B.

25. In particular, D'Ambrosio 1998 and Ascher 1991; 2002. There also several useful surveys and position statements in Selin 2000.

26. See Bernhardsson 2005 on the role of archaeology in the formation of the modern Iraqi state.

CHAPTER TWO: BEFORE THE MID-THIRD MILLENNIUM

1. §§2.2 and 2.3 are based on Robson (forthcoming 2); §2.4 grew out of a paper called 'Shape, space, and symmetry in Old Babylonian mathematics and beyond', given to the 46e Rencontre Assyriologique Internationale held in Paris in July 2000. For an excellent in-depth study of accounting in this period see Nissen, Damerow, and Englund 1993. For general historical background, see Kuhrt 1995, 19–44; Van De Mieroop 2004b, 19–38.

2. Matthews 2000, 13–22.

3. The siglum 'bc' indicates uncalibrated radiocarbon dates.

4. Matthews 2000, 52.

5. Matthews 2000, 51–4.

6. On the so-called Uruk phenomenon see most recently Algaze 2004.

7. Algaze 2004, 141.

8. Robson 1999, 7.

9. See Robson 1999, 13, for discussion of this question in relation to 'model' documents in Old Babylonian scribal schooling.

10. See table A.1 for an overview of the late fourth-millennium metrologies discussed in this chapter.

11. See most conveniently Nissen, Damerow, and Englund 1993, 6.

12. The only substantial assemblage of published cuneiform tablets from the early third millennium (Early Dynastic I–II) comprises around 300 administrative records found below the famous 'royal cemetery' of Ur (Burrows 1935). They have rarely been studied since, though their metrology has recently been the subject of a short study (Chambon 2003).

13. Nissen 1988 and Pollock 1999 give good general overviews of the Early Dynastic period.

14. See Cooper 2004.

15. Visicato 2000, 19–21.

16. Visicato 2000, 21–50.

17. For a convenient summary of the archive's contents, see Pettinato 1991, 45–64. The latest evidence on the date of the destruction of Ebla is given by Archi and Biga 2003.

18. See Gelb 1981; Archi 1987.

19. See the DCCLT database <http://cdli.ucla.ecu/dcclt> under 'Archaic Word List D (Food/Grain)' and 'Early Dynastic Food', lines 1–18; and, in more detail, Civil 1982, 1–8.

20. See most recently Edzard 2005; the most detailed study is still Powell 1971.

21. Both stances have been thoroughly critiqued. The most important reviews of Schmandt-Besserat's work (1992) are by Friberg 1994; Michalowski 1993; Oates 1993; Powell 1994; and Zimansky 1993. Glassner's book (2003) has been reviewed by Dalley 2005; Englund 2005; George 2005b; Postgate 2005; and Robson 2006, amongst others. The best general overview remains Nissen, Damerow, and Englund 1993, with more detailed, technical coverage in Englund 1998.

22. Tokens were presumably in use at this time in northern Iraq too; the lack of evidence may be a function of the restricted opportunities for excavation there over the last few decades since the significance of tokens was first recognised. See Quenet 2005 for a concise survey of the origins of writing in northern Iraq. For an excellent overview of counters and proto-literate accounting in Elam (southwest Iran) see Englund 2004b.

23. Akkermans and Verhoeven 1995; Akkermans 1996.

24. Akkermans and Verhoeven 1995, 25.

25. Friberg 1994, 484.

26. Strommenger 1980.

27. Schmandt-Besserat 1992, 88–91.

28. Strommenger 1980, 64 fig. 58; Schmandt-Besserat 1992, 116, n13.

29. Glassner 2003, 117, 180–3, points out that the numerical tablets and sealed clay packets of accounting tokens were found exclusively in private houses in Habuba (as also in contemporaneous Susa and elsewhere). It may well be that they served as the non-literate counterparts to written legal contracts between institutions and private entrepreneurs. If so, then writing itself—the systematically symbolic visual representation of assets and transactions—belonged firmly in the institutional domain but semi-literate representations of the numbers involved were accessible to and interpretable by individual non-literate parties to legally binding contractual relationships between, for instance, temples and merchants, which are so clearly documented for the later third millennium (e.g., Neumann 1999).

30. Englund 1998, 51.

31. See Pollock 1999, 173–94.

32. Englund 1998, 41.

33. See Englund 1998, 51–5, for more detail.

34. There is no room, on this view, for Schmandt-Besserat's 'complex' or 'decorated' tokens (which, with rare exceptions, are not archaeologically attested in administrative contexts) as the precursors of proto-cuneiform word signs. See Englund 1998, 55.

35. Englund 1998, 61; Nissen, Damerow, and Englund 1993, 25–9.

36. Englund 1998, 215.

37. Englund 1998, 61–64.

38. If this seems unnecessarily complicated it is useful to remember that there are 20 fluid ounces in an Imperial pint and 8 pints in a gallon; almost all pre-metric weights and measures use a variety of number bases, usually based on socially accepted standards of storage containers and the like.

39. These two tablets are discussed in more detail by Nissen, Damerow, and Englund 1993, 43–6, and Englund 1998, 192.

40. On the *umbisağ* in archaic Uruk, see Visicato 2000, 3–4, with further bibliography.

41. Englund 2004a, 28.

42. Powell 1995 imaginatively reconstructs the ways in which clay accounting tokens could have been manipulated as calculating aids.

43. This text is discussed in more detail by Englund 1998, 192–3. The other metrological list, IM 23426, has been dealt with most recently by Englund 2001, 13–7; see also the bibliographies in table B.2.

44. Veldhuis 2006, 194.

45. See Veldhuis 2006 for more detail.

46. Melville 2002; Friberg 2005b.

47. The grain metrology of Early Dynastic Ebla is summarised in table A.2.

48. Archi 1989, 2 n1; Milano 1987, 185.

49. Rubio 2003, 205. Thus some scholars have argued that 'Word list C' is not an elementary scribal exercise but an esoteric literary work (e.g., J.G. Westenholz 1998).

50. Civil 1987, 37.

51. Rubio 2003, 201.

52. SF 34, 75, 76, 77 (Deimel 1923, nos. 34, 75–7) and TSŠ 973 (Jestin 1937, no. 973), all from Šuruppag; OIP 99, 47 and 282 (Biggs 1974, nos. 47, 282), both from Abu Salabikh. They would benefit from analysis along the lines of Ascher 1991, 30–6.

53. The key works in ethno-mathematics are Ascher 1991; 2002; A.B. Powell and Frankenstein 1997; D'Ambrosio 1998; Eglash 2000.

54. E.g., Nemet-Nejat 1993; cf. Robson 1995.

55. For more on symmetry in Mesopotamian decorative arts, see Robson 2000b.

56. Washburn and Crowe 1988; 2004.

57. Ziegler 1999, no. 37. Reference courtesy of Jon Taylor.

58. Until recently this tablet was thought to originate in Early Dynastic Šuruppag; see Krebernik 2006 for the correct attribution. Reference courtesy of Niek Veldhuis.

59. Kilmer 1990; see also Robson 1999, 34–56; 2007d.

60. Each side of the tunic (i.e., front and back) consists of 156 such squares, arranged 12 across and 13 down (Boone 1996, 457–65).

61. Powell 1987–90, 508; Krebernik 1998, 305. But note that the still badly understood 'EN' metrological system of the Uruk period may have recorded weight (Nissen, Damerow, and Englund 1993, 28).

62. On Uruk-period fields see Friberg 1997–8, 19–27. There are no field lengths recorded in extant documents from Early Dynastic Šuruppag or Abu Salabikh,

but plenty of areas (Krebernik 1998, 304). For later third-millennium fields see Liverani 1990.

63. See Kilmer 1990, 84–6; Robson 2000b, 95–101, on the relationship between early Mesopotamian mathematics and the decorative arts; Albenda 1998 and Guralnick 2004 examine pattern in Assyrian design. For the continuation of motifs into the first millennium CE, see for instance Moorey 1978, pls. K, L, N, in Sasanian Kiš; also observed by the author under door lintels at Roman-Parthian Hatra (site visit 2001) and Palmyra (site visit 2006).

64. Lave and Wenger 1991.

CHAPTER THREE: THE LATER THIRD MILLENNIUM

1. For general historical background see Kuhrt 1995, 45–73; Van De Mieroop 2004b, 59–79. For more detail see Westenholz 1999; Sallaberger 1999. §3.3 began life as a paper titled 'Mesopotamian human resource management: changing techniques of labour accounting in early city states' at a conference called Les Mathématiques et l'État, Centre International de Rencontres Mathématiques, Luminy, France, in October 2001; another version, under the title 'Men at work: quantified methods of labour control in early Mesopotamia', was presented to the University of Oxford's Classical Archaeology Seminar, Technology, Economy, and Power in the Ancient World, in the same month. §3.2 is derived from 'Shape, space, and symmetry in Old Babylonian mathematics and beyond', a paper presented to the 46e Rencontre Assyriologique Internationale held in Paris in July 2000.

2. Thus this chapter differentiates between Old Akkadian, used to designate the earliest attested phase of the Akkadian language, and Sargonic, used to designate the political period c. 2340–2200 BCE.

3. Westenholz 1999, 19.

4. E.g., Foster and Robson 2004, 7; Friberg 2005b, §4.6.1.

5. None of the names (see table B.5) appear in Visicato's list of scribes from Sargonic Ĝirsu (Visicato 2000, table 13). Ur-Ištaran is named as scribe on one Sargonic tablet of unknown provenance now in the Iraq Museum (Visicato 2000, table 21).

6. Steinkeller 1991, 16.

7. Cuneiform Digital Library Initiative (CDLI): <http://cdli.ucla.edu>; accessed 15 October 2006.

8. Apart from the 3000 unprovenanced tablets, there are also small quantities from several identified locations, including Adab (74 tablets), Mari (60), Mazyad (30), Susa (66), and Tell Wilayah (70).

9. Robson 1999, 169–71 gives a preliminary account of Ur III school mathematics. The tablets CBS 9044 (Chiera 1916–9, III no. 51) and YBC 3892 (Clay 1915, no. 21), listed by Robson 1999, 13 as possibly Ur III model documents, are in fact Old Babylonian. The first is a fragment of a Type II elementary school tablet from Nippur (not Umma) with a list of personal names on the obverse and a model account on the reverse (see §4.2); the latter is a rough school diagram, as also suggested there. Also from that list, HSM 7500 (Edzard 1962, 81) and YBC 3895 (Clay 1915, no. 25) should be reclassified as genuine Ur III field plans (Liverani 1990,

nos. 29, 11); see §3.2. Similarly VAT 7031 (Schneider 1930, no. 504) is probably a genuine building plan (Dolce 2000). Neugebauer 1935–7, III 50, also assigned to the Ur III period the three times multiplication table MDP 27, 61, from Susa (van der Meer 1935, no. 61). It is, however, probably early Old Babylonian (see §6.1).

10. See, for instance, Høyrup 2002a; 2002b, 314–5.

11. Høyrup 2002a, 141.

12. Powell 1999 makes this point particularly clearly.

13. Veldhuis 1997, 17.

14. Zettler 1996, 93; earlier publication by Falkowitz 1983–4, 41–2.

15. Rubio 2000, 203–4.

16. The sole exception is Ist Ni 4254, bearing *Šulgi hymn A* (ETCSL 2.4.2.01) <http://etcsl.orinst.ox.ac.uk>, which has nothing to say about mathematics. See most recently Rubio 2000, 215–7.

17. In particular, *Šulgi hymn B* (ETCSL 2.4.2.02), lines 16–7, and *Šulgi hymn C* (ETCSL 2.4.2.03), lines 46–7, quoted, for instance, by Robson 1999, 170, and Høyrup 2002b, 375. Jacob Klein's study of 'archaic' (i.e., Ur III) orthography in the hymns to Šulgi categorises Šulgi Hymn B as 'modernised' to OB orthography and grammar and Šulgi C as containing 'mixed' orthographies, both modernising and archaic (Klein 2000, 135 n2). Thus they are hardly reliable witnesses of Ur III praxis.

18. Høyrup 2002b, 375–6.

19. Høyrup 2002b, 377; Høyrup's emphasis.

20. Two of the Sumerian strategies for expressing a condition are:

(a) changing the modality of the verb, as discussed in detail by Miguel Civil (2000). For instance, the verbal infix $ḫe_2$- may mark conditionality in Sumerian: *a-ra$_2$ ḫe$_2$-bi$_2$-šid zag-bi-še$_3$ nu-e-zu* 'Should the multiplication table have to be recited, you do not know it completely' [*Dialogue 1* (ETCSL 5.4.1), line 57; Civil (2000, 32 no. 10)].

(b) using the subordinating conjunction *tukum-bi* 'if' with an indicative verb: *tukum-bi ğe$_{26}$-e-gin$_7$-nam lugal si-sa$_2$-am$_3$* 'If he is a just king, like me' [*Šulgi hymn B* (ETCSL 2.4.2.01), line 285]. But as *tukum-bi* is written with five signs (ŠU.NIĞ$_2$. TUR.LAL.BI) it is hardly a useful logogram for *aššum*, which needs just two signs *aš-šum*.

To express a cause, Sumerian typically uses a case-marked noun phrase containing a complement clause. For example, *ğeštug$_2$-ğeštug$_2$-ga dirig-ga-ğu$_{10}$-še$_3$ um-mi-a-gin$_7$ mu libir-ra-ğu$_{10}$-še$_3$* 'Because of my extraordinary wisdom and my ancient name as a master' [*Šulgi hymn B* (ETCSL 2.4.2.02), lines 305–6]. The subordinating conjunctions *mu* and *bar* more occasionally precede such clauses (Thomsen 1984, 248–9).

21. Høyrup 2002b, 377–8 n464.

22. Winter 1999; Collon 2000.

23. A good general overview is Millard 1987; more detailed studies include Heinrich and Seidl 1967; Röllig 1980–83; Liverani 1990; Robson 1996; 1999, 148–52; Dolce 2000.

24. SMN 4172 (Meek 1935, no. 1).

25. The most recent published study is Liverani 1990, 148–54, which also surveys the 'implicit plans' in 70 or so purely textual administrative documents drawn up in the Ur III province of Lagaš between Šulgi year 27 and Ibbi-Suen year 3 (c. 2068–2026) in order to predict the yield from state-owned fields shortly before harvest (Liverani 1990, 155–73).

26. MIO 1107 (Thureau-Dangin 1897; Dunham 1986) and Wengler 36 (Deimel 1922, no. 26; Nissen, Damerow, and Englund 1993, figs. 58–9). Both have been restudied by Quillien 2003.

27. Itinerary descriptions of long journeys also appear from the Ur III period; they may quantify travelled time (in days and months) or distance (Edzard 1976–80). In fact the 'league' was also a measure of time, being a sixth of a seasonal day or night (and thus variable over the year).

28. Ist Ni 2464, lines i 17–ii 23 (Kraus 1955; Postgate 1992, 151; Frayne 1997, 51).

29. It may be that the problem was a more complex one—perhaps to find the length of the middle line such that it bisects the area into two equal halves of ½ sar each (Friberg 1987–90, 541). But that would involve two unknowns—the length of the line itself and its distance from either end of the figure—and thus it would be solvable only with a cut-and-paste method equivalent to simultaneous equations (see §§4.1, 4.3). The likelihood is remote, given the simple level of contemporary mathematical problems (see §3.1). Another possibility is that the middle line was meant to be 13 cubits long, thus satisfying the property—known from Old Babylonian mathematics—that $13^2 = (17^2 + 7^2)/2$ (Damerow 2001, 264). Attractive as this interpretation is, it remains speculative without the relevant quantitative data on the tablet.

30. Explained first in Robson 1999, 47–60; see also Robson 2001a; 2002c.

31. Haddad 104 problem (3), lines 26–8 (Al-Rawi and Roaf 1984, 188–95, 214).

32. Three was a standard approximation of the ratio between diameter and circumference in Old Babylonian mathematics: although it was known to be inaccurate it was easy to calculate with. In certain circumstances the much more accurate 3;07 30, or 3⅛, was used. The 'constant of a circle', 0;05, was listed in most known Mesopotamian coefficient lists (Robson 1999, 34–8).

33. Visual representations of three-dimensional objects, like the log, are rare in Old Babylonian mathematics: perhaps a dozen have been published. There are three on BM 85196 (table 4.4), out of seventeen problems about three-dimensional objects. One gives a bird's-eye view of a triangular pier in a river, while the second shows a composite view of the top and two ends of a ramp. The last is not a visual representation at all, but a logical arrangement of the variables in the problem: the depth and width of the cross-section of a canal, and the amounts by which they are to be enlarged.

34. Some ideas relating to angles were in use, however. Gradients measured the external slope of walls and ramps in formulations like, 'for every 1 cubit depth the slope (of the canal) is ½ cubit' (YBC 4666, rev. 25: Neugebauer and Sachs 1945, text K). There was also a rough distinction made between right angles and what we might call 'wrong angles', with probably a 10°–15° leeway (Robson 1997).

35. Measuring: extract from DP 641 (Lugalanda year 2: Allotte de la Fuÿe 1912, no. 641); assigning work: extract from VAT 4713 (Urukagina year 1: Förtsch 1916,

no. 130; Maeda 1984, 34–5); surveying completed work: extract from VAT 4851 (Urukagina year 1: Förtsch 1916, no. 187; Nissen, Damerow, and Englund 1993, 82–3).

36. Westenholz 1999, 41, n129.

37. Nik 2 64 (Powell 1976b; Foster 1982d, 26–7; Monaco 1985, 311–4).

38. Foster 1982a, 115–6; 1982b, 22–4; 1983.

39. Foster 1982b, 23.

40. Steinkeller (1991; 2003; 2004) and Van De Mieroop (2004a) usefully discuss the principles and practicalities of Ur III accounting; Englund 1991; 2003 and Van De Mieroop 1999–2000 present detailed studies of individual balanced accounts.

41. Genouillac 1921, no. 5676.

42. Englund 1991, 280. But in the Ur III period it was common to pay off both private and public debt through the debtor's labour, or that of his dependents or guarantor (Steinkeller 2001). And Heimpel 1997 shows that 'the state often allowed years for repayment of such debt'.

43. Neumann 1999; 2000; Garfinkle 2003.

44. AO 5499 (Gelb 1967; Nissen, Englund, and Damerow 1993, 97–102).

45. Gomi 1980.

46. This analysis is based on Robson 1999, 105–6.

47. Robson 1999, 160–1.

48. Robson 1999 examines in great detail quantity-surveying constants in Ur III administration and Old Babylonian mathematics, and the relationships between them; for a rather different interpretation, see Friberg 2001.

49. According to the CDLI database <http://cdli.ucla.edu>, searched on 1 November 2006, the shekel ($gi\breve{g}_4$) is attested 22 times at ED IIIa Šuruppag, always in respect of silver. The mina (*ma-na*) occurs 69 times at ED IIIa Šuruppag and Adab, used of copper, wool, and textiles as well as silver. There are 327 attestations of the mina and 280 of the shekel in ED IIIb accounts. The grain (*še*) is not attested as a metrological unit in Early Dynastic records.

50. Powell 1987–90, 508. For more details of early Mesopotamian weight metrology, with an important methodological discussion, see Powell 1979. For archaeologically attested weights, see especially Ascalone and Peyronel 2001.

51. CDLI, on 1 November 2006, listed seven such instances, in four land accounts from ED IIIb Ǧirsu.

52. An ED III writing of *ma-na* or $gi\breve{g}_4$ with sa_{10} 'to buy' is discussed most recently by Friberg 1999b, 133; 2005b, §2.2.8. The 'small mina' (*ma-na tur*) is used four times in Sargonic metal accounts and the 'small shekel' ($gi\breve{g}_4$ *tur*) three times, according to the CDLI on 1 November 2006. These rare phrases do not survive the Sargonic period.

53. Powell 1987–90, 508. For the details of the metrology and mathematics of bricks, see Robson 1999, 57–73; and rather differently Friberg 2001.

54. For instance, there are over 500 instances of $ku\breve{s}_3$ 'cubit', 44 of $\breve{s}u$-du_3-*a* 'double-hand', and 9 of *šu-si* 'finger' in ED IIIb documents, and none from earlier periods in the CDLI database (accessed 1 November 2006).

55. The only exception was 1–9 talents, written with horizontal wedges rather than vertical ones. The sign $gun_{(2)}$ 'talent' was always explicitly given though.

56. Archaeological remains at places such as the Abu Salabikh ash tip, a rubbish dump that probably served an ED III temple or administrative complex, strongly suggest that accounting tokens were used well into the third millennium: some 70 accounting tokens were found there, as well as nearly 100 crudely filed pottery discs some 2–4 cm in diameter, which may also have served as counters (Green 1993, I 125–34, II figs. 5:1–6:3). Lieberman 1980 argues that third-millennium impressed numerals are representations of accounting tokens, and thus counted numbers, whereas incised, cuneiform numerals represent non-computed numbers such as year counts. Powell 1995 proposes various concrete ways in which counters could have been used in arithmetical operations. Both Proust 2000 and Høyrup 2002a have argued, from numerical errors in Old Babylonian mathematical problems, that a place-marked abacus board with five columns was in use by the early second millennium if not earlier. This topic is in desperate need of careful and extended study, drawing on both archaeological and textual evidence, as well as Reviel Netz's stimulating study of calculi and early Greek numeracy (Netz 2002a).

57. On 1 November 2006 CDLI listed 571 instances of ½, five occurrences of ⅓, and three attestations of ⅔ in ED IIIa documents; and just two attestations of ⅚ in Sargonic Ĝirsu. All notations are ubiquitous in Ur III accounts.

58. According to CDLI on 1 November 2006, la_2 'minus' is attested 51 times in ED IIIa.

59. Attested in 17 documents according to CDLI on 1 November 2006. The examples are from DP 519, rev. iii 1; DP 429, obv. ii 1 (Allotte de la Fuÿe 1912, nos. 519, 429).

60. Suggested already in Sargonic mathematical exercises and administrative records, where it was not compulsory to write the unit 'rod' (see §3.1).

61. Compare Imperial height measures in feet versus length measures in yards.

62. Powell 1976a. The CDLI (1 November 2006) has six attestations of 'writing boards' le-um in the Ur III period: three from Ĝirsu, two from Umma and one from Ur. Ist Um 2870 (Gomi and Yildiz 1997, no. 2870), from Umma, is the earliest of the three that are dated.

63. HSM 6388 (Fish and Lambert 1963, 95 no. 8; Owen 1982, no. 91).

64. Sexagesimal numbers are are also written on a blank space on the obverse of Erlenmayer 152 (Englund 2003), an annual labour account from Umma for the year 2036 BCE. The informal totals at the bottom of the columns are in discrete notation, however.

65. E.g., Ist L 5225 (Genouillac 1912, no. 5225), from Ĝirsu, Šulgi year 44 (2051 BCE); YBC 964 (Sigrist 2000, no. 441), from Umma, Šulgi year 45 (2050 BCE). The model account Walker 47a (§3.3) does not record large areas of hoed land in sexagesimal multiples of the sar, however—another reason to interpret it as a badly executed training exercise.

66. Foster and Robson 2004; see also Friberg 2005b for a different interpretation (which, however, entails misreading the text at three points).

67. Robson 2003–4.

68. Lave and Wenger 1991. On Sargonic scribal education, see Westenholz 1974–7; Foster 1982c. There is no general overview of its Ur III counterpart.

69. De Certeau 1984, 118–22. I have repeated this experiment in three consecutive years with my undergraduate students (almost all academically trained scientists),

asking them to describe the department in which we work. Interestingly, in each trial nearly half of them choose the 'map' approach—as opposed to the 3 percent of participants in Linde and Labov's study.

CHAPTER FOUR: THE EARLY SECOND MILLENNIUM

1. For background history see Kuhrt 1995, 74–117; Van De Mieroop 2004b, 80–8, 104–18; more detailed political history in Charpin 2004 and socio-economic history in Stol 2004. Parts of §§4.1–2 are based on Robson 2007e; I discuss House F in more detail in Robson 2001c; 2002b. Much more on §4.3 can be found in Robson 2001a; 2002a; 2002c. §4.4 is based on Robson 2007a; forthcoming 2.

2. Veldhuis 1997; 2004.

3. For instance, Neugebauer and Sachs 1945, 1.

4. On the non-tabular character of almost all mathematical 'tables', see Robson 2002b, 361; 2003, 31–3.

5. First discussed by Robson 1999, 7–15, where tables of published exemplars are given.

6. Neugebauer and Sachs 1945, text H; my translation.

7. See Høyrup 2002b for a detailed exposition of this archetypically Old Baby-lonian mathematical technique.

8. See Civil 1980; Faivre 1995; Tanret 2002, 143–53.

9. Al-Rawi and Roaf 1984.

10. About forty mathematical tablets are now known from the kingdom of Ešnuna; see most recently Isma'el forthcoming.

11. Goetze 1945; Høyrup 2002b, 317–61.

12. Høyrup 2002b, 358.

13. Potentially single-author clusters of (anonymous) word problem tablets are signalled by Høyrup 2002b, 317–61 passim. For clusters of anonymous calcula-tions and diagrams, see for instance Nemet-Nejat 2002, nos. 1–5, 7–8, 9–10. Scribes are frequently named on a variety of school tablets (including Type III met-rological and arithmetical tables) that are probably from Larsa and now in the Ashmolean Museum, the Louvre, and the Yale Babylonian Collection. A major study needs to be undertaken.

14. Goetze 1945, 150; Høyrup 2002b, 329–31, Groups 6A, 6B (n383).

15. The tablets with BM numbers in the range 85194–85210 (see §8.2) were part of a mixed lot of 787 pieces that the British Museum brought from the dealer I. Elias Gejou in Paris and accessioned on 14 April 1899. The lot comprises tablets from the Sargonic to the Achaemenid period; the OB administrative records come variously from Kisurra, Larsa, Sippar, and Uruk but most are unprovenanced (Sig-rist, Zadok, and Walker 2006, ix, 248–76). Others were acquired from the dealers Shamhiry and Shamas in 1902 and from Naaman in 1906. I am very grateful to Jon Taylor of the British Museum for his help with tracing the provenance of these tablets.

16. Høyrup 2002b, 332.

17. In BM 78476 (Ammi-ṣaduqa year 14) he successfully sues for a share of a pay-ment in silver. (Richardson 2002, Text 7.98). In BM 79761 (reign of Ammi-ṣaduqa),

he receives barley from the temple of Šamaš (Richardson 2002, I 177–8, II 208–10). In BM 81310 (Ammi-ṣaduqa year 17), he lends silver for the purchase of barley (Richardson 2002, Text 7.102). I am very grateful to Seth Richardson for his indefatigable help in the hunt for Iškur-mansum.

18. Friberg 2000; Robson 2001c; 2002b; 2007a; Tanret 2002.

19. Civil 1979, 5–7. For further elaborations of the Type IV typology see also Gordon 1959, 7–8.

20. On the association between typology and pedagogical function, see Robson 2002b, 339–45.

21. Veldhuis 1997. For the details of the elementary curriculum in House F, see Robson 2001c, 45–50.

22. OB Nippur Ura 01, lines 279–87, on Niek Veldhuis's *Digital corpus of cuneiform lexical texts* <http://cuneiform.ucla.edu/dcclt/>.

23. The preceding paragraphs are a précis of Robson 2002b, 330–46.

24. Lion 2001.

25. Tanret 2002; 2004.

26. The multiplication table shown in figure 4.2 was written by a different Bēlānum, who probably lived in Larsa around 1800 BCE.

27. Cohen 2005.

28. ETCSL 3.3.18, lines 1–17 (translation after Black et al. 2004, 281–2).

29. See Robson 2002b, 352–9, for a comparison of calculations in House F and the schools of Ur.

30. See Tanret 2003.

31. Høyrup 2002b, 385–6, gives a useful summary (although he is mistaken about the provenance of Plimpton 322; see below). For more detail see Høyrup 1999; Damerow 2001.

32. Høyrup 2002b, 268–72, 297–8.

33. See Robson 1999 for an in-depth study.

34. Neugebauer and Sachs 1945, text Aa. See Fowler and Robson 1998 for more details.

35. Neugebauer and Sachs 1945, text Ue line 10; Robson 1999, list E. Both YBC 7289 and YBC 7243 are unprovenanced but likely to be from Larsa, like many of the mathematical tablets in the Yale Babylonian Collection.

36. Some examples are given in Fowler and Robson 1998.

37. See for instance Robson 1999, 173–4, on variant constants and problems about the volume of a grain pile, known from Iškur-mansum's compilations and from a tablet from Susa.

38. Discussed by Robson 1999, 9–10; see also tables B.11 and B.17.

39. Robson 1999, 252–9; Friberg 2000, 122–35.

40. Neugebauer and Sachs 1945, text A. See Robson 2001a for some of the voluminous subsequent bibliography.

41. On Banks and Plimpton see Robson 2002a. Plimpton bequeathed his whole collection of historical mathematical books and artefacts to Columbia University in the mid-1930s along with a large number of personal effects.

42. Neugebauer and Sachs 1945, text Ua; Høyrup 2002b, 55–8.

43. Bruins 1949; 1955, confirmed by three independent publications in the early 1980s: Buck 1980; Friberg 1981a; Schmidt 1980. There have been three other

major interpretations of the tablet's function since it was first published. First, it has popularly been seen as a form of trigonometric table: if columns II and III contain the short sides and diagonals of right-angled triangles, then the values in the first column are *tan²* or $1/cos^2$—and the table is arranged so that the acute angles of the triangles decrease by approximately 1° from line to line. But as there is strong evidence against a concept of measured angle in early Mesopotamia (§3.2), this theory can instantly be dismissed. It is purely accidental that the reciprocal pairs chosen by the author of Plimpton 322 lead to a series of triangles which so closely approximate modern angles. Similarly, any notion that Plimpton 322 might be connected with observational astronomy can also be summarily disposed of. Although some simple records of the movements of the moon and Venus *may* have been made for divination in the early second millennium, the accurate and detailed programme of astronomical observations for which Mesopotamia is rightly famous began a thousand years later, in the eighth century BCE (§8.1). Finally, Otto Neugebauer (1952, 36–40), and Asger Aaboe following him (1964, 30–1), argued that the table was generated like this:

If *p* and *q* take on all whole values subject only to the conditions

(1) $p > q > 0$,
(2) *p* and *q* have no common divisor (save 1),
(3) *p* and *q* are not both odd,

then the expressions

$$x = p^2 - q^2$$
$$y = 2pq$$
$$z = p^2 + q^2$$

will produce all reduced Pythagorean number triples, and each triple only once.

The quest has then been to find how *p* and *q* were chosen. But if the missing columns at the left of the tablet had listed *p* and *q*, they would not have been in descending numerical order and would thus have violated the standard principles of Mesopotamian tabulation (§6.3). Nor, under this theory, has anyone satisfactorily explained the presence of column I in the table. There are other good reasons to reject it; I deal with them in Robson 2001a.

44. For more details of this argument, and a discussion of the errors in the table, see Robson 2001a; 2002c.

45. ETCSL 5.1.3, lines 1–8: <http://etcsl.orinst.ox.ac.uk>.

46. ETCSL 5.1.3, lines 36–41.

47. The Sumerian literary contents of House F are described by Robson 2001c, 50–9.

48. Robson 2007a.

49. Tinney 1999, 162–8.

50. ETCSL 2.5.5.2, lines 18–24.

51. Winter 1994.

52. ETCSL 4.16.1, line 1.

53. ETCSL 4.16.1, lines 29–31. Enki was the great god of wisdom and creation.

54. ETCSL 2.5.5.2, lines 25–7. Enlil, father of the gods, oversaw human affairs too.

55. Robson 2001c, 53–5.

56. Robson 2001c, 56–7.

57. ETCSL 1.1.3, lines 412–7.

58. ETCSL 1.2.2, lines 165–8.

59. ETCSL 1.4.1, lines 154–8.

60. Editions and translations are given by Roth 1995.

61. Laws of Ur-Namma, lines A iii 135–iv 139, C i 11–21 (Roth 1995, 16; my translation).

62. BM 11265 (Finkel 1987).

63. Laws of Hammurabi, line xlvii 59 (Roth 1995, 133).

64. Gap ¶t; §109 (Roth 1995, 97, 101). See also, for instance, Gap ¶¶u–x (Roth 1995, 98).

65. And see Ascalone and Peyronel 2001 for archaeological finds of weights in various Old Babylonian temples of Šamaš, from Ebla to the kingdom of Ešnuna.

66. For the most recent treatments see Spycket 2000; Hallo 2005; Slanski 2007.

67. Collon 2005, 30–1, 39–45.

68. Jacobsen 1987.

69. ETCSL 5.4.1, lines 30–32; Vanstiphout 1997, 589.

70. Ziegler 1999, 91–2, 106.

71. Lion 2001.

72. Lion and Robson 2006.

73. Høyrup 1990d; 2002b, 362–85.

CHAPTER FIVE: ASSYRIA

1. For political background, see Kuhrt 1995, 81–108, 348–65, 473–546; Van De Mieroop 2004b, 89–98, 169–74, 216–52. Many ideas in this chapter were first tested in 'Did mathematics count in ancient Assyria?' a paper presented to the 49[e] Rencontre Assyriologique Internationale in London, July 2003. Parts of §5.3 are explored in more depth in Robson 2001b; 2004b.

2. Throughout this book, the transcription Assur is used for the city, Aššur for the city god. In cuneiform writing they were distinguished by their determinatives (see §1.2), not by their spellings.

3. Dates follow Veenhof 2003, 83.

4. Many of the Mari letters are now conveniently available in English translation (Heimpel 2003). On the close relationship between Mari and Ešnuna scribal culture see most recently Charpin 2004, 140 n628.

5. The most recent translation is by Foster 2005, 298–317.

6. Some Neo-Assyrian tablets have been wrongly identified as mathematical: the 'mathematical exercises' on the reverse of an Assyrian eponym list from Huzirīna (Gurney and Finkelstein 1957, no. 47) are in fact a tabular account of cattle, while the fragments of mathematical 'vocabularies' (Neugebauer 1935–7, I 28–30) are simply parts of the standard school vocabularies or lexical lists *Ana ittišu*

(Tablet IV ii 43–57) (Landsberger 1937, 58) and *Ur₅-ra = ḫubullu* (Tablet I 328–34) (Landsberger 1957, 34).

7. See Charpin 1995.

8. Parrot 1956, 188–9 and pls. XLI–XLII; Aaboe 1964, 13 fig. 1.4; correct interpretation by Margueron 1986, 144.

9. Adapted from the translation by Robins and Shute 1987, 56. The nature of this Mari tablet was first recognised and explained by Proust 2002.

10. Neugebauer 1935–7, I 77–8.

11. For more details see Høyrup 1990c.

12. See for instance Soubeyran 1984, 33–4; Guichard 1997, 317–9.

13. Unfortunately the exact findspots of the Kaneš tablets remain unpublished, while most the tablets from Assur were found in secondary context, in a Middle Assyrian house.

14. Hecker 1993; Michel 1998, 253–6.

15. Veenhof 2003, 96.

16. Ass 13058e, Ass 13058i, Kt a/k 178 (table B.13).

17. See Dercksen 1996, 35–6, for copper-silver exchange rates; Larsen 1977 for gold. Translation after Larsen 1977, 125.

18. Michel 1998, 249.

19. Neugebauer 1935–7, I 287–9; Neugebauer and Sachs 1945, 18.

20. This problem is related to two tabular calculations on Type R tablets: YBC 7353 and YBC 11125 (Neugebauer and Sachs 1945, 17; Nemet-Nejat 2002, nos. 7, 8; see Friberg 1987–90, §5.2h).

21. See, for instance, De Odorico 1995.

22. A.0.77.2, lines 5–13, 21–24, after Grayson 1987, 189. The order of eponyms in Shalmaneser's reign is not known.

23. Esarhaddon building inscription Ass A, lines iii 16–41 (Borger 1956, 1–6); my translation.

24. Eponym list: Millard 1994; king list: Grayson 1980–3, 101–15.

25. Erišum's inscriptions: Grayson 1987, 19–37, A.0.33.1–14; Samsi-Addu's inscriptions: Grayson 1987, 47–63, A.0.39.1, 9, 11.

26. Sargon's Eighth Campaign, lines 346–52, 367–9, 405 (Mayer 1979, 273–80); my translation.

27. Fuchs 1994, 113–5, lines 149–60; my translation. A third account, found in three copies of the so-called 'Display inscription', also on the walls of the Fort Sargon palace, has very little quantification and is not treated here. See Fuchs 1994, 214–5, lines 72–6.

28. Curiously the text and images of Fort Sargon do not fit well together: the *Annals* account of Muṣaṣir, on Slabs 13–14 in Room 2, accompanies an image of the sack of the town of Kinda'u in Media (northwest Iran) the year before, while the image is on Slab 4 in Room 13 in a different part of the palace with an *Annals* account of the military campaigns of Sargon's 11th regnal year (Albenda 1986, pls. 118, 133; Fuchs 1994, 82, 131–6, 325–6). On this problem, see Russell 1999, 111–23.

29. This is the only surviving fragment of the relief (Salvini 1995, 149 fig. 1b); the rest (as many of the other Fort Sargon reliefs) is known only from Flandin's drawings, made *in situ* in the 1840s.

30. K 1987 + K 7384 (Lanfranchi and Parpola 1990, no. 206 lines 1–10).

31. K 833, undated (Fales and Postgate 1995, no. 167).

32. See Robson 2004a, 138.

33. Gurney and Finkelstein 1957, no. 47.

34. See, for instance, Fales 1986; 1995.

35. K 1276 (Fales and Postgate 1992, no. 1).

36. Published by Starr 1990; Hunger 1992; Parpola 1993; now all online at <http://cdl.museum.upenn.edu/saa/>.

37. See Gurney and Finkelstein 1957; Gurney and Hulin 1964 for the contents of the tablets; and Gurney 1997 for the names of the scribes. The archaeological data are given by Lloyd and Gökçe 1953, and the whole is summarised by Pedersén 1998, 178–80.

38. See most recently Cooper 2000.

39. Library M2/N1 (Pedersén 1985–6, I 31–42, II 12–24; 1998, 83–4).

40. Library N4 (Pedersén 1985–6, II 41–76; 1998, 135).

41. Klengel-Brandt 1975. A cuneiform inscription on the cover of one set of ivory boards, found down a well in Aššur-nāṣirpal's palace at Kalḫu, shows that the celestial omen series *Enūma Anu Ellil* had been copied for king Sargon for his palace at Fort Sargon—and presumably removed from there after his death (Oates and Oates 2001, 99).

42. Hunger 1968, no. 308.

43. On Nabû-zuqup-kēna see Lieberman 1987, 204–17; Frahm 1999; 2005; Guinan 2002.

44. *Examenstext A*, line 27 (Sjöberg 1975).

45. K 3258, obv. 1–10 (after Livingstone 1989, 26). I thank Karen Radner for this reference.

46. L⁴, lines i 13–8 (Streck 1916, II 254–6; Frahm 2004, 45).

47. Lieberman 1987; Brown 2000a, 126–30.

48. Koch 2005, 56–65, 447–79. All fifteen Neo-Assyrian exemplars are very fragmentary and difficult to interpret. Nos. 91 and 93 were originally published in Neugebauer and Sachs 1945, texts V and W.

49. Koch 2005, 451 no. 93 lines A1–4 (my translation).

50. Koch 2005, 460 no. 97 obv. 10–11 (my translation). The tablet's colophon states that it was copied and collated from an original from the city of Assur, by Šamaš-zēru-iddina, *bārû* ('sacrificial diviner'), son of Šamaš-šumu-lēšir, *bārû*, in a year whose name is now missing.

51. MUL.APIN Tablet II, lines ii 41–2, iii 13–15 (Hunger and Pingree 1989, 101, 107–8).

52. BM 17175+17285 (Hunger and Pingree 1989, 163–4). The same scheme is tabulated in *Enūma Anu Ellil* Tablet 14 Table C (Al-Rawi and George 1991–2). These schemes are discussed by Brown, Fermor, and Walker 1999–2000.

53. Discussed in detail by Brown 2000a, 113–20.

54. Qurdi-Nergal: Gurney and Hulin 1964, nos. 331–8; Kişir-Aššur: Pedersén 1985–6, II 71 no. 478; Nabû-zuqup-kēna: Lieberman 1987, 206–7.

55. Parpola 1983; Fales and Postgate 1992, nos. 49–56. See also Fincke 2003–4; Frame and George 2005.

56. See Livingstone 1986, 22–3; list of constants discussed by Robson 1999, list J.

57. Babylon inscription A (Borger 1956, 15); my translation.

58. 'Secrets of Extispicy': K 8705 (+) K 9843 (Koch 2005, no. 93 mss B, C); timekeeping: K 2077+ (Pingree and Reiner 1974–7).

59. There is, of course, always the possibility that all mathematics and quantified astronomy were written on long-perished writing boards; but in that case one would expect to find some reference to them on the clay library records.

CHAPTER SIX: THE LATER SECOND MILLENNIUM

1. More detail of the historical background can be found in Kuhrt 1995, 194–204, 300–29, 332–48, 365–80; Van De Mieroop 2004b, 115–40, 154–69, 174–94. I am grateful to Kathryn Slanski for her comments on §6.3.

2. Sassmannshausen 1999, 420.

3. Veldhuis 2000a.

4. For instance Hilprecht 1906, nos. 1–6, 8–19, 26, 31; followed by Neugebauer 1935–7, I 10–3 and *passim*.

5. 'Kassite': Kilmer 1960, 273; Duchesne-Guillemin 1984, 6; 'Neo-Babylonian': Kilmer 1980–3, 575; 1992, 101.

6. For English translations of the diplomatic correspondence, see Moran 1992; for the scholarly tablets, see Izre'el 1997. Both books have introductions summarising the circumstances of the find. This cannot have been the only such diplomatic archive in existence; similar tablets have been found at Hattusas (Beckman 1999) and must have been kept at Babylon and other regional capitals too.

7. EA 3 (Moran 1992, 7); EA 5 (Moran 1992, 11).

8. Ben-Tor 1997; Horowitz and Wasserman 2000.

9. Horowitz 1997, 195, where the spelling *šanabi-bi-bi* 'two-thirds of it' is also identified on ARM 9, 299, from Mari (table B.12). The same spelling is further found on Ash 1924.590 from OB Kiš (table B.9).

10. Goren 2000, 34–5. The letter is EA 228 (Moran 1992, 289–90).

11. Goren 2000, 41.

12. There is an excellent recent overview of Ugarit by Yon 2006; on scribal education in the city, see van Soldt 1995a; 1999.

13. van Soldt 1995a.

14. Grammar: RS 20.214A; *Ur$_5$-ra*: RS 20.245 (Tablet 2), RS 21.08A (Tablets 14–15), 20.201A+ (Tablets 20–21a), RS 22.343 (Tablet 23) (Van Soldt 1995a, 197–203).

15. See Negahban 1991, 104 plans 5–6 pls. 53–55; Pedersén 1998, 120–3.

16. De Mequenem et al. 1943, 41–62; Tanret 1986, 140. Atta-ḫušu was a contemporary of king Gungunum of Larsa (dated source: 1917 BCE) and Sumu-abum of Babylon (dated source: 1894 BCE) (Potts 1999, 163).

17. TS B V 68: Tablet IV of OB *Ur$_5$-ra* (Tanret 1986, 145–6).

18. Stève, Gasche, and De Meyer 1980, 132; Tanret 1986, 141–8.

19. Bruins and Rutten 1961, 2; Tanret 1986, 141 n3. The latest dateable Old Babylonian references to Elam are from 1658 and 1646 BCE (Potts 1999, 171).

20. Bruins and Rutten 1961, no. 4.

21. Bruins and Rutten 1961.

22. Perhaps during the reign of Iddin-Dagan of Isin (r. 1794–1754), whose daughter married an Elamite ruler in 1973 BCE; or during the reign of Gungunum of Larsa (r. 1932–1906), who may have conquered Elam in 1927 BCE (Potts 1999, 162).

23. CBS 3359 (Clay 1906, no. 136).

24. HS 154 (Bernhardt 1976, no. 23).

25. CBS 10744 (Sassmannshausen 2001, no. 120).

26. The following ideas are explored in greater detail in Robson 2003; 2004a.

27. Ash 1910.759 (Grégoire 1996, pl. 17; Robson 2004a, 118) is less securely attributable to the Ur III period, but it shares many of the same features. Once again, the row labels are on the right, the column labels at the bottom. Both sets of labels appear to be single-sign abbreviations. The entries are in proportion 2 : 1 across the columns and 2 : 3 : 5 : 5 : 15 down the rows. All are conspicuously round and regular, especially the columnar totals. The numeral 3, written on the left edge, may be the grand total, as Neugebauer 1935–7, I 82, suggests.

28. Sigrist 1984b, no. 56.

29. Sigrist 1984b, no. 63; date missing.

30. As suggested by Neugebauer 1935–7, I 82.

31. Published in Sigrist 1984a.

32. YBC 4721 (Grice 1919, no. 103).

33. For examples, see Robson 2004a, 133–5.

34. Summarised in Robson 2004a, 138.

35. Brinkman 1980–3 provides a useful general overview. Important recent studies include Seidl 1989; Slanski 2000; 2003a; 2003b, and Charpin 2002.

36. L. W. King 1912a, no. 4; Slanski 2003a, 75–9.

37. Sometimes traditional area metrology was used, as in the case of king Marduk-apal-iddina I's (r. 1172–1159) grant of 18 *bur* 2 *eše* of land to Marduk-zākir-šumi, descendant of Arad-Ea (BM 90850: L. W. King 1912a, no. 5); see further below.

38. See Powell 1987–90, 481–2.

39. There are some exceptions, however. The fragmentary entitlement stele AS 6018 (Scheil 1905, 39–42; Borger 1970, 13–6), from the reign of Marduk-apal-iddina I, quantifies the borders of land, apparently in the context of a problem that was resolved by the king appointing a new land surveyor:

[......] north;
38 rods, long side facing west, as far as the bridge on the royal road;
38 rods, long side on the bank of the Radānu canal, east;
11½ rods, short side facing south;

When measuring that field, Marduk-bānī, the mayor of Bīt-Pir'i-Amurru, [. . .] gold, 10 shekels in weight [. . .] shekels of extra gold, the gift of silver [. . .]. Marduk-apal-iddina heard him and [. . .] he wrote to Ina-Esağila-zēra-ibni, son of Arad-Ea, and *he* measured the field and [. . .].

40. For early third-millennium land sales, see Gelb, Steinkeller, and Whiting 1991; for the Ur III period Steinkeller 1989, 122–8 (houses and orchards only); for the Old Babylonian period, e.g., Dekiere 1994–7.

41. *Mašāḫu* is first attested in Middle Babylonian dialect (Oppenheim et al. 1956–2006, M I 352–3); *rēš eqli našû*: Sb 26 III 8, reign of Marduk-apal-iddina I

(Scheil 1905, 31–8), BE 1/2 149, I 20, reign of Marduk-ahhe-eriba (r. 1045) (Hinke 1907, 188–99); *palāku*: UM 29-20-1, III 13, reign of Nebuchadnezzar I (Hinke 1907); *taràşu*: Tehran *kudurru* I 16 (Borger 1970, 1–11). The verb *madādu* 'to measure', while often used of land surveying in other periods, is used only of measuring grain in the later second millennium (Oppenheim et al. 1956–2006, M I 7). The family name *Mādidu* 'measurer' is known from the patronyms of two women on Kassite-period administrative documents from Nippur and Dur-Kurigalzu (Brinkman 2006).

42. 2N-T 356 = UM 55-1-62, lines iii 2'–7' (Sassmannshausen 1994, 452).

43. Old title deeds: Charpin 2002, 178–80; entitlement stelae as copies of clay tablets: Slanski 2003a, 117–8.

44. Slanski 2003a, 133.

45. Walker and Collon 1980, 101 no. 48. Slanski 2003a, 305–8, gives a complete list of entitlement stelae found in archaeological context.

46. Another example is the very fragmentary BM 90940 (L. W. King 1912a, no. 13). Reference courtesy of Kathryn Slanksi.

47. Hinke 1907; Hurowitz 1997.

48. Analysed in detail by Hurowitz 1997.

49. On Arad-Ea, see most recently Lambert 2005, XIV–XV.

50. One [Ap]laya, son of Arad-Ea, *bēl pīḫāti*, is amongst the witnesses of a land grant of Aššur-nādin-šumī (r. 699–694) (Brinkman and Dalley 1988, 82 line iii 12).

51. Sb 26, lines ii 2–3 (Scheil 1905, 31–8).

52. Duck weight: Scheil 1905, 48; legal dispute: IM 85512 (Gurney 1974, no. 41; 1983, no. 41).

53. Gadd 1921, pl. 5; Limet 1971, 85 no. 6.10.

54. Brinkman 1993.

55. Dašpi: Limet 1971, no. 2.21; Kiribti-Marduk: Limet 1971, no. 2.26; Qīšti-Marduk: Ward 1910, no. 517. See also Brinkman 1993; Sassmannshausen 2001, 52.

56. Steinkeller 1989, 100–2; over 300 Ur III attestations in the CDLI database <http://cdli.ucla.edu> on 17 November 2006.

57. B 1062 (Matthews 1992, no. 54); see also Sassmannshausen 2001, 52. This is perhaps the individual after whom the province of Bīt-Pir'i-Amurru is named (see BM 90829 and AS 6018 above, figure 6.5; note 39).

58. Bau: *An adab to Bau for Išme-Dagan*, lines 18, 28 (ETCSL 2.5.4.02); Nisaba: *A praise poem of Enlil-bāni*, lines 43, 65 (ETCSL 2.5.8.1); and *Nisaba hymn A*, line 12 (ETCSL 4.1.6.1); *An adab to Ninisina*, lines 13, 24 (ETCSL 4.22.05) <http://etcsl.orinst.ox.ac.uk>.

59. Lines 41–3 (Lambert 1967; Foster 2005, 583–91).

60. For instance, Wilcke 1998–2001 makes no mention of Kassite evidence in his description of Nin-sumun.

61. The definitive scholarly edition of *Gilgameš* is by Andrew George 2003; there are more accessible translations in George 2000; Dalley 2000, 39–135; and Foster 2001.

62. This section was written without the benefit of access to Edzard's 1991 study of quantifications and measurements in *Gilgameš*. There are many other interesting uses of number and measure in Mesopotamian literature. In Sumerian, see for

instance *The Farmer's Instructions* (Civil 1994; ETCSL 5.6.3), which quantifies the tasks of the agricultural year, and *The Herds of Nanna* (Black et al. 2004, 145–7; ETCSL 4.13.06), in which the goddess Nisaba calculates the size of the moon-god's vast herd of starry cows. See <http://etcsl.orinst.ox.ac.uk>.

63. After George 2003, 539.

64. George 2003, 782.

65. See George 2003, 784, for further discussion of the complexities and complications of this passage.

66. And see the commentary by George 2003, 843.

67. Also see George 2003, 797–8.

68. Discussed in detail in George 2003, 817–8. These first two lines are also used to describe Gilgameš's eventual journey home to Uruk (Tablet XI, lines 301–2, 319–20).

69. In Tablet X, line 180 (only 1 ms preserved), I suggest that Gilgameš runs out of punting poles *ina* 5! (ms has: 2) GEŠ$_2$ 'At 5 sixties', instead of George's 'At one hundred and twenty double-furlongs' (2003, 689, 872–3). This interpretation depends only on correcting a single numeral—numbers are notoriously unstable in manuscript transmission—whereas, as George notes, his interpretation assumes a missing metrological unit of unknown size.

70. And see George's (2003, 882) comments on this passage.

71. See Høyrup 2002b, 27–32.

72. And see Høyrup 1993.

73. See, for instance, Beckman 1983 on Hattusas; Civil 1989 on Emar.

74. Sassmannshausen 1999, 416.

CHAPTER SEVEN: THE EARLY FIRST MILLENNIUM

1. For historical background, see Joannès 2004, 112–202. I am grateful to Cornelia Wunsch for casting her expert eye over this chapter, thereby contributing many small improvements.

2. In 686 and 685 BCE: Sachs and Hunger 1988–, V 395, no. 3 obv. ii' 1–6, 1'–4'.

3. In 650 BCE: Sachs and Hunger 1988–, V 395, no. 3 obv. iv' 5–8.

4. Waerzeggers 2003–4.

5. Two tables of squares previously assigned to the first millennium are probably Old Babylonian: CBS 1535 (Collection Khabaza; Neugebauer and Sachs 1945, 34) and VAT 8492 ('Larsa'; Neugebauer 1935–7, I 69, 76).

6. Discussed in more detail in Robson 2007d.

7. BM 78822 (table B.18). The dating follows Jursa 1993–4, 71.

8. The scribe of course had no means of marking final zeros in the sexagesimal place value system (see further §7.2). They are shown here in parentheses for clarity.

9. Wunsch 2000, 26–8.

10. Wunsch 2000, 24–5; BM 30627 (Wunsch 2000, no. 11) is a beautiful plan of eight adjacent fields owned by the Egibi family.

11. Nemet-Nejat 1982, no. 2.

12. Cole 1996; school tablets: Cole 1997, nos. 89, 114–24, 128.

13. IM 77077 (Cole 1997, no. 114). On Vocabulary S^b B see Cavigneaux 1981, 623–4; Gesche 2000, 66–74.

14. Another tablet (Cole 1997, no. 128) contains the only known Babylonian copy of a composition now called 'Advice to a Prince', otherwise attested only in Assyria. In fact, this carefully constructed literary work reads more like a declaration of Babylonian rights under Assyrian occupation and may have been part of the governor's political archive. If it is a school tablet it was written by a student much more advanced than the scribe of the other exercises found with it. On the 'Advice to a Prince' see for instance Hurowitz 1998; Biggs 2004.

15. Cole 1997, no. 124, lines 12–8.

16. Gesche 2000, 44–7.

17. Syllabary and vocabularies: Hallock et al. 1955; the 'Weidner' god list: Cavigneaux 1981, 82–99; Ur$_5$-ra I–III: Landsberger 1957.

18. CBS 11019, CBS 11032; BM 29440 (table B.18).

19. BM 53939 (table B.18).

20. UET 7, 155 (table B.18).

21. BM 57537 (table B.18).

22. Cavigneaux 1980; 1981; 1985; Ishaq 1985; George 1986, 12–6.

23. 79.B.1.59+ (table B.18).

24. BM 60185+ (table B.18).

25. Langdon 1924, 87–9; Moorey 1978, 48–9; Robson 2004c, 46–9.

26. There is a large literature on early Mesopotamian orality and later Mesopotamian library culture. See Robson 2007c for a convenient overview with many references to earlier publications.

27. Robson 2004c, 62.

28. Robson 2004c, no. 28. Separation signs had sporadically been used to disambiguate adjacent sexagesimal places since the Old Babylonian period: for examples see Friberg 1987–90, 536; Høyrup 2002b, 293–4. On zeros in Seleucid mathematical astronomy see Neugebauer 1941; 1955, II 511.

29. Joannès 1989; on Aḫušunu and Lurindu's marriage, see also Joannès 2004, 165–70.

30. Roth 1989–90, 10–2, usefully discusses real estate in Neo-Babylonian dowries.

31. See, for instance, Vargyas 2000.

32. Kaşirtu's quantifications of her measuring vessels are highly unusual; Roth lists no other capacity measures specified amongst the household goods given in Neo-Babylonian dowries (Roth 1989–90, 23–8).

33. Miglus 1999, 206.

34. AO 17648 (Durand 1981, pl. 52; Joannès 1982, no. 24, 81–84). Reference courtesy of Heather Baker, to whom I also owe the explanation that in the context of Neo-Babylonian house-plans the cardinal directions refer to the direction each room or wing faces, not its orientation with respect to the central courtyard.

35. See Miglus 1999, pl. 88.

36. See, for instance, Dandemaev 1985.

37. Nemet-Nejat 1982, no. 24.

38. For a city plan of Babylon see, for instance, George 1992, 24 fig. 4.

39. Nemet-Nejat 1982, no. 3.

CHAPTER EIGHT: THE LATER FIRST MILLENNIUM

1. For background history, see Joannès 2004, 203–60; for more detail of sources, Oelsner 1986; Boiy 2004. A shortened version of §8.3 appears in Robson 2007c; another treatment of the scholars of Seleucid Uruk (§8.4), which focuses on the writings of Šamaš-ēṭir of the Ekur-zākir family, is given in Robson forthcoming 3. I warmly thank John Steele for reading and critiquing an early draft of this chapter.

2. Kuhrt and Sherwin-White 1991.

3. Waerzeggers 2003–4.

4. Jursa 2005, 1–2; Geller 1997.

5. Sachs 1976.

6. See Macginnis 2002; Clancier 2005.

7. W 22656/02 from Uruk, provisionally labelled as a mathematical problem in its primary publication (von Weiher 1983–98, IV no. 178), turns out to be a ragment of a building ritual. A 1555 (Holt 1911, 212), which Boiy 2004, 37, tentatively attributes to Hellenistic Babylon, is in fact a standard OB multiplication table for 50, provenance unknown (Neugebauer and Sachs 1945, 20 no. 65).

8. There is some uncertainty in dating unprovenanced first-millennium mathematical tablets (see §7.1). Especially some of those listed in table B.19 may well be Neo-Babylonian rather than Late Babylonian.

9. See Friberg 2005a for details.

10. See Friberg 1993 for details.

11. Late Babylonian measurements of Esaĝila: George 1995; 2005a; Late Babylonian measurements of Anu's temple Rēš in Uruk: Van Dijk and Mayer 1980, no. 96; Late Babylonian measurements of Šamaš's temple in Sippar: George 1992, nos. 36–7; Neo-Assyrian measurements of Esaĝila and Ezida in Borsippa: George 1992, no. 14; Middle Babylonian areas (in *sar* and *iku*) of temples at Nippur: Bernhardt and Kramer 1975; Old Babylonian temples: Charpin 1982; 1983.

12. AO 6555 obv. 16–9 (table B.20).

13. The rather garbled metrological table at the end of the tablet is little help, as 'whoever drew up this tablet was not familiar with the older metrological system employed in the text alongside the Neo-Babylonian system, and thus [. . .] the table is not as old as the text it tries to explain' (George 1992, 434).

14. Schmid 1995; see also Keetman 2005.

15. Quote taken from Robson 1999, 7.

16. The developments are described very clearly by Britton 1993. About a dozen tabulations of observed or calculated solar, lunar, and planetary phenomena can be dated in the two centuries before the Seleucid Era (Brown 2000a, 262). The earliest-known astronomical observations are from the mid-seventh century: positions of Saturn (Walker 1998) and Mars (Britton 2004); lunar eclipses (Sachs and Hunger 1988–, V no. 1).

17. All 300-odd ephemerides, precepts, and auxiliary tables then known were published in Neugebauer 1955; about 30 more have been discovered since (see for instance Brown 2000a, 263; Hunger and Pingree 1999, 243).

18. The best introductory description of the mathematical tools used by Babylonian astronomy is Britton and Walker 1996, 52–60, where the quotes used here

come from (55, 57). The observational basics are given by Aaboe 2001, 1–23. Dates of development and places of attestation are laid out by Hunger and Pingree 1999, 221–4, who also discuss two 'proto-procedure texts' which give non-standard methods for calculating lunar and solar motion (Hunger and Pingree 1999, 210–2). Excellent general overviews of celestial divination are Koch-Westenholz 1995; Rochberg 2004; the genres of observational astronomy are explained by Hunger 1998.

19. See, for instance, Robson 1999, 129–31; Brown 2000a, 249.

20. The diaries have been edited by Sachs and Hunger 1988– and are most usefully summarised by Hunger 1993, who also published the three Uruk tablets just mentioned (1976a; 1976b, no. 93). The market prices in the diaries are analysed by Slotsky 1997; van der Spek and Mandemakers 2003. The observational and computational rules are on AO 6455 (table 8.3: 3) and the zodiacal-medical calendars have been surveyed by Brack-Bernsen and Steele 2004. See Brown 2000b; Robson 2004b for more general surveys of scholarly measurement and conceptualisation of time in the first millennium BCE.

21. Bab 54099 = VAT 19211: multiplication table × 24; Bab 58813 (tablet not identified; excavation photos survive): multiplication table × 9; Bab 66746 = VAT 19275b: 'mathematical' (Pedersén 2005, 295–6 nos. 70, 79, 87).

22. Hormuzd Rassam could be considered the first Iraqi archaeologist. His story is a sad one. He was a well-educated Christian from Mosul, first employed by Layard to help in the excavation of Nineveh, and who later became a British museum employee. Certain other members of staff treated him very shabbily and the museum certainly exploited him; his tale is told sympathetically in Reade 1993. Details of his discoveries in Babylonia are presented in Reade 1986. The 1880-6-17 collection comprises tablets with BM numbers in the range 36277–38098; the Sp II tablets have BM numbers 34529–35494.

23. Reade 1986, xv.

24. Data on theoretical astronomical tablets taken from Neugebauer 1955; Sachs and Neugebauer 1956; Neugebauer and Sachs 1967; 1969; Aaboe and Sachs 1969; Aaboe et al. 1991; Britton and Walker 1991. Data from the observational tablets come from Sachs and Hunger 1988–. There are also over 350 fragments of various types of school tablets in the 1880-6-17 collection, but they are not immediately relevant to our problem, as none of them contain metrological, arithmetical, or mathematical exercises and only two have colophons, neither of which is legible (Gesche 2000, 671–86).

25. Oelsner 2000, 802–10. See also Britton and Walker 1991, 110–2; Slotsky 1997, 99–101. Rochberg 2000, 371, erroneously identifies Marduk-šāpik-zēri with the astronomer of that name attested in a legal record from 119 BCE; however, that individual's patronym is not Bēl-apla-iddina but Bēl-bullissu.

26. Itti-Marduk-balāṭu and his son Bēl-aḫḫe-uṣur (table 8.1: 13, 14) must be the offspring of a later Iddin-Bēl of the same family, as they can be quite firmly dated to c. 120 BCE, some two centuries later (Oelsner 2000, 805; George 2003, 740 n11). A solitary ration list for the fourteen astronomers of Esagila, dating to the second half of the fourth century, does not give patronyms for any of the nine extant names; thus none can be identified with any known members of the Mušēzib family (Beaulieu 2006).

27. As already Britton and Walker 1991. The tablets concerned are BM 36301 = 1880-6-17, 27 (Neugebauer and Sachs 1967, text C); BM 36580 = 1880-6-17, 590 (Aaboe et al. 1991, text G/S); BM 36651+ = 1880-6-17, 383 + 452 + 776 + 797 (Aaboe et al. 1991, texts E/M and E/L); BM 36712 = 1880-6-17, 445 + 820 + 919 (Sachs and Neugebauer 1956); BM 36731 = 1880-6-17, 464 (Aaboe and Sachs 1969, text A); and BM 37151+ = 1880-6-17, 901 + 1003 (Aaboe and Huber 1977).

28. Pedersén 2005, 275–85, esp. N19, N20. The Mušallim-Bēl who wrote a scribal exercise on the names of trees, found in the Amran area (Pedersén 2005, 284 N21 (2)), is the son of one Ea-iddin, and thus not the Mušallim-Bēl of the Mušēzib family (Maul 1998).

29. BM 35399 obv. 1–4. The procedure continues until the end of line 13. See Neugebauer 1955, 226–35, for a full discussion of this text and its astronomical content.

30. BM 34568 obv. ii 1–5 (table B.21). Most recent edition by Høyrup 2002b, 393, 395–6, on which these paragraphs are based.

31. AO 6484 (tables B.20, 8.4: 26). See most recently Høyrup 2002b, 390–1.

32. See Oelsner 1983; 1993.

33. There are not enough archaeological data on the remaining 80 tablets to assign them to either family. The stratigraphy is rather confused because of over 40 later coffins buried within the walls of the house (see J. Schmidt et al. 1979, pl. 71).

34. von Weiher 1983–98, IV nos. 222A & B; V nos. 231, 283, 284, 286–94, 299, 300–2.

35. Hunger 1976b, nos. 90, 98, 128; von Weiher 1983–98, II no. 38; III no. 97; IV nos. 162, 170; V nos. 308, 310–1. The earlier tablets (625–419 BCE) are edited by Hunger 1976b, no. 86; von Weiher 1983–98, II nos. 55–6; III no. 86; IV no. 221; V nos. 295, 307. I have attributed the tablets found in Level III around the recycling areas outside the house (of which more below) to the Ekur-zākirs rather than to the Šangû-Ninurtas for the following two reasons: four tablets dated to 312–300 BCE and none dated earlier; four tablets owned by Iqīšâ of the Ekur-zākir family and none owned by the Šangû-Ninurtas.

36. A detailed description is given in J. Schmidt et al. 1979, 49–50.

37. W 23281-x, lines iv 38–40 (table B.20; Friberg, Hunger, and al-Rawi 1990, 529–30).

38. For LB bricks c. 31–33.5 cm square see Walker 1981, nos. 115–7; for LB lists of brick constants see Robson 1999, lists K and L.

39. W 23273, lines iv 2–8 (table B.20).

40. W 23281, lines iii 6–11, iv 26–31 (table B.20).

41. W 23021 (table B.20) table G, which matches line 35 of the reciprocal table. The other calculations are for the reciprocals of 1 08 20 27 30 (line 26), [1 09 07 12] (lines 27 and 29), 1 09 26 40 (line 28), 1 10 18 45 (line 30), 1 11 11 29 03 45 (line 32), 1 12 (line 33), and 1 13 14 31 52 30 (line 37) Cf. Friberg 1999a, who restores the second (his third) reciprocal pair differently, on the basis of a much later reciprocal table (see §8.4). W 23016 (table B.20) is an unfinished calculation of line 13 of the table.

42. W 22307/81 (Hunger 1976b, no. 100); W 23009, and W 23293/13 (von Weiher 1983–98, V nos. 266, 268).

43. J. Schmidt et al. 1979, 28–9.

44. 322 BCE: two commentaries on the celestial omen series *Enūma Anu Ellil* (Tablets 20 and 56), copied for Iqīšâ by Anu-aba-uṣur, son of Anu-mukīn-apli, descendant of Kurî; 318 BCE: one tablet of the birth omen series *Šumma Izbu* and one of the terrestrial omen series *Šumma Alu*, the latter copied by Iqīšâ's son Ištar-šuma-ēreš (Hunger 1976b, no. 90; von Weiher 1983–98, II no. 38; III no. 97; IV no. 162).

45. W 22226/0 lines 1–14, 20–1 (Hunger 1976b, no. 128). The omitted lines contain a damaged list of witnesses. The tablet was sealed by the legal parties and witnesses.

46. 'Secrets of Extispicy': W 22656/10a (von Weiher 1983–98, IV no. 158; Koch 2005, no. 95); calendars: W 22074 and W 22619/6+22 (von Weiher 1983–98, III 104–5); astrological scheme: W 22246a (Hunger 1976b, no. 94). Another tablet from 'Secrets of Extispicy', on calculating the duration of a liver omen's validity, was copied for a descendant of Iqīšâ in 229 BCE and found in the same house (W 22839+: von Weiher 1983–98, IV no. 157; Koch 2005, no. 99).

47. Table W 22646; descriptions W 22554/6; calculations W 22925 (von Weiher 1983–98, II no. 43; V no. 269; IV no. 168; Hunger 1991).

48. Tablets W 22801 (+) 22805, unplaced within the stratigraphy of this house, use a very similar scheme to determine the dates of the summer solstice during the reigns of the Babylonian kings Nabopolassar to Neriglissar (late seventh to early sixth centuries BCE) (von Weiher 1983–98, IV no. 169; Hunger 1991).

49. W 22755/3 (von Weiher 1983–98, IV no. 170). W 22342 (Hunger 1976b, no. 98), a fragmentary ephemeris for full moons according to System A, was also found in the house. It is either for SE 40 (271 BCE) or for SE 265 (46 BCE), and thus almost certainly belongs to a later phase of the house's occupation, which had eroded away long before excavation. That makes it either one of the earliest-known System A lunar ephemerides or one of the latest: cf. BM 36611+ (Neugebauer 1955, no. 18a), dated SE 41–44 or SE 266–269 from Babylon. The other extant lunar ephemerides from Uruk all date to the 100s–140s SE (c. 200–160 BCE); Urukean planetary ephemerides date to the 80s–150s SE (c. 225–155 BCE) (Neugebauer 1955, 7–9).

50. But, as references in colophons to perishable objects such as waxed writing boards suggest, Late Babylonian scholarly libraries probably never consisted solely of cuneiform tablets. It is entirely possible that the Ekur-zākir family wrote mathematics on boards, papyri, or parchment and/or that the Šangû-Ninurtas wrote astronomy on similarly ephemeral media.

51. Planetary ephemeris: Ist U 163 (Neugebauer 1955, no. 651 O); planetary precept: Ist U 150 (Neugebauer 1955, no. 803 N).

52. Kümmel 1979, 157; McEwan 1981, 34.

53. Witnessed by Nidinti-Anu: division of inherited house and land in Uruk, SE 92 (A 3680: Weisberg 1991, no. 45), sales of land, SE 104 and undated (NCBT 2036, NCBT 1974: Wallenfels 1994, no. 330). Witnessed by Ina-qibīt-Anu: quit-claim, in which the right to temple income is surrendered, SE 118 (MLC 2131: Clay 1913, no. 32). Witnessed by Anu-uballiṭ: sale of shares in the Temple Enterer's prebend, SE 110–119 (Ash 1923.78: McEwan 1982, no. 57), sale of temple-owned land in Uruk, SE 137 (A 2518: Weisberg 1991, no. 13).

54. Wallenfels 1993.

55. See most recently Beaulieu 2000; Pearce and Doty 2000; Rochberg 1993; 2000.

56. Old Babylonian predecessor for problems (1)–(2): see BM 85196 problem (13) (table 4.4); on the OB length and volume-capacity conversion coefficients used in problems (9)–(13) see Robson 1999, 112–8; for problems (14)–(17) see Høyrup 2002b, 390–1.

57. On Hellenistic *kalûs* and *Kalûtu*, see the excellent study Linssen 2004.

58. Beaulieu and Rochberg 1996, 91. The horoscopes are edited by Rochberg 1998.

59. Or perhaps seven years earlier in SE 112 (tablet 9, where the scribe's name is broken).

60. The two lunar auxiliary tables with damaged colophons, VAT 7852 and A 3419+, also dated SE 124 (Neugebauer 1955, no. 161 V; no. 174 W), should probably be ascribed to Anu-aba-utēr (owner) and Anu-uballiṭ (scribe) too.

61. McEwan 1981, 118–20.

62. Kümmel 1979, 122, 130–1; Beaulieu 2000, 5–6.

63. Doty 1988, 100; Boiy 2005a; 2005b.

64. Doty 1988, 97. Anu-uballiṭ/Kephalon Aḫi'ūtu was the brother of Antiochus's father Ina-qibīt-Anu, and of Anu-bēlšunu, who taught Anu-bēlšunu Sîn-lēqi-unninni in 229 BCE (table 8.4: 3).

65. Downey 1988, 17–28.

66. Downey 1988, 25; Van Dijk and Mayer, 1980.

67. Linssen 2004, 100–8, 283–305.

68. Linssen 2004, 109–15, 306–20.

69. Linssen 2004, 92–9, 252–82.

70. Beaulieu and Britton 1994.

71. If this explanation seems unlikely—the mathematical sophistication of the astronomy is far beyond the basic needs for predicting lunar eclipses, after all—a useful parallel might be Mamluk Islam, where piety became mathematised in a similar way. During the thirteenth and fourteenth centuries CE the mosques of Damascus and Cairo sponsored increasingly complex mathematical methods for timing the five daily calls to prayer and for correctly orienting the direction of prayer towards Mecca (D. A. King 1998).

72. Clancier 2005.

73. For instance HSM 913.1.1 (7309) (Wallenfels 1998, no. II) records the boundaries of a house in Uruk, measured in cubits, sold in SE 72.

74. Cocquerillat 1965. Although this move away from sexagesimal fractions may indicate a Hellenisation of numerate practices, unit fractions had been a feature of cuneiform administrative numeracy since the mid-third millennium (§3.4) and also played a role in Old Babylonian mathematics (Høyrup 1990b).

75. See Høyrup 2002b, 387–99, on the characteristics of Late Babylonian 'algebra'.

76. BM 36859+; Ist U 91+ (tables B.19, B.20).

77. Neugebauer 1955, II 575–6 sub *igi*. There are no entries in that index for $ib_2\text{-}si_{(8)}$ or *miṭḫartum* 'square (side)'.

78. See Robson forthcoming 3 for a comparison of the language of Šamaš-ēṭir's rituals and precepts.

79. For instance, Høyrup 2002b, 399, on BM 34568 (table B.21): 'We have no evidence as to the time or the place where the new methods and problems were created: . . . indirect evidence speaks vaguely in favour of a non-Mesopotamian cradle'.

80. van der Spek 2000, 439; Verbrugghe and Wickersham 1996, 13–91.

81. Geller 1997.

82. Robson forthcoming 3.

CHAPTER NINE: EPILOGUE

1. Parts of this chapter are revisions of parts of Robson 2005b; forthcoming 1. Other parts are based on two conference papers: 'Babylon, Mesopotamia, or Iraq? Locations and *appropriations of an ancient mathematical culture*', presented at the workshop 'Sciences in Asia: Representations and Historiography' at the Needham Research Institute, Cambridge, in January 2005, and 'What counts as mathematics? A re-examination of the cuneiform corpus', given to the 51ᵉ Rencontre Assyriologique Internationale, Chicago, in July 2005.

2. Van De Mieroop 1999, 9; Bohrer 2003, 1.

3. Kline 1972, I 3.

4. Waerden 1954, 5.

5. Neugebauer 1952, 147.

6. For the deep and multi-faceted influence of Mesopotamia and Persia on ancient Greece see for instance Burkert 1992; 2004; West 1997.

7. De Kuyper 1993; Kuhrt 1982.

8. Wiesehöfer 1996, 100–1; Frame 1997.

9. No systematic study has yet been made of the Classical, Late Antique, and European images of the Magi and Chaldaeans as mathematicians.

10. Larsen 1996; Kuklick 1996.

11. For instance Scheil 1902b, 48–54; 1915; 1916; Thureau-Dangin 1912; 1918; 1921.

12. Hilprecht 1906.

13. Hilprecht 1906, 11–3; Robson 2002a, 252 n7.

14. Robson 2002a.

15. Smith 1907; 1923–5.

16. The first publication was Eisenlohr 1877. For a critique of the historiography of ancient Egyptian mathematics, see Imhausen 2003, 5–32.

17. For an overview of the *Elements* in English, see Barrow-Green 2006.

18. Heiberg and Menge 1883–1916; Heath 1926.

19. Neugebauer 1935–7; Thureau-Dangin 1938; Neugebauer and Sachs 1945.

20. Neugebauer 1952; 1934.

21. Aaboe 1964.

22. Neugebauer and Sachs 1945, 42; Neugebauer 1952, 35; Aaboe 1964, 25–7; Fowler and Robson 1998.

23. Baqir 1950a; Neugebauer 1952, 52; Høyrup 2002b, 231–4. Waerden (1954) discussed some fourteen mathematical problems that had been published in Neugebauer 1935–7, but these were not replicated in the popular literature nearly so frequently. However, two of the three cuneiform tablets illustrated in photographic plates are Plimpton 322 and YBC 7289 (Waerden 1954, pls. 8a–b).

24. See Cuomo 2001 for a marvellous redressing of the balance.

25. Drucker 1993; Dold-Samplonius 1997.

26. Swerdlow 1993.

27. See, for instance, Balaguer 1998.

28. Aaboe 1964, 1.

29. The most striking instance is Friberg 1987–90, an impressively detailed account of the internal workings of Mesopotamian mathematics in an Assyriological encyclopedia.

30. Said 1978, 204.

31. Bahrani 1998, 162.

32. See even Høyrup 2002b, 388.

33. Høyrup 1990d.

34. Robson 1999, 118–22, 173–4.

35. Høyrup 2002b, 400–5.

36. Tybjerg 2005, 567.

37. Netz 2004, 3.

38. Høyrup 2002b, 14–5.

39. Waerden 1954, 63.

40. Høyrup 2002b, 169–70.

41. Neugebauer 1935–7, III 20–1.

42. Høyrup 2002b, 396–7. He uses [] to represent geometrical multiplication and □ to stand for geometrical squaring.

43. Heath 1926, I 383.

44. Unguru 1975, 68: 'As to the goal of these so-called "historical" studies, it can easily be stated in one sentence: to show how past mathematicians hid their modern ideas and procedures.'

45. Netz 1999.

46. Quoted in Taisbak 1999.

47. The one index fossil that Høyrup identifies—a shared concept of 'squareness' in Akkadian *mithartum* and Greek *dýnamis* (Høyrup 1990c; 2002b, 402–3) has been challenged by Fowler 1992 for its lack of cultural specificity.

48. Fowler 1999a, 150–1.

49. See already Unguru 1975 for the first cogent attack on the 'crisis of commensurability' and Fowler 1999b for a detailed alternative early history.

50. Fowler 1999a, 151. For Hypsicles see Bulmer-Thomas 1972.

51. Cuomo 2001, 50–61. Diogenes Laertius (*Lives* VIII.3) and Porphyry (*Life of Pythagoras*, §§8, 11–2) tell similar tales.

52. Iamblichus/Clark 1989, 8 §19.

53. Burkert 1962.

54. O'Meara 1989. And for an excellent overview of the historicity of Pythagoras, see Burnyeat 2007.

55. Cuomo 2001, 4–38; Netz 1999, 272–7; Fowler 1999b, 3–29.

56. Cuomo 2001, 79.

57. Fowler 1999b, 7.

58. Netz 1999, 289.

59. Netz 1999, 285; 2002b.

60. Netz 1999, 291–2.

61. For instance, Høyrup 2002b, 405.

62. Kuhrt 1982.

63. van der Spek 1985; 1987; 2001.

64. For concrete examples of Babylonian influence on Greco-Egyptian astronomy, see Jones 1991; 1993; 1996. On Hipparchus see Toomer 1988.

65. Verbrugghe and Wickersham 1996.

66. That is, D'Orville 301, copied in Byzantium in 888 CE and now owned by the Bodleian Library, Oxford.

67. Used in the colophons of collections of word problems: Robson 1999, 176.

68. Nemet-Nejat 1993, 184; on the writings of Sîn-iqīšam see Robson 1999, 219–45.

69. Brown 2000a, 127–9.

70. Robson 2004c, no. 17.

71. Robson 1999, 7.

BIBLIOGRAPHY

Aaboe, A. 1964. *Episodes from the early history of mathematics.* Washington, DC: Mathematical Association of America.

Aaboe, A. 1965. Some Seleucid mathematical tables (extended reciprocals and squares of regular numbers). *Journal of Cuneiform Studies* 19: 79–86.

Aaboe, A. 1998. A new mathematical text from the astronomical archive in Babylon: BM 36849. In *Ancient astronomy and celestial divination*, ed. N. M. Swerdlow, 179–86. Cambridge, MA: MIT Press.

Aaboe, A. 2001. *Episodes from the early history of astronomy.* Berlin: Springer.

Aaboe, A., et al. 1991. Saros cycle dates and related Babylonian astronomical texts. *Transactions of the American Philosophical Society* 81/6: 1–75.

Aaboe, A., and P. Huber. 1977. A text concerning subdivision of the synodic motion of Venus from Babylon: BM 37151. In *Essays on the ancient Near East in memory of Jacob Joel Finkelstein*, ed. M. Ellis, 1–4. New Haven: Connecticut Academy of Sciences.

Aaboe, A., and A. J. Sachs. 1969. Two lunar texts of the Achaemenid period from Babylon. *Centaurus* 14: 1–22.

Akkermans, P.M.M.G. (ed.). 1996. *Tell Sabi Abyad: the Late Neolithic settlement.* Leiden and Istanbul: Nederlands Historisch-Archaeologisch Instituut te Istanbul.

Akkermans, P.M.M.G., and M. Verhoeven. 1995. An image of complexity: the Burnt Village at late Neolithic Sabi Abyad, Syria. *American Journal of Archaeology* 99: 5–32.

Albenda, P. 1986. *The palace of Sargon king of Assyria* (Synthèse 22). Paris: Éditions Recherche sur les Civilisations.

Albenda, P. 1998. *Monumental art of the Assyrian Empire: dynamics of composition styles* (Monographs on the Ancient Near East 3/1). Malibu: Undena.

Al-Fouadi, A. H. 1979. *Lenticular exercise school texts,* vol.1 (Texts in the Iraq Museum 10/1). Baghdad: Republic of Iraq Ministry of Culture and Arts.

Algaze, G. 2004. *The Uruk world system: the dynamics of expansion of early Mesopotamian civilization,* 2nd ed. Chicago: University of Chicago Press.

Al-Jadir, W. 1987. Découverte en Irak d'une bibliothèque babylonienne. *Archeologia* 224: 23–5.

Allotte de la Fuÿe, M. F. 1912. *Documents présargoniques.* Paris: Leroux.

Al-Rawi, F.N.H., and S. M. Dalley. 2000. *Old Babylonian texts from private houses at Abu Habbah, ancient Sippir: Baghdad University excavations* (É-dubba-a 7). London: Nabu.

Al-Rawi, F.N.H., and A. R. George. 1991–2. Enūma Anu Enlil XIV and other early astronomical tables. *Archiv für Orientforschung* 38–9: 52–73.

Al-Rawi, F.N.H., and M. Roaf. 1984. Ten Old Babylonian mathematical problems from Tell Haddad, Himrin. *Sumer* 43: 175–218.

Alster, B. 1997. *Proverbs of ancient Sumer: the world's oldest proverb collections.* Bethesda: CDL.

Archi, A. 1980. Un testo matematico d'età protosiriana. *Studi Eblaiti* 3: 63–4.

Archi, A. 1987. More on Ebla and Kiš. In *Eblaitica: essays on the Ebla archives and the Eblaite language,* vol. 1, ed. C. H. Gordon and G. A. Rendsburg, 125–40. Winona Lake: Eisenbrauns.

Archi, A. 1989. Tables de comptes eblaïtes. *Revue d'Assyriologie* 83: 1–6.

Archi, A., and M. G. Biga. 2003. A victory over Mari and the fall of Ebla. *Journal of Cuneiform Studies* 55: 1–44.

Arnaud, D. 1994. *Texte aus Larsa: Die epigraphischen Funde der I. Kampagne in Senkereh-Larsa 1933* (Berliner Beitrage zum Vorderen Orient. Texte 3). Berlin: Reimer.

Ascalone, E., and L. Peyronel. 2001. Two weights from Temple N at Tell Mardikh-Ebla, Syria: a link between metrology and cultic activities in the second millennium BC? *Journal of Cuneiform Studies* 53: 1–12.

Ascher, M. 1991. *Ethnomathematics: a multicultural view of mathematical ideas.* Pacific Grove: Brooks/Grove.

Ascher, M. 2002. *Mathematics elsewhere: an exploration of ideas across cultures.* Princeton: Princeton University Press.

Bahrani, Z. 1998. Conjuring Mesopotamia: imaginative geography and a world past. In *Archaeology under fire: nationalism, politics and heritage in the eastern Mediterranean and Middle East,* ed. L. Meskell, 159–74. London: Routledge.

Balaguer, M. 1998. *Platonism and anti-Platonism in mathematics.* Oxford: Oxford University Press.

Baqir, T. 1950a. An important mathematical problem text from Tell Harmal (on a Euclidean theorem). *Sumer* 6: 39–54.

Baqir, T. 1950b. Another important mathematical text from Tell Harmal. *Sumer* 6: 130–48.

Baqir, T. 1951. Some more mathematical texts from Tell Harmal. *Sumer* 7: 28–45.

Baqir, T. 1962. Tell Dhiba'i: new mathematical texts. *Sumer* 18: 11–4.

Barrow-Green, J. E. 2006. 'Much necessary for all sortes of men': 450 years of Euclids in English. *BSHM Bulletin* 21: 2–25.

Beaulieu, P.-A. 2000. The descendants of Sîn-lēqi-unninni. In *Assyriologica et Semitica: Festschrift für Joachim Oelsner* (Alter Orient und Altes Testament 252), ed. J. Marzahn and H. Neumann, 1–16. Münster: Ugarit-Verlag.

Beaulieu, P.-A. 2006. The astronomers of the Esagil temple in the fourth century BC. In *If a man builds a joyful house: Assyriological studies in honor of Erle Verdun Leichty* (Cuneiform Monographs 31), ed. A. K. Guinan et al., 5–22. Leiden: Brill.

Beaulieu, P.-A., and J. Britton. 1994. Rituals for an eclipse possibility in the 8th year of Cyrus. *Journal of Cuneiform Studies* 46: 73–86.

Beaulieu, P.-A., and F. Rochberg. 1996. The horoscope of Anu-bēlšunu. *Journal of Cuneiform Studies* 48: 89–94.

Beckman, G. 1983. Mesopotamians and Mesopotamian learning at Hattuša. *Journal of Cuneiform Studies* 35: 97–114.

Beckman, G. 1999. *Hittite diplomatic texts* (Writings from the Ancient World 7). Atlanta: Scholars Press.

Ben-Tor, A. 1997. Hazor. In *The Oxford encyclopedia of archaeology in the Near East*, vol. 3, ed. E. M. Meyers, 1–5. Oxford: Oxford University Press.

Berggren, J. L. 1986. *Episodes in the mathematics of medieval Islam*. Berlin: Springer.

Bernhardsson, M. 2005. *Reclaiming a plundered past: archaeology and nation building in modern Iraq*. Austin: University of Texas Press.

Bernhardt, I. 1976. *Sozialökonomische Text und Rechtsurkunden aus Nippur zur Kassitenzeit* (Texte und Materialen der Frau Professor Hilprecht Sammlung. Neue Folge 5). Berlin: Akademie-Verlag.

Bernhardt, I., and S. N. Kramer. 1975. Die Tempel und Götterschreine von Nippur. *Orientalia* 44: 96–102.

Biggs, R. D. 1974. *Inscriptions from Tell Abu Salabikh* (Oriental Institute Publications 99). Chicago: University of Chicago Press.

Biggs, R. D. 2004. The Babylonian Fürstenspiegel as a political forgery. In *From the upper sea to the lower sea: studies on the history of Assyria and Babylonia in honour of A. K. Grayson*, ed. G. Frame and L. Wilding, 1–7. Leiden: Nederlands Instituut voor het Nabije Oosten.

Birot, M. 1960. *Textes administratifs de la salle 5* (Archives Royales de Mari 9). Paris: Geuthner.

Black, J. A. 2007. Sumerian. In *Languages of Iraq, ancient and modern*, ed. J. N. Postgate, 4–30. London: British School of Archaeology in Iraq.

Black, J. A., G. Cunningham, E. Robson, and G. Zólyomi. 2004. *The literature of ancient Sumer*. Oxford: Oxford University Press.

Bloor, D. 1976. *Knowledge and social imagery*. London: Routledge & Kegan Paul.

Bohrer, F. N. 2003. *Orientalism and visual culture: imagining Mesopotamia in nineteenth-century Europe*. Cambridge: Cambridge University Press.

Boiy, T. 2004. *Late Achaemenid and Hellenistic Babylon* (Orientalia Lovaniensia Analecta 136). Leuven: Peeters.

Boiy, T. 2005a. Akkadian-Greek double names in Hellenistic Babylonia. In *Ethnicity in ancient Mesopotamia: papers read at the 48th Rencontre Assyriologique Internationale, Leiden, 1–4 July 2002*, ed. W. H. van Soldt, 47–60. Leiden: Nederlands Instituut voor het Nabije Oosten.

Boiy, T. 2005b. The fifth and sixth generation of the Nikarchos = Anu-uballiṭ family. *Revue d'Assyriologie* 99: 105–10.

Boone, E. H. 1996. *Andean art at Dumbarton Oaks*. Washington, DC: Dumbarton Oaks.

Borger, R. 1956. *Die Inschriften Asarhaddons Königs von Assyrien* (Archiv für Orientforschung, Beiheft 9). Osnabrück: Biblio-Verlag.

Borger, R. 1970. Vier Grenzsteinurkunden Merodachbaladans I. von Babylon: Der Teheran Kudurru, SB 33, SB 169 und SB 26. *Archiv für Orientforschung* 23: 1–26.

Bos, H.J.M., and H. Mehrtens. 1977. The interactions of mathematics and society in history: some exploratory remarks. *Historia Mathematica* 4: 7–30.

Brack-Bernsen, L., and H. Hunger. 2002. *TU* 11: a collection of rules for the prediction of lunar phases and of month lengths. *SCIAMVS—Sources and Commentaries in Exact Sciences* 3: 3–90.

Brack-Bernsen, L., and O. Schmidt. 1990. Bisectable trapezia in Babylonian mathematics. *Centaurus* 33: 1–38.

Brack-Bernsen, L., and J. M. Steele. 2004. Babylonian mathemagics: two mathematical astronomical-astrological texts. In *Studies in the history of the exact sciences in honour of David Pingree*, ed. C. Burnett et al., 95–125. Leiden: Brill.

Brinkman, J. A. 1980–3. Kudurru. A. Philologisch. In *Reallexikon der Assyriologie*, vol. 6, ed. D. O. Edzard, 268–74. Berlin: de Gruyter.

Brinkman, J. A. 1993. A Kassite seal mentioning a governor of Dilmun. *NABU* 1993: no. 106.

Brinkman, J. A. 2006. The use of occupation names as patronyms in the Kassite period: a forerunner of Neo-Babylonian ancestral names? In *If a man builds a joyful house: Assyriological studies in honor of Erle Verdun Leichty* (Cuneiform Monographs 31), ed. A. K. Guinan et al., 23–43. Leiden: Brill.

Brinkman, J. A., and S. M. Dalley. 1988. A royal *kudurru* from the reign of Aššur-nādin-šumi. *Zeitschrift für Assyriologie* 78: 76–98.

Britton, J. P. 1991–3. A table of 4th powers and related texts from Seleucid Babylon. *Journal of Cuneiform Studies* 43–5: 71–87.

Britton, J. P. 1993. Scientific astronomy in pre-Seleucid Babylonia. In *Die Rolle der Astronomie in den Kulturen mesopotamiens* (Grazer Morgenländische Studien 3), ed. H. D. Galter, 61–76. Graz: rm-Druck- & Vergesellschaft mbH.

Britton, J. P. 2004. An early observation text for Mars: HSM 1899.2.112 (= HSM 1490). In *Studies in the history of the exact sciences in honour of David Pingree*, ed. C. Burnett et al., 33–55. Leiden: Brill.

Britton, J. P., and C.B.F. Walker. 1991. A 4th century Babylonian model for Venus: BM 33552. *Centaurus* 34: 97–118.

Britton, J. P., and C.B.F. Walker. 1996. Astronomy and astrology in Mesopotamia. In *Astronomy before the telescope*, ed. C.B.F. Walker, 42–67. London: British Museum Press.

Brown, D. R. 2000a. *Mesopotamian planetary astronomy-astrology* (Cuneiform Monographs 18). Groningen: Styx.

Brown, D. R. 2000b. The cuneiform conception of celestial space and time. *Cambridge Archaeological Journal* 10: 103–22.

Brown, D. R., J. Fermor, and C.B.F. Walker. 1999–2000. The water-clock in Mesopotamia. *Archiv für Orientforschung* 46–7: 130–48.

Bruins, E. M. 1949. On Plimpton 322. Pythagorean numbers in Babylonian mathematics. *Koninklijke Nederlandse Akademie van Wetenschappen. Proceedings* 52: 629–32.

Bruins, E. M. 1953. Revision of the mathematical texts from Tell Harmal. *Sumer* 9: 241–53.

Bruins, E. M. 1954. Some mathematical texts. *Sumer* 10: 55–61.

Bruins, E. M. 1955. Pythagorean triads in Babylonian mathematics: the errors on Plimpton 322. *Sumer* 11: 117–21.

Bruins, E. M. 1962. Interpretation of cuneiform mathematics. *Physis* 4: 277–341.

Bruins, E. M. 1971. Computation in the Old Babylonian period. *Janus* 58: 222–67.

Bruins, E. M., and M. Rutten. 1961. *Textes mathématiques de Suse* (Mémoires de la Mission Archéologique en Iran 34). Paris: Geuthner.

Buck, R. C. 1980. Sherlock Holmes in Babylon. *American Mathematical Monthly* 87: 335–45.

Bulmer-Thomas, I. 1972. Hypsicles of Alexandria. In *Dictionary of scientific biography*, vol. 6, ed. C. C. Gillespie, 616–7. New York: Scribner.

Burkert, W. 1962. *Weisheit und Wissenschaft: Studien zu Pythagoras, Philoloas und Platon.* Nürnberg: Carl.

Burkert, W. 1992. *The orientalizing revolution: Near Eastern influence on Greek culture in the early archaic age*, trans. M. E. Pinder and W. Burkert. Cambridge, MA: Harvard University Press.

Burkert, W. 2004. *Babylon, Memphis, Persepolis: eastern contexts of Greek culture.* Cambridge, MA: Harvard University Press.

Burnyeat, M. 2007. Other lives. *London Review of Books* 29/4: 3–6.

Burrows, E. R. 1935. *Archaic texts* (Ur Excavations. Texts 2). London: Trustees of the British Museum and of the Museum of the University of Pennsylvania.

Canby, J. V. 2001. *The 'Ur-Nammu' stela* (University Museum Monographs 110). Philadelphia: University of Pennsylvania Museum of Archaeology and Anthropology.

Cavigneaux, A. 1980. Le temple de Nabû *ša harê*—rapport préliminaire sur les textes cunéiformes. *Sumer* 37: 118–26.

Cavigneaux, A. 1980–3. Lexikalische Listen. In *Reallexikon der Assyriologie* vol. 6, ed. D. O. Edzard, 609–41. Berlin: de Gruyter.

Cavigneaux, A. 1981. *Textes scolaires du temple de Nabû ša harê.* Baghdad: Ministry of Culture and Information State Organization of Antiquities and Heritage.

Cavigneaux, A. 1982. Schültexte aus Warka. *Baghdader Mitteilungen* 13: 21–30.

Cavigneaux, A. 1985. Nabû *ša harê* temple and cuneiform texts. *Sumer* 41: 27–9.

Cavigneaux, A. 1996. *Uruk: altbabylonische Texte aus dem Planquadrat Pe XVI-4/5 nach Kopien von Adam Falkenstein* (Ausgrabungen in Uruk-Warka. Endberichte 23). Mainz: Von Zabern.

Cavigneaux, A. 1999. A scholar's library in Meturan? With an edition of the tablet H 72 (Textes de Tell Haddad VII). In *Mesopotamian magic: textual, historical, and interpretative perspectives* (Ancient Magic and Divination 1), ed. T. Abusch and K. van der Toorn, 253–73. Groningen: Styx.

Chambon, G. 2002. Trois documents pédagogiques de Mari. In *Florilegium Marianum VI: Recueil d'études à la mémoire d'André Parrot* (Mémoires de NABU 7), ed. D. Charpin and J.-M. Durand, 497–503. Paris: Société pour l'étude du Proche-orient ancien.

Chambon, G. 2003. Archaic metrological systems from Ur. *Cuneiform Digital Library Journal* <http://cdli.ucla.edu/pubs/cdlj/2003/cdlj2003_005.html>.

Charpin, D. 1982. Le temple de Kahat d'après un document inédit de Mari. *MARI, Annales de Recherches Interdisciplinaires* 1: 137–47.

Charpin, D. 1983. Temples à découvrir en Syrie du Nord d'après des documents inédits de Mari. *Iraq* 45: 56–63.

Charpin, D. 1986. *Le clergé d'Ur au siècle d'Hammurabi (XIXe–XVIIIe siècles av. J.-C.)* (Hautes Études Orientales 22). Geneva: Droz.

Charpin, D. 1993. Données nouvelles sur la poliorcétique à l'époque paléo-babylonienne. 1) Le plan d'une fortification (M.288). *MARI, Annales de Recherches Interdisciplinaires* 7: 193–7.

Charpin, D. 1995. La fin des archives dans le palais de Mari. *Revue d'Assyriologie* 89: 29–40.

Charpin, D. 2002. La commemoration d'actes juridiques: à propos des *kudurrus* babyloniens. *Revue d'Assyriologie* 96: 169–91.

Charpin, D. 2004. Histoire politique du Proche-orient Amorrite (2002–1595). In *Mesopotamien: die altbabylonische Zeit* (Orbis Biblicus et Orientalis 106/4), by D. Charpin, D. O. Edzard, and M. Stol, 25–480. Fribourg and Göttingen: Academic Press and Vandenhoeck & Ruprecht.

Chiera, E. 1916–9. *Lists of personal names from the temple school of Nippur*, 3 vols. (Publications of the Babylonian Section 11). Philadelphia: University of Pennsylvania Museum of Archaeology and Anthropology.

Civil, M. 1979. *Ea A* = nâqu, *Aa A* = nâqu, *with their forerunners and related texts* (Materials for the Sumerian Lexicon 14). Rome: Pontifical Biblical Institute.

Civil, M. 1980. Les limites de l'information textuelle. In *L'archéologie de l'Iraq du début de l'époque néolithique à 333 avant notre ère: perspectives et limites de l'interprétation anthropologique des documents* (Colloques Internationaux du Centre National de la Recherche Scientifique 580), ed. M. Barrelet, 225–32. Paris: Éditions du Centre National de la Recherche Scientifique.

Civil, M. 1982. Studies on Early Dynastic lexicography. *Oriens Antiquus* 21: 1–26.

Civil, M. 1987. Feeding Dumuzi's sheep: the lexicon as a source of literary inspiration. In *Language, literature, and history: philological and historical studies presented to Erica Reiner* (American Oriental Series 67), ed. F. Rochberg-Halton, 37–55. New Haven: American Oriental Society.

Civil, M. 1989. The texts from Meskene-Emar. *Aula Orientalis* 7: 11–20.

Civil, M. 1994. *The Farmer's Instructions: a Sumerian agricultural manual* (Aula Orientalis Supplementa 5). Barcelona: Editorial AUSA.

Civil, M. 2000. Modal prefixes. *Acta Sumerologica* 22: 29–42.

Clancier, P. 2005. Les scribes sur parchemin du temple d'Anu. *Revue d'Assyriologie* 99: 85–104.

Clay, A. T. 1906. *Documents from the temple archives of Nippur dated in the reigns of Cassite rulers (complete dates)* (Babylonian Expedition of the University of Pennsylvania. Series A: Cuneiform Texts 14). Philadelphia: Department of Archaeology, University of Pennsylvania.

Clay, A. T. 1913. *Legal records dated in the Seleucid Epoch* (Babylonian Records in the Library of J. Pierpont Morgan 2). New Haven: Yale University Press.

Clay, A. T. 1915. *Miscellaneous inscriptions in the Yale Babylonian Collection* (Yale Oriental Series, Babylonian Texts 1). New Haven: Yale University Press.

Clay, A. T. 1923. *Epics, hymns, omens, and other texts* (Babylonian Records in the Library of J. Pierpont Morgan 4). New Haven: Yale University Press.

Cocquerillat, D. 1965. Les calculs pratiques sur les fractions à l'époque séleucide. *Bibliotheca Orientalis* 22: 239–42.

Cohen, Y. 2005. Feet of clay at Emar: a happy end? *Orientalia* 74: 165–70.

Cole, S. J. 1996. *Nippur in Late Assyrian times, c. 755–612 BC* (State Archives of Assyria Studies 4). Helsinki: Neo-Assyrian Text Corpus Project.

Cole, S. J. 1997. *The early Neo-Babylonian governor's archive from Nippur* (Oriental Institute Publications 114). Chicago: Oriental Institute of the University of Chicago.

Collon, D. 2000. Early landscapes. In *Landscapes: territories, frontiers and horizons in the ancient Near East,* vol. 3 (History of the Ancient Near East Monographs III/3), ed. L. Milano, S. de Martino, F. M. Fales, and G. B. Lanfranchi, 15–22. Padova: Sargon.

Collon, D. 2005. *The queen of the night.* London: British Museum Press.

Contenau, G. 1929. *Contrats néo-babyloniens,* vol. 2 (Textes Cunéiformes du Louvre 13). Paris: Geuthner.

Cooper, J. S. 2000. Right writing: talking about Sumerian orthography and texts. *Acta Sumerologica* 22: 43–52.

Cooper, J. S. 2004. Babylonian beginnings: the origin of the cuneiform writing system in comparative perspective. In *The first writing: script invention as history and process,* ed. S. Houston, 71–99. Cambridge: Cambridge University Press.

Corò, P. 2005. *Prebende templari in età Seleucide* (History of the Ancient Near East. Monographs 8). Padova: Sargon.

Cuomo, S. 2001. *Ancient mathematics.* London: Routledge.

Curtis, J. E., and J. E. Reade. 1995. *Art and empire: treasures from Assyria in the British Museum.* London: British Museum Press.

Dalley, S. M. 2000. *Myths from Mesopotamia: Creation, the Flood, Gilgamesh, and others,* 2nd ed. Oxford: Oxford World's Classics.

Dalley, S. M. 2005. Review of Glassner 2003. *Technology and Culture* 46: 408–9.

Dalley, S. M., and N. Yoffee. 1991. *Old Babylonian texts in the Ashmolean Museum: texts from Kish and elsewhere* (Oxford Editions of Cuneiform Texts 13). Oxford: Clarendon.

D'Ambrosio, U. 1998. *Ethnomathematics: the art or technique of explaining and knowing,* trans. P. B. Scott. Las Cruces, NM: NMSU/ISGEm.

Damerow, P. 2001. Kannten der Babylonier den Satz des Pythagoras? Epistemologische Anmerkungen zur Natur der babylonischen Mathematik. In *Changing views on ancient Near Eastern mathematics* (Berliner Beiträge zum Vorderen Orient 19), ed. J. Høyrup and P. Damerow, 219–310. Berlin: Reimer.

Dandemaev, M. 1985. Review of Nemet-Nejat 1982. *Orientalische Literaturzeitung* 80: 27–9.

De Certeau, M. 1984. *The practice of everyday life,* trans. S. Rendall. Berkeley: University of California Press.

De Kuyper, J. 1993. Mesopotamian astronomy and astrology as seen by Greek literature: the Chaldaeans. In *Die Rolle der Astronomie in den Kulturen mesopotamiens* (Grazer Morgenländische Studien 3), ed. H. D. Galter, 135–8. Graz: rm-Druck- & Vergesellschaft mbH.

De Mecquenem, R., et al. 1943. *Archéologie Susienne* (Mémoires de la Mission Archéologique en Perse 29). Paris: Presses Universitaires de France.

De Odorico, M. 1995. *The use of numbers and quantifications in the Assyrian royal inscriptions* (State Archives of Assyria Studies 3). Helsinki: Neo-Assyrian Text Corpus Project.

Deimel, A. 1922. Miszellen. *Orientalia* 5: 42–63.

Deimel, A. 1923. *Die Inschriften von Fara, II: Schultexte aus Fara.* Leipzig: Hinrichs.

Dekiere, L. 1994–7. *Old Babylonian real estate documents from Sippar in the British Museum* (Mesopotamian History and Environment III. Texts 2/1–6). Ghent: University of Ghent.

Delaporte, L. 1911. Document mathématique de l'époque des rois d'Our. *Revue d'Assyriologie* 8: 131–3.

Delaporte, L. 1912. *Inventaire des tablettes de Tello conservées au Musée Impérial Ottoman, IV: Textes de l'époque d'Ur.* Paris: Leroux.

Dercksen, J. G. 1996. *The Old Assyrian copper trade in Anatolia.* Istanbul: Nederlands Historisch-Archaeologisch Instituut te Istanbul.

Dolce, R. 2000. Some architectural drawings on clay tablets: examples of planning activity or sketches? In *Proceedings of the First International Conference on the Archaeology of the Ancient Near East*, ed. P. Matthiae, A. Enea, L. Peyronel, and F. Pinnock, 365–93. Rome: Università degli Studi di Roma 'La Sapienza'.

Dold-Samplonius, Y. 1997. In memoriam: Bartel Leendert van der Waerden (1903–1996). *Historia Mathematica* 24: 125–30.

Donbaz, V. 1985. More Old Assyrian tablets from Aššur. *Akkadica* 42: 1–23.

Donbaz, V., and B. R. Foster. 1982. *Sargonic texts from Telloh in the Istanbul archeological museums* (Occasional Publications of the Babylonian Fund 5). Philadelphia: University of Pennsylvania Museum of Archaeology and Anthropology.

Dossin, G. 1927. *Autres textes sumériens et accadiens* (Mémoires de la Mission archéologique en Iran 18). Paris: Leroux.

Doty, L. T. 1977. *Cuneiform archives from Hellenistic Uruk.* PhD diss., Yale University.

Doty, L. T. 1988. Nikarchos and Kephalon. In *A scientific humanist: studies in memory of Abraham Sachs*, ed. M. Ellis et al., 95–111. Philadelphia: University of Pennsylvania Museum of Archaeology and Anthropology.

Downey, S. 1988. *Mesopotamian religious architecture: Alexander through the Parthians.* Princeton: Princeton University Press.

Drucker, T. 1993. Morris Kline. *Modern Logic* 3: 156–8.

Duchesne-Guillemin, M. 1984. *A Hurrian musical score from Ugarit: the discovery of Mesopotamian music* (Sources from the Ancient Near East 2/2). Malibu: Undena Publications.

Dunham, S. 1982. Bricks for the temples of Šara and Ninurra. *Revue d'Assyriologie* 76: 27–41.

Dunham, S. 1986. Sumerian words for foundation. *Revue d'Assyriologie* 80: 31–64.

Durand, J.-M. 1981. *Textes babyloniens d'époque récente* (Recherche sur les Grands Civilisations. Cahier 6). Paris: Éditions ADPF.

Edzard, D. O. 1962. Texts and fragments. *Journal of Cuneiform Studies* 16: 78–81.

Edzard, D. O. 1969. Eine altsumerische Rechentafel (OIP 14, 70). In *lišān mitḫurti, Festschrift Wolfram Freiherr von Soden* (Alter Orient und Altes Testament 1), ed. W. Röllig, 101–4. Kevelaer/Neukirchen-Vluyn: Verlag Butzon und Bercker/Neukirchener Verlag des Erziehungsvereins.

Edzard, D. O. 1976–80. 'Itinerare'. In *Reallexikon der Assyriologie*, vol. 5, ed. D. O. Edzard, 216–20. Berlin: de Gruyter.

Edzard, D. O. 1981. *Verwaltungstexte verschiedenen Inhalts aus dem Archiv L. 2769* (Archivi reali di Ebla. Testi 2). Roma: Missione archeologica italiana in Siria.

Edzard, D. O. 1991. Zahlen, Zählen und Messen im Gilgameš-Epos. In *Texte, Methode und Grammatik*, ed. W. Gross, H. Irsigler, and T. Seidl, 57–66. St Ottilien: EOS.

Edzard, D. O. 2000. Wann ist Sumerisch als gesprochene Sprache ausgestorben? *Acta Sumerologica* 22: 53–70.

Edzard, D. O. 2005. Sumerian one to one hundred twenty revisited. In *An experienced scribe who neglects nothing: ancient Near Eastern studies in honor of Jacob Klein*, ed. Y. Sefati, P. Artzi, C. Cohen, B. Eichler, and V. Hurowitz, 98–107. Bethesda: CDL.

Eglash, R. 2000. Anthropological perspectives on ethnomathematics. In *Mathematics across cultures: the history of non-Western mathematics*, ed. H. Selin, 13–22. Dordrecht: Kluwer.

Eisenlohr, A. 1877. *Ein mathematisches Handbuch der alten Aegypter*. Leipzig: Hinrichs.

Englund, R. K. 1991. Hard work: where will it get you? Labor management in Ur III Mesopotamia. *Journal of Near Eastern Studies* 50: 255–80.

Englund, R. K. 1994. *Archaic administrative texts from Uruk: the early campaigns* (Archaische Texte aus Uruk 5). Berlin: Mann.

Englund, R. K. 1996. *The proto-cuneiform texts from diverse collections* (Materialen zu den frühen Schriftzeugnissen des Vorderen Orients 4). Berlin: Mann.

Englund, R. K. 1998. Texts from the Late Uruk period. In *Mesopotamia: Späturuk-Zeit und Frühdynastische Zeit* (Orbis Biblicus et Orientalis 160/1), by J. Bauer, R. K. Englund, and M. Krebernik, 15–233. Göttingen: Vandenhoek & Ruprecht.

Englund, R. K. 2001. Grain accounting practices in archaic Mesopotamia. In *Changing views on ancient Near Eastern mathematics* (Berliner Beiträge zum Vorderen Orient 19), ed. J. Høyrup and P. Damerow, 1–35. Berlin: Reimer.

Englund, R. K. 2003. The year: 'Nissen returns joyous from a distant land'. *Cuneiform Digital Library Journal* <http://cdli.ucla.edu/pubs/cdlj/2005/cdlj2003_001.html>.

Englund, R. K. 2004a. Proto-cuneiform account-books and journals. In *Creating economic order: record-keeping, standardization, and the development of accounting in the ancient Near East*, ed. M. Hudson and C. Wunsch, 23–46. Bethesda: CDL.

Englund, R. K. 2004b. The state of decipherment of proto-Elamite. In *The first writing: script invention as history and process*, ed. S. Houston, 100–49. Cambridge: Cambridge University Press.

Englund, R. K. 2005. Review of Glassner 2003. *Journal of the American Oriental Society* 125: 113–6.

Englund, R. K., and H. J. Nissen. 2001. *Archaische Verwaltungstexte aus Uruk: die Heidelberger Sammlung* (Archaische Texte aus Uruk 7). Berlin: Mann.

Faivre, X. 1995. Le recyclage des tablettes cunéiformes. *Revue d'Assyriologie* 89: 57–66.

Fales, F. M. 1986. *Aramaic epigraphs on clay tablets of the Neo-Assyrian period* (Studi Semitici, Nuove Serie 2). Rome: Università degli Studi 'La Sapienza'.

Fales, F. M. 1995. Assyro-Aramaica: the Assyrian lion-weights. In *Immigration and emigration within the ancient Near East: Festschrift E. Lipiński* (Orientalia Lovaniensia Analecta 65), ed. K. van Lerberghe and A. Schoors, 33–56. Leuven: Peeters.

Fales, F. M., and J. N. Postgate. 1992. *Imperial administrative records, I: palace and temple administration* (State Archives of Assyria 7). Helsinki: Helsinki University Press.

Fales, F. M., and J. N. Postgate. 1995. *Imperial administrative records, II: provincial and military administration* (State Archives of Assyria 11). Helsinki: Helsinki University Press.

Falkowitz, R. S. 1983–4. Round Old Babylonian school tablets from Nippur. *Archiv für Orientforschung* 29–30: 18–45.

Farber, W. 1989. Lamaštu, Enlil, Anu-ikşur: Streiflichter aus Uruks Gelehrtenstuben. *Zeitschrift für Assyriologie* 79: 223–41.

Fincke, J. C. 2003–4. The Babylonian texts of Nineveh: report on the British Museum's *Aššurbanipal Library Project*. *Archiv für Orientforschung* 50: 111–49.

Finkel, I. L. 1987. An issue of weights from the reign of Amar-Sin. *Zeitschrift für Assyriologie* 77: 192–3.

Fish, T., and M. Lambert. 1963. 'Verification' dans la bureaucratie sumérienne. *Revue d'Assyriologie* 57: 93–7.

Förtsch, W. 1916. *Altbabylonische Wirtschaftstexte aus der Zeit Lugalandas und Urukaginas* (Vorderasiatische Schriftdenkmäler der Staatlichen Museen zu Berlin 14). Leipzig: Hinrichs.

Foster, B. R. 1982a. *Administration and use of institutional land in Sargonic Sumer* (Mesopotamia 9). Copenhagen: Akademisk Forlag.

Foster, B. R. 1982b. Archives and record-keeping in Sargonic Mesopotamia. *Zeitschrift für Assyriologie* 72: 1–26.

Foster, B. R. 1982c. Education of a bureaucrat in Sargonic Sumer. *Archív Orientální* 50: 238–41.

Foster, B. R. 1982d. *Umma in the Sargonic period* (Memoirs of the Connecticut Academy of Arts & Sciences 20). Hamden: Archon Books.

Foster, B. R. 1983. Ebla and the origins of Akkadian accountability. *Bibliotheca Orientalis* 40: 298–305.

Foster, B. R. 2001. *The epic of Gilgamesh: a new translation, analogues, criticism.* New York: Norton.

Foster, B. R. 2005. *Before the muses: an anthology of Akkadian literature*, 2nd ed. Bethesda: CDL.

Foster, B. R., and E. Robson. 2004. A new look at the Sargonic mathematical corpus. *Zeitschrift für Assyriologie* 94: 1–15.

Fowler, D. H. 1992. *Dýnamis, mithartum*, and square. *Historia Mathematica* 19: 418–9.

Fowler, D. H. 1999a. Inventive interpretations. *Revue d'Histoire des Mathématiques* 5: 149–53.

Fowler, D. H. 1999b. *The mathematics of Plato's academy: a new reconstruction*, 2nd ed. Oxford: Oxford University Press.

Fowler, D. H., and E. Robson. 1998. Square root approximations in Old Babylonian mathematics: YBC 7289 in context. *Historia Mathematica* 25: 366–78.

Frahm, E. 1999. Nabû-zuqup-kēna, das Gilgameš-Epos und der Tod Sargons II. *Journal of Cuneiform Studies* 51: 73–90.

Frahm, E. 2004. Royal hermeneutics: observations on the commentaries from Aššurbanipal's libraries at Nineveh. *Iraq* 66: 45–50.

Frahm, E. 2005. Nabû-zuqup-kēna, Gilgameš XII, and the rites of Du'uzu. *NABU* 2005: no. 5.

Frame, G. 1997. Chaldeans. In *The Oxford encyclopedia of archaeology in the Near East*, ed. E. Meyers, 482–4. Oxford: Oxford University Press.

Frame, G., and A. R. George. 2005. The royal libraries of Nineveh: new evidence for their formation. *Iraq* 67: 265–84.

Frayne, D. R. 1997. *Ur III period (2112–2004 BC)* (The Royal Inscriptions of Mesopotamia. Early Periods 3/2). Toronto: University of Toronto Press.

Freydank, H. 1991. *Beiträge zur mittelassyrischen Chronologie und Geschichte* (Schriften zur Geschichte und Kultur des alten Orients 21). Berlin: Akademie-Verlag.

Friberg, J. 1981a. Methods and traditions of Babylonian mathematics: Plimpton 322, Pythagorean triples and the Babylonian triangle parameter equations. *Historia Mathematica* 8: 277–318.

Friberg, J. 1981b. Methods and traditions of Babylonian mathematics, II: an Old Babylonian catalogue text with equations for squares and circles. *Journal of Cuneiform Studies* 33: 57–64.

Friberg, J. 1983. On the big 6-place tables of reciprocals and squares from Seleucid Babylon and Uruk and their Old Babylonian and Sumerian predecessors. *Sumer* 42: 81–7.

Friberg, J. 1986. Three remarkable texts from ancient Ebla. *Vicino Oriente* 6: 3–25.

Friberg, J. 1987–90. Mathematik. In *Reallexikon der Assyriologie,* vol. 7, ed. D. O. Edzard, 531–85. Berlin: de Gruyter.

Friberg, J. 1993. On the structure of cuneiform metrological texts from the −1st millennium. In *Die Rolle der Astronomie in den Kulturen mesopotamiens* (Grazer Morgenländische Studien 3), ed. H. D. Galter, 383–405. Graz: rm-Druck- & Vergesellschaft mbH.

Friberg, J. 1994. Preliterate counting and accounting in the Middle East: a constructively critical review of Schmandt-Besserat's *Before Writing. Orientalistische Literaturzeitung* 89: 477–502.

Friberg, J. 1997. 'Seed and reeds' continued: another metro-mathematical topic text from Late Babylonian Uruk. *Baghdader Mitteilungen* 28: 251–365.

Friberg, J. 1997–8. Round and almost round numbers in proto-literate metro-mathematical field texts. *Archiv für Orientforschung* 44–5: 1–58.

Friberg, J. 1999a. A Late Babylonian factorisation algorithm for the computation of reciprocals of many-place regular sexagesimal numbers. *Baghdader Mitteilungen* 30: 139–61.

Friberg, J. 1999b. Counting and accounting in the proto-literate Middle East: examples from two new volumes of proto-cuneiform texts. *Journal of Cuneiform Studies* 51: 107–37.

Friberg, J. 2000. Mathematics at Ur in the Old Babylonian period. *Revue d'Assyriologie* 94: 97–188.

Friberg, J. 2001. Bricks and mud in metro-mathematical cuneiform texts. In *Changing views on ancient Near Eastern mathematics* (Berliner Beiträge zum Vorderen Orient 19), ed. J. Høyrup and P. Damerow, 61–154. Berlin: Reimer.

Friberg, J. 2005a. Nos. 72–77: mathematical texts. In *Literary and scholastic texts of the first millennium BC* (Cuneiform Texts in the Metropolitan Museum of Art 2), ed. I. Spar and W. G. Lambert, 288–314. New York: Metropolitan Museum of Art.

Friberg, J. 2005b. On the alleged counting with sexagesimal place value numbers in mathematical cuneiform texts from the third millennium BC. *Cuneiform Digital Library Journal* <http://cdli.ucla.edu/pubs/cdlj/2005/cdlj2005_002.html>.

Friberg, J., H. Hunger, and F.N.H. Al-Rawi. 1990. 'Seed and reeds': a metro-mathematical topic text from Late Babylonian Uruk. *Baghdader Mitteilungen* 21: 483–557.

Fuchs, A. 1994. *Die Inschriften Sargons II. aus Khorsabad*. Göttingen: Cuvillier.

Gadd, C. J. 1921. *Cuneiform texts from Babylonian tablets in the British Museum*, vol. 36. London: Trustees of the British Museum.

Garfinkle, S. 2003. SI.A-a and family: the archive of a 21st century (BC) entrepreneur. *Zeitschrift für Assyriologie* 93: 161–98.

Gelb, I. J. 1967. Growth of a herd of cattle in ten years. *Journal of Cuneiform Studies* 21: 64–9.

Gelb, I. J. 1970a. *Sargonic texts in the Louvre Museum* (Materials for the Assyrian Dictionary 4). Chicago: University of Chicago Press.

Gelb, I. J. 1970b. *Sargonic texts in the Ashmolean Museum, Oxford* (Materials for the Assyrian Dictionary 5). Chicago: University of Chicago Press.

Gelb, I. J. 1981. Ebla and the Kiš civilization. In *La lingua di Ebla*, ed. L. Cagni, 9–73. Napoli: Istituto Universitario Orientali.

Gelb, I. J., P. Steinkeller, and R. Whiting. 1991. *Earliest land tenure systems in the Near East: ancient kudurrus* (Oriental Institute Publications 104). Chicago: Oriental Institute of the University of Chicago.

Geller, M. J. 1997. The last wedge. *Zeitschrift für Assyriologie* 87: 43–95.

Genouillac, H. de. 1910. *Inventaire des tablettes de Tello conservées au Musée Impérial Ottoman, II: Textes de l'époque d'Agadé, de l'époque d'Ur*. Paris: Leroux.

Genouillac, H. de. 1912. *Inventaire des tablettes de Tello conservées au Musée Impérial Ottoman, III: Textes de l'époque d'Ur*. Paris: Leroux.

Genouillac, H. de. 1921. *Inventaire des tablettes de Tello conservées au Musée Impérial Ottoman, V: Époque présargonique, époque d'Agadé, époque d'Ur*. Paris: Leroux.

George, A. R. 1986. The topography of Babylon reconsidered. *Sumer* 44: 7–24.

George, A. R. 1992. *Babylonian topographical texts* (Orientalia Lovaniensia Analecta 40). Leuven: Peeters.

George, A. R. 1995. The bricks of E-sagil. *Iraq* 57: 173–98.

George, A. R. 2000. *The epic of Gilgamesh: a new translation*. Harmondsworth: Penguin.

George, A. R. 2003. *The Babylonian Gilgamesh epic: introduction, critical edition and cuneiform texts*. Oxford: Oxford University Press.

George, A. R. 2005a. No. 68: measurements of the interior of the Esagil temple. In *Literary and scholastic texts of the first millennium BC* (Cuneiform Texts in the Metropolitan Museum of Art 2), ed. I. Spar and W. G. Lambert, 267–77. New York: Metropolitan Museum of Art.

George, A. R. 2005b. Review of Glassner 2003. *Bulletin of the School of Oriental and African Studies* 68: 107–9.

George, A. R. 2007. Babylonian and Assyrian: a history of Akkadian. In *Languages of Iraq, ancient and modern*, ed. J. N. Postgate, 31–71. London: British School of Archaeology in Iraq.

Gesche, P. D. 2000. *Schulunterricht in Babylonien im ersten Jahrtausend v. Chr.* (Alter Orient und Altes Testament 275). Münster: Ugarit-Verlag.

Glassner, J.-J. 2003. *The invention of cuneiform: writing in Sumer*, trans. Z. Bahrani and M. Van De Mieroop. Baltimore: John Hopkins University Press.

Goetze, A. 1945. The Akkadian dialects of the Old-Babylonian mathematical texts. In Neugebauer and Sachs 1945, 146–51.

Gomi, T. 1980. On dairy productivity at Ur in the late Ur III period. *Journal of the Economic and Social History of the Orient* 23: 1–42.

Gomi, T., and F. Yildiz. 1997. *Die Umma-Texte aus den Archäologischen Museen zu Istanbul*, vol. 4. Bethesda: CDL.

Gordon, E. I. 1959. *Sumerian proverbs: glimpses of everyday life in ancient Mesopotamia* (Museum Monographs 19). Philadelphia: University of Pennsylvania Museum of Archaeology and Anthropology.

Goren, Y. 2000. Provenance study of the cuneiform texts from Hazor. *Israel Exploration Journal* 20: 29–42.

Grayson, A. K. 1980–3. Königslisten und Chroniken. B. Akkadisch. In *Reallexikon der Assyriologie*, vol. 6, ed. D. O. Edzard, 86–135. Berlin: de Gruyter.

Grayson, A. K. 1987. *Assyrian rulers of the third and second millennia BC (to 1115 BC)* (Royal Inscriptions of Mesopotamia. Assyrian Periods 1). Toronto: University of Toronto Press.

Green, A. R. 1993. *The 6G ash-tip and its contents: cultic and administrative discard from the temple?* (Abu Salabikh Excavations 4). London: British School of Archaeology in Iraq.

Greengus, S. 1979. *Old Babylonian tablets from Ishchali and vicinity* (Uitgaven van het Nederlands Historisch-Archaeologisch Instituut te Istanbul 44). Istanbul: Nederlands Historisch-Archaeologisch Instituut te Istanbul.

Grégoire, J.-P. 1996. *Archives administratives et inscriptions cuneiforms de l'Ashmolean Museum et de la Bodleian Collection d'Oxford*, vol. 1/1. Paris: Geuthner.

Grice, E. M. 1919. *Records from Ur and Larsa dated in the Larsa dynasty* (Yale Oriental Series. Babylonian Texts 5). New Haven: Yale University Press.

Guichard, M. 1997. Présages fortuits à Mari (copies et ajouts à *ARMT* XXVI/1). *MARI, Annales de Recherches Interdisciplinaires* 8: 305–28.

Guinan, A. 2002. A severed head laughed: stories of divinatory interpretation. In *Magic and divination in the ancient world* (Ancient Magic and Divination 2), ed. L. Ciraolo and J. Seidel, 7–40. Groningen: Styx.

Guralnick, E. 2004. Assyrian patterned fabrics. *Iraq* 66: 221–32.

Gurney, O. R. 1974. *Middle Babylonian legal documents and other texts* (Ur Excavations Texts 7). London: British Museum Publications.

Gurney, O. R. 1983. *The Middle Babylonian legal and economic texts from Ur.* London: British School of Archaeology in Iraq.

Gurney, O. R. 1986. Lexical texts in the Ashmolean Museum. In *Materials for the Sumerian Lexicon. Supplementary Series*, vol. 1, by M. Civil, O. R. Gurney, and D. A. Kennedy, 45–71. Rome: Pontifical Biblical Institute.

Gurney, O. R. 1989. *Literary and miscellaneous texts in the Ashmolean Museum* (Oxford Editions of Cuneiform Texts 11). Oxford: Clarendon.

Gurney, O. R. 1997. Scribes at Huzirīna. *NABU* 1997: no. 17.

Gurney, O. R., and J. J. Finkelstein. 1957. *The Sultantepe tablets*, vol. 1. London: British Institute of Archaeology at Ankara.

Gurney, O. R., and P. Hulin. 1964. *The Sultantepe tablets*, vol. 2. London: British Institute of Archaeology at Ankara.

Hackman, G. G. 1958. *Sumerian and Akkadian administrative texts from Predynastic times to the end of the Akkad dynasty* (Babylonian Inscriptions in the Collection of James B. Nies 8). New Haven: Yale University Press.

Hallo, W. W. 2005. Sumerian history in pictures: a new look at the 'Stele of the Flying Angels'. In *An experienced scribe who neglects nothing: ancient Near Eastern studies in honor of Jacob Klein*, ed. Y. Sefati, P. Artzi, C. Cohen, B. Eichler, and V. Hurowitz, 142–62. Bethesda: CDL.

Hallock, R. T., et al. 1955. *Das Syllabar A, das Vokabular Sᵃ, das Vokabular Sᵇ* (Materialen zum Sumerischen Lexikon 3). Rome: Pontifical Biblical Institute.

Harper, P. O., J. Aruz, and F. Tallon. 1996. *The royal city of Susa: ancient Near Eastern treasures in the Louvre*. New York: Metropolitan Museum of Art.

Heath, T. L. 1921. *A history of Greek mathematics*. Oxford: Clarendon.

Heath, T. L. 1926. *The thirteen books of Euclid's Elements*. Cambridge: Cambridge University Press.

Hecker, K. 1993. Schültexte vom Kültepe. In *Aspects of art and iconography: Anatolia and its neighbours. Studies in honor of Nimet Özgüç*, ed. M. J. Mellink, E. Porada, and T. Özgüç, 281–91. Ankara: Türk Tarih Kurumu Basımevi.

Hecker, K. 1996. Schültexte aus Kültepe: ein Nachtrag. *NABU* 1996: no. 30.

Heiberg, J. L., and H. Menge. 1883–1916. *Euclidis opera omnia*. Leipzig: Teubner.

Heimerdinger, J. 1979. *Sumerian literary fragments from Nippur* (Occasional Publications of the Babylonian Fund 4). Philadelphia: University of Pennsylvania Museum of Archaeology and Anthropology.

Heimpel, W. 1997. Disposition of households of officials in Ur III and Mari. *Acta Sumerologica* 19: 63–82.

Heimpel, W. 2003. *Letters to the king of Mari: a new translation, with historical introduction, notes, and commentary* (Mesopotamian Civilizations 12). Winona Lake: Eisenbrauns.

Heinrich, E., and U. Seidl. 1967. Grundrißzeichnungen aus dem Alten Orient. *Mitteilungen der Deutschen Orientgesellschaft* 98: 24–45.

Herréro, P., and J.-J. Glassner. 1991. Haft-Tépé: choix de textes, II. *Iranica Antiqua* 26: 39–80.

Herréro, P., and J.-J. Glassner. 1996. Haft-Tépé: choix de textes, IV. *Iranica Antiqua* 31: 51–82.

Hilprecht, H. V. 1896. *Old Babylonian inscriptions, chiefly from Nippur*, vol. 2 (The Babylonian Expedition of the University of Pennsylvania. Series A: Cuneiform Texts 1/2). Philadelphia: University of Pennsylvania.

Hilprecht, H. V. 1906. *Mathematical, metrological and chronological tablets from the temple library of Nippur* (The Babylonian Expedition of the University of Pennsylvania. Series A: Cuneiform Texts 20/1). Philadelphia: University of Pennsylvania.

Hinke, W. J. 1907. *A new boundary stone of Nebuchadrezzar I from Nippur* (The Babylonian Expedition of the University of Pennsylvania. Series D: Researches and Treatises 4). Philadelphia: University of Pennsylvania.

Hinke, W. J. 1911. *Selected Babylonian kudurru inscriptions* (Semitic Study Series 14). Leiden: Brill.

Holt, I. L. 1911. Tablets from the R. Campbell Thompson Collection in Haskell Oriental Museum, the University of Chicago. *American Journal of Semitic Literatures* 27: 193–232.

Horowitz, W. 1997. A combined multiplication table on a prism fragment from Hazor. *Israel Exploration Journal* 47: 190–7.

Horowitz, W. 1998. *Mesopotamian cosmic geography* (Mesopotamian Civilizations 8). Winona Lake: Eisenbrauns.

Horowitz, W., T. Oshima, and S. Sanders. 2006. *Cuneiform in Canaan: cuneiform sources from the land of Israel in ancient times*. Jerusalem: Israel Exploration Society and the Hebrew University of Jerusalem.

Horowitz, W., and O. Tammuz. 1998. A multiplication table for 40 in the Israel Museum. *Israel Exploration Journal* 48: 262–4.

Horowitz, W., and N. Wasserman. 2000. An Old Babylonian letter from Hazor with mention of Mari and Ekallātum. *Israel Exploration Journal* 50: 169–74.

Høyrup, J. 1982. Investigations of an early Sumerian division problem, c. 2500 BC. *Historia Mathematica* 9: 19–36.

Høyrup, J. 1990a. Algebra and naive geometry. An investigation of some basic aspects of Old Babylonian mathematical thought. *Altorientalische Forschungen* 17: 27–69, 262–354.

Høyrup, J. 1990b. *Dýnamis*, the Babylonians, and *Theaetetus* 147c7–148d7. *Historia Mathematica* 17: 201–22.

Høyrup, J. 1990c. On parts of parts and ascending continued fractions. An investigation of the origins and spread of a peculiar system. *Centaurus* 33: 293–324.

Høyrup, J. 1990d. Sub-scientific mathematics: observations on a pre-modern phenomenon. *History of Science* 28: 63–86.

Høyrup, J. 1993. 'Remarkable numbers' in Old Babylonian mathematical texts: a note on the psychology of numbers. *Journal of Near Eastern Studies* 52: 281–6.

Høyrup, J. 1994. *In measure, number, and weight: studies in mathematics and culture*. Albany: State University of New York.

Høyrup, J. 1996. Changing trends in the historiography of Mesopotamian mathematics: an insider's view. *History of Science* 34: 1–32.

Høyrup, J. 1999. Pythagorean 'rule' and 'theorem': mirror of the relation between Babylonian and Greek mathematics. In *Babylon: Focus mesopotamischer Geschichte, Wiege früher Gelehrsamkeit, Mythos in der Moderne*, ed. J. Renger, 292–407. Berlin: Deutsche Orient-Gesellschaft / Saarbrücken: SDV Saarbrücker Druckerei und Verlag.

Høyrup, J. 2002a. How to educate a Kapo, or reflections on the absence of a culture of mathematical problems in Ur III. In *Under one sky: astronomy and mathematics in the ancient Near East* (Alte Orient und Altest Testament 297), ed. J. M. Steele and A. Imhausen, 121–45. Münster: Ugarit-Verlag.

Høyrup, J. 2002b. *Lengths, widths, surfaces: a portrait of Old Babylonian algebra and its kin*. Berlin: Springer.

Høyrup, J. 2002c. A note on Old Babylonian computational techniques. *Historia Mathematica* 29: 193–8.

Huehnergard, J., and C. Woods. 2003. Akkadian and Eblaite. In *The Cambridge encyclopedia of the world's ancient languages*, ed. R. J. Woodard, 218–87. Cambridge: Cambridge University Press.

Hulin, P. 1963. A table of reciprocals with Sumerian spellings. *Journal of Cuneiform Studies* 17: 72–6.

Hunger, H. 1968. *Babylonische und assyrische Kolophone* (Alter Orient und Altes Testament 2). Kevelaer/Neukirchen-Vluyn: Butzon & Bercker/Neukirchener Verlag.

Hunger, H. 1976a. Astrologische Wettervorhersagen. *Zeitschrift für Assyriologie* 66: 234–60.

Hunger, H. 1976b. *Spätbabylonische Texte aus Uruk*, vol. 1 (Ausgrabungen der Deutschen Forschungs-gemeinschaft in Uruk-Warka. Endberichte 9). Berlin: Mann.

Hunger, H. 1991. Schematische Berechnungen der Sonnenwenden. *Baghdader Mitteilungen* 22: 513–9.

Hunger, H. 1992. *Astrological reports to Assyrian kings* (State Archives of Assyria 8). Helsinki: Helsinki University Press.

Hunger, H. 1993. Astronomische Beobachtungen in neubabylonischer Zeit. In *Die Rolle der Astronomie in den Kulturen mesopotamiens* (Grazer Morgenländische Studien 3), ed. H. D. Galter, 139–48. Graz: rm-Druck- & Vergesellschaft mbH.

Hunger, H. 1998. Non-mathematical astronomical texts and their relationships. In *Ancient astronomy and celestial divination*, ed. N. M. Swerdlow, 77–98. Cambridge, MA: MIT Press.

Hunger, H., and D. Pingree. 1989. *MUL.APIN: an astronomical compendium in cuneiform* (Archiv für Orientforschung. Beiheft 24). Horn: Berger.

Hunger, H., and D. Pingree. 1999. *Astral sciences in Mesopotamia* (Handbuch der Orientalistik 1/44). Leiden: Brill.

Hurowitz, V. A. 1997. *Divine service and its rewards: ideology and poetics in the Hinke* kudurru (Beer-Sheva Studies by the Department of Bible and Ancient Near East 10). Beer-Sheva: Ben-Gurion University of the Negev.

Hurowitz, V. A. 1998. Advice to a prince: a message from Ea. *State Archives of Assyria Bulletin* 12: 39–53.

Iamblichus of Chalcis. 1989. *On the Pythagorean life*, trans. G. Clark. Liverpool: Liverpool University Press.

Imhausen, A. 2003. *Ägyptische Algorithmen: Eine Untersuchung zu den mittelägyptischen mathematischen Aufgabentexten* (Ägyptologische Abhandlungen 65). Wiesbaden: Harrassowitz.

Ishaq, D. 1985. The excavations at the southern part of the procession street and Nabû *ša harê* temple. *Sumer* 41: 30–3.

Isma'el, K. S. 1999a. A new table of square roots. *Akkadica* 112: 18–26.

Isma'el, K. S. 1999b. A new mathematical text in the Iraqi Museum. *Akkadica* 113: 6–12.

Isma'el, K. S. Forthcoming. Arithmetical tablets from Iraqi excavations in the Diyala. In *Your praise is sweet: memorial volume for Jeremy Black from students, colleagues and friends*, ed. H. D. Baker, E. Robson, and G. G. Zólyomi.

Izre'el, S. 1997. *The Amarna scholarly tablets* (Cuneiform Monographs 9). Groningen: Styx.

Jacobsen, T. 1987. Pictures and pictorial language (the Burney Relief). In *Figurative language in the ancient Near East*, ed. M. Mindlin, M. J. Geller, and J. E. Wansborough, 1–11. London: School of Oriental and African Studies.

Jestin, R. 1937. *Tablettes sumériennes de Shuruppak conservées au Musée de Stamboul*. Paris: de Boccard.

Joannès, F. 1982. *Textes économiques de la babylonie récente*. Paris: Éditions Recherche sur les Civilisations.

Joannès, F. 1989. *Archives de Borsippa: la famille Ea-ilûta-bâni. Étude d'un lot d'archives familiales en Babylonie du VIIe au Ve siècle av. J.-C.* (Hautes Études Orientales 25). Geneva: Droz.

Joannès, F. 2004. *The age of empires: Mesopotamia in the first millennium* BC. Edinburgh: Edinburgh University Press.

Jones, A. 1991. The adaptation of Babylonian methods in Greek numerical astronomy. *Isis* 82: 441–53.

Jones, A. 1993. The evidence for Babylonian arithmetical schemes in Greek astronomy. In *Die Rolle der Astronomie in den Kulturen Mesopotamien* (Grazer Morgenländische Studien 3), ed. H. D. Galter, 77–94. Graz: rm-Druck- & Vergesellschaft mbH.

Jones, A. 1996. Babylonian astronomy and its Greek metamorphoses. In *Tradition, transmission, transformation: proceedings of two conferences on pre-modern science held at the University of Oklahoma*, ed. F. J. Ragep and S. Ragep, 139–55. Leiden: Brill.

Jones, T. B., and J. W. Snyder. 1961. *Sumerian economic texts from the third Ur dynasty: a catalogue and discussion of documents from various collections*. Minneapolis: University of Minnesota Press.

Jordan, J. 1928. *Uruk-Warka nach den Ausgrabungen durch die Deutsche Orient-Gesellschaft*. Leipzig: Hinrichs.

Jursa, M. 1993–4. Zweierlei Maß. *Archiv für Orientforschung* 40–1: 71–3.

Jursa, M. 2005. *Neo-Babylonian legal and administrative documents: typology, contents and archives* (Guides to the Mesopotamian Textual Record 1). Münster: Ugarit-Verlag.

Jursa, M., and K. Radner. 1995–6. Keilschrifttexte aus Jerusalem. *Archiv für Orientforschung* 42–3: 89–108.

Keetman, J. 2005. Wie hoch war die 6. Stufe von Etemenanki? Rekonstruktion. *Revue d'Assyriologie* 99: 77–84.

Kilmer, A. D. 1960. Two new lists of key numbers for mathematical operations. *Orientalia* 29: 273–308.

Kilmer, A. D. 1980–3. Leier. A. Philologisch. In *Reallexikon der Assyriologie und vorderasiatischen Archäologie*, vol. 6, ed. D. O. Edzard, 571–6. Berlin: de Gruyter.

Kilmer, A. D. 1990. Sumerian and Akkadian names for designs and geometrical shapes. In *Investigating artistic environments in the ancient Near East*, ed. A. C. Gunter, 83–91. New York: Smithsonian Institution.

Kilmer, A. D. 1992. Musical practice at Nippur. In *Nippur at the centennial: papers read at the 35ᵉ RAI, Philadelphia 1988* (Occasional Publications of the Samuel Noah Kramer Fund 14), ed. M. de J. Ellis, 101–12. Philadelphia: University of Pennsylvania Museum of Archaeology and Anthropology.

King, D. A. 1998. Mamluk astronomy and the institution of the *muwaqqit*. In *The Mamluks in Egyptian politics and society*, ed. T. Philipp and U. Haarmann, 153–62. Cambridge: Cambridge University Press.

King, L. W. 1912a. *Babylonian boundary-stones and memorial tablets in the British Museum*. London: Trustees of the British Museum.

King, L. W. 1912b. *Cuneiform texts from Babylonian tablets in the British Museum*, vol. 33. London: Trustees of the British Museum.

Klein, J. 2000. The independent pronouns in the Šulgi hymns. *Acta Sumerologica* 22: 135–52.

Klengel-Brandt, E. 1975. Ein Schreib-Tafel aus Assur. *Altorientalische Forschungen* 3: 169–71.

Kline, M. 1972. *Mathematical thought from ancient to modern times*. Oxford: Oxford University Press.

Koch, J. 2001. Neue Überlegungen zu einigen astrologischen und astronomischen Keilschrifttexten. *Journal of Cuneiform Studies* 53: 69–81.

Koch, U. 2005. *Secrets of extispicy: the chapter* multābiltu *of the Babylonian extispicy series and* nişirti bārûti *texts mainly from Aššurbanipal's library* (Alter Orient und Altes Testament 326). Münster: Ugarit-Verlag.

Koch-Westenholz, U. 1995. *Mesopotamian astrology: an introduction to Babylonian and Assyrian celestial divination* (Carsten Niebuhr Institute Publications 19). Copenhagen: Museum Tusculanum Press.

Kraus, F. R. 1955. Provinzen des neusumerischen Reiches von Ur. *Zeitschrift für Assyriologie* 51: 45–75.

Krebernik, M. 1998. Die Texte aus Fara und Abu Salabikh. In *Mesopotamia: Späturuk-Zeit und Frühdynastische Zeit* (Orbis Biblicus et Orientalis 160/1), by J. Bauer, R. K. Englund, and M. Krebernik, 317–427. Göttingen: Vandenhoek & Ruprecht.

Krebernik, M. 2006. Neues zu den Fara-Texten. *NABU* 2006: no. 15.

Kuhrt, A. 1982. Assyrian and Babylonian traditions in classical authors: a critical synthesis. In *Mesopotamien und seine Nachbarn: Politische und kulturelle Wechselbeziehungen im Alten Vorderasien vom 4. Bis 1. Jahrtausend v. Chr.* (Berliner Beiträge zum Vorderen Orient 1), ed. H. Kühne, H.-J. Nissen, and J. Renger, 539–54. Berlin: Reimer.

Kuhrt, A. 1995. *The ancient Near East, c. 3000–300 BC*. London: Routledge.

Kuhrt, A., and S. Sherwin-White. 1991. Aspects of Seleucid royal ideology: the cylinder of Antiochus I from Borsippa. *Journal of Hellenic Studies* 111: 71–86.

Kuklick, B. 1996. *Puritans in Babylon: the ancient Near East and American intellectual life, 1880–1930*. Princeton: Princeton University Press.

Kümmel, H. M. 1979. *Familie, Beruf und Amt im spätbabylonischen Uruk: prosopographische Untersuchungen zu Berufsgruppen des 6. Jahrhunderts v. Chr. in Uruk* (Abhandlungen der Deutschen Orient-Gesellschaft 20). Berlin: Mann.

Lambert, W. G. 1957. Ancestors, authors, and canonicity. *Journal of Cuneiform Studies* 11: 1–14.

Lambert, W. G. 1967. The Gula hymn of Bullutsa-rabi. *Orientalia* 36: 105–32.

Lambert, W. G. 2005. Introduction: the transmission of the literary and scholarly texts. In *Literary and scholastic texts of the first millennium BC* (Cuneiform

Texts in the Metropolitan Museum of Art 2), ed. I. Spar and W. G. Lambert, xi–xix. New York: Metropolitan Museum of Art.

Landsberger, B. 1937. *Die Serie* ana ittišu (Materialen zum Sumerischen Lexikon 1). Rome: Pontifical Biblical Institute.

Landsberger, B. 1957. *The series* ḪAR-ra = ḫubullu, *tablets I–IV* (Materialen zum Sumerischen Lexikon 5). Rome: Pontifical Biblical Institute.

Lanfranchi, G. B., and S. Parpola. 1990. *The correspondence of Sargon II, part II: letters from the northern and north-eastern provinces* (State Archives of Assyria 5). Helsinki: Helsinki University Press.

Langdon, S. H. 1924. *Excavations at Kish: the Herbert Weld (for the University of Oxford) and Field Museum of Natural History (Chicago) Expedition to Mesopotamia, I: 1923–1924*. Paris: Geuthner.

Largement, R. 1957. Contribution à l'étude des astres errants dans l'astrologie chaldéenne. *Zeitschrift für Assyriologie* 52: 235–64.

Larsen, M. T. 1977. Partnerships in the Old Assyrian trade. *Iraq* 39: 119–45.

Larsen, M. T. 1996. *The conquest of Assyria: excavations in an antique land, 1840–1860*. London: Routledge.

Lave, J., and E. Wenger. 1991. *Situated learning: legitimate peripheral participation*. Cambridge: Cambridge University Press.

Legrain, L. 1935. Quelques textes anciens, II. *Revue d'Assyriologie* 32: 128–9.

Leichty, E., and C.B.F. Walker. 2004. Three Babylonian chronicle and scientific texts. In *From the upper sea to the lower sea: studies on the history of Assyria and Babylonia in honour of A. K. Grayson*, ed. G. Frame and L. Wilding, 203–212. Leiden: Nederlands Instituut voor het Nabije Oosten.

Lieberman, S. J. 1980. Of clay pebbles, hollow clay balls, and writing: a Sumerian view. *American Journal of Archaeology* 84: 339–58.

Lieberman, S. J. 1987. A Mesopotamian background for the so-called *aggadic* 'measures' of Biblical hermeneutics? *Hebrew Union College Annual* 58: 157–225.

Limet, H. 1971. *Les legends des sceaux cassites* (Académie royale de Belgique. Mémoires 8/40). Bruxelles: Palais des Académies.

Limet, H. 1973. *Étude de documents de la période d'Agadé appartenant à l'Université de Liège* (Bibliothèque de la Faculté de Philosophie et Lettres de l'Université de Liège 206). Paris: Société d'Éditions 'Les Belles Lettres'.

Linssen, M.J.H. 2004. *The cults of Uruk and Babylon: the temple ritual texts as evidence for Hellenistic cult practices* (Cuneiform Monographs 25). Leiden: Styx.

Lion, B. 2001. Dame Inana-amamu, scribe à Sippar. *Revue d'Assyriologie* 95: 7–32.

Lion, B., and E. Robson. 2006. Quelques textes scolaires paléo-babyloniens rédigés par des femmes. *Journal of Cuneiform Studies* 57: 37–53.

Liverani, M. 1990. The shape of Neo-Sumerian fields. *Bulletin on Sumerian Agriculture* 5: 147–86.

Livingstone, A. 1986. *Mystical and mythological explanatory works of Assyrian and Babylonian scholars*. Oxford: Clarendon.

Livingstone, A. 1989. *Court poetry and literary miscellanea* (State Archives of Assyria 3). Helsinki: Helsinki University Press.

Lloyd, S. 1980. *Foundations in the dust: the story of Mesopotamian exploration*, 3rd ed. London: Thames & Hudson.

Lloyd, S., and N. Gökçe. 1953. Sultantepe: Anglo-Turkish joint excavations, 1952. *Anatolian Studies* 3: 27–51.

Luckenbill, D. D. 1930. *Inscriptions from Adab* (Oriental Institute Publications 14). Chicago: University of Chicago Press.

Macginnis, J. 2002. The use of writing boards in the Neo-Babylonian temple administration at Sippar. *Iraq* 64: 217–36.

Maeda, T. 1984. Work concerning irrigation canals in pre-Sargonic Lagaš. *Acta Sumerologica* 6: 33–54.

Margueron, J. 1986. Quelques remarques concernant les archives retrouvées dans le palais de Mari. In *Cuneiform archives and libraries: papers read at the 30ᵉ Rencontre Assyriologique Internationale*, ed. K. R. Veenhof, 141–52. Istanbul: Nederlands Historisch-Archaeologisch Instituut te Istanbul.

Márquez Rowe, I. 1998. Dos tables de multiplicar paleo-babilónicas en el Museu Bíblic de Montserrat. *Aula Orientalis* 16: 268–8.

Martin, H. 1988. *Fara*. Birmingham: Chris Martin.

Matthews, D. M. 1992. *Kassite glyptic of Nippur* (Orbis Biblicus et Orientalis 116). Freiburg and Göttingen: Universitätsverlag Freiburg and Vandenhoeck & Ruprecht.

Matthews, R. 2000. *The early prehistory of Mesopotamia: 500,000 to 4,500 BC*. Turnhout: Brepols.

Maul, S. M. 1998. *Tikip santakki mala bašmu . . .* Anstelle eines Vorwortes. In *Festschrift für Rykle Borger zu seinem 65. Geburtstag am 24. Mai 1994* (Cuneiform Monographs 10), ed. S. M. Maul, vii–xvii. Groningen: Styx.

Mayer, W. 1978. Seleukidische Rituale aus Warka mit Emesal-Gebeten. *Orientalia* 47: 431–58.

Mayer, W. 1979. Die Finanzierung einer Kampagne (TCL 3, 346–410). *Ugarit-Forschungen* 11: 571–96.

McCown, D. E., and R. C. Haines. 1967. *Nippur I: Temple of Enlil, scribal quarter, and soundings* (Oriental Institute Publications 78). Chicago: Oriental Institute of the University of Chicago.

McEwan, G. 1981. *Priest and temple in Hellenistic Babylonia* (Freiburger Altorientalische Studien 4). Wiesbaden: Steiner.

McEwan, G. 1982. *Texts from Hellenistic Babylonia in the Ashmolean Museum* (Oxford Editions of Cuneiform Texts 9). Oxford: Clarendon.

Meek, T. J. 1935. *Excavations at Nuzi, III: Old Akkadian, Sumerian, and Cappadocian texts from Nuzi* (Harvard Semitic Series 10). Cambridge, MA: Harvard University Press.

Meissner, B. 1893. *Beiträge zum altbobylonishen Privatrecht* (Assyriologische Bibliothek, 11). Leipzig: Hinrichs.

Melville, D. 2002. Ration computations at Fara: Multiplication or repeated addition? In *Under one sky: mathematics and astronomy in the ancient Near East* (Alter Orient und Altes Testament 297), ed. J. M. Steele and A. Imhausen, 237–52. Münster: Ugarit-Verlag.

Michalowski, P. 1993. Tokenism. *American Anthropologist* 95: 996–9.

Michalowski, P. 2000. The life and death of the Sumerian language in comparative perspective. *Acta Sumerologica* 22: 177–202.

Michalowski, P. 2003. Sumerian. In *The Cambridge encyclopedia of the world's ancient languages*, ed. R. J. Woodard, 19–59. Cambridge: Cambridge University Press.

Michel, C. 1998. Les marchands et les nombres: l'exemple des Assyriens à Kanis. In *Intellectual life of the ancient Near East: papers presented at the 43rd Rencontre assyriologique internationale*, ed. J. Prosecky, 249–67. Prague: Oriental Institute of the University of Chicago.

Michel, C. 2003. *Old Assyrian bibliography of cuneiform texts, bullae, seals and the results of the excavations at Aššur, Kültepe/Kaniš, Acemhöyük, Alişar and Boğazköy* (Old Assyrian Archives. Studies 1). Leiden: Nederlands Instituut voor het Nabije Oosten.

Miglus, P. 1999. *Städtische Wohnarchitektur in Babylonien und Assyrien* (Baghdader Forschungen 22). Mainz: von Zabern.

Milano, L. 1987. Barley for rations and barley for sowing (*ARET* II 51 and related matters). *Acta Sumerologica* 9: 177–201.

Millard, A. 1987. Cartography in the ancient Near East. In *The history of cartography, I: cartography in prehistoric, ancient, and medieval Europe and the Mediterranean*, ed. J. B. Harley and D. Woodward, 107–16. Chicago: University of Chicago Press.

Millard, A. R. 1994. *The eponyms of the Assyrian empire, 910–612 BC* (State Archives of Assyria Studies 2). Helsinki: Neo-Assyrian Text Corpus Project.

Monaco, S. 1985. Review of Foster 1982d. *Oriens Antiquus* 24: 310–5.

Moore, E. W. 1939. *Neo-Babylonian documents in the University of Michigan collection*. Ann Arbor: University of Michigan Press.

Moorey, P.R.S. 1978. *Kish excavations, 1923–1933*. Oxford: Clarendon.

Moran, W. L. 1992. *The Amarna letters*. Baltimore: Johns Hopkins University Press.

Negahban, E. O. 1991. *Excavations at Haft Tepe, Iran* (University Museum Monographs 70). Philadelphia: University of Pennsylvania Museum of Archaeology and Anthropology.

Nemet-Nejat, K. R. 1982. *Late Babylonian field plans in the British Museum* (Studia Pohl Series Maior 11). Rome: Pontifical Biblical Institute.

Nemet-Nejat, K. R. 1993. *Cuneiform mathematical texts as a reflection of everyday life in Mesopotamia* (American Oriental Series 75). New Haven: American Oriental Society.

Nemet-Nejat, K. R. 2001a. A Late Babylonian mathematical text. *NABU* 2001: nos. 10, 44.

Nemet-Nejat, K. R. 2001b. Lucky thirteen. *Bibliotheca Orientalis* 58: 59–72.

Nemet-Nejat, K. R. 2002. Square tablets in the Yale Babylonian Collection. In *Under one sky: mathematics and astronomy in the ancient Near East* (Alter Orient und Altes Testament 297), ed. J. M. Steele and A. Imhausen, 253–82. Münster: Ugarit-Verlag.

Nemet-Nejat, K. R., and R. Wallenfels. 1994. A sealed mathematical tablet. *NABU* 1994: no. 91.

Netz, R. 1999. *The shaping of deduction in Greek mathematics: a study in cognitive history* (Ideas in Context 51). Cambridge: Cambridge University Press.

Netz, R. 2002a. Counter culture: towards a history of Greek numeracy. *History of Science* 40: 321–51.

Netz, R. 2002b. Greek mathematicians: a group picture. In *Science and mathematics in ancient Greek culture*, ed. C. J. Tuplin and T. E .Rihll, 196–216. Oxford: Oxford University Press.

Netz, R. 2004. *The two books* On the Sphere and the Cylinder (The Works of Archimedes. Translation and Commentary 1). Cambridge: Cambridge University Press.

Neugebauer, O. 1934. *Vorgriechische Mathematik*. Berlin: Springer.

Neugebauer, O. 1935–7. *Mathematische Keilschrift-Texte*, 3 vols. (Quellen und Studien zur Geschichte der Mathematik, Astronomie und Physik A3). Berlin: Springer.

Neugebauer, O. 1941. On a special use of the sign 'zero' in cuneiform astronomical texts. *Journal of the American Oriental Society* 61: 213–5.

Neugebauer, O. 1952. *The exact sciences in antiquity*. Princeton: Princeton University Press.

Neugebauer, O. 1955. *Astronomical cuneiform texts*, 3 vols. Princeton: Institute for Advanced Study.

Neugebauer, O., and A. J. Sachs. 1945. *Mathematical cuneiform texts* (American Oriental Series 29). New Haven: American Oriental Society.

Neugebauer, O., and A. J. Sachs. 1967. Some atypical astronomical cuneiform texts, I. *Journal of Cuneiform Studies* 21: 183–218.

Neugebauer, O., and A. J. Sachs. 1969. Some atypical astronomical cuneiform texts, II. *Journal of Cuneiform Studies* 22: 92–113.

Neugebauer, O., and A. J. Sachs. 1984. Mathematical and metrological texts. *Journal of Cuneiform Studies* 36: 243–51.

Neumann, H. 1999. Ur-Dumuzida and Ur-DUN. Reflections on the relationship betwen state-initiated foreign trade and private economic activity in Mesopotamia towards the end of the third millennium BC. In *Trade and finance in ancient Mesopotamia* (MOS Studies 1), ed. J. G. Dercksen, 43–53. Leiden: Nederlands Historisch-Archaeologisch Instituut te Istanbul.

Neumann, H. 2000. Staatliche Verwaltung und privates Handwerk in der Ur III-Zeit: die Auftragstätigkeit der Schmiede von Girsu. In *Interdependency of institutions and private entrepreneurs* (MOS Studies 2), ed. A.C.M. Bongenaar, 119–33. Leiden: Nederlands Historisch-Archaeologisch Instituut te Istanbul.

Nies, J. B., and C. E. Keiser. 1920. *Historical, religious and economic texts and antiquities* (Babylonian Inscriptions in the Collection of James B. Nies 2). New Haven: Yale University Press.

Nissen, H. J. 1988. *The early history of the ancient Near East, 9000–2000 BC*. Chicago: University of Chicago Press.

Nissen, H. J., P. Damerow, and R. K. Englund. 1993. *Archaic bookkeeping: early writing and techniques of administration in the ancient Near East*. Chicago: Chicago University Press.

Nougayrol, J. 1968. *Nouveaux textes accadiens, hourrites et ugaritiques des archives et bibliothèques privées d'Ugarit* (Ugaritica 5). Paris: Imprimerie Nationale.

Oates, J. 1986. *Babylon*, rev. ed. London: Thames & Hudson.

Oates, J. 1993. Review of Schmandt-Besserat 1992. *Cambridge Archaeological Journal* 3: 149–53.

Oates, J., and D. Oates. 2001. *Nimrud: an Assyrian imperial city revealed*. London: British School of Archaeology in Iraq.

Oelsner, J. 1983. Review of Hunger 1976b. *Orientalische Literaturzeitung* 78: 246–50.

Oelsner, J. 1986. *Materialen zur Babylonischen Gesellschaft und Kultur in Hellenistischer Zeit*. Budapest: Eötvös Loránd Tudományegeyetem.

Oelsner, J. 1993. Aus dem Leben babylonischer Priester in der 2. Hälfte des 1. Jahrtausends v. Chr. (am Beispiel der Funde aus Uruk). In *Šulmu IV: Everyday life in ancient Near East*, ed. J. Zabłocka and S. Zawadzki, 235–42. Poznan: Adam Mickiewicz University Press.

Oelsner, J. 2000. Von Iqīšâ und einigen anderen spätgeborenen Babyloniern. In *Studi su vicino Oriente antico dedicati alla memoria di Luigi Cagni*, ed. S. Graziani, 797–813. Napoli: Istituto Universitario Orientale.

Oelsner, J. 2001. HS 201—eine Reziprokentabelle der Ur III-Zeit. In *Changing views on ancient Near Eastern mathematics* (Berliner Beiträge zum Vorderen Orient 19), ed. J. Høyrup and P. Damerow, 53–9. Berlin: Reimer.

Oelsner, J. 2006. Ein mathematische Tabellentext aus kassitischer Zeit, gefunden in Babylon. *NABU* 2006: no. 60.

O'Meara, D. J. 1989. *Pythagoras revived: mathematics and philosophy in Late Antiquity*. Oxford: Oxford University Press.

Oppenheim, A. L., et al. 1956–2006. *The Assyrian dictionary of the Oriental Institute of the University of Chicago*. Chicago: Oriental Institute of the University of Chicago.

Owen, D. 1982. *Selected Ur III texts from the Harvard Semitic Museum* (Materiali per il Vocabolario Neosumerico 11). Rome: Multigrafica Editrice.

Page, S. 1967. A new boundary stone of Merodach-baladan I. *Sumer* 23: 45–67.

Parpola, S. 1983. Assyrian library records. *Journal of Near Eastern Studies* 42: 1–29.

Parpola, S. 1993. *Letters from Assyrian and Babylonian scholars* (State Archives of Assyria 10). Helsinki: Helsinki University Press.

Parrot, A. 1956. *Le palais: l'architecture* (Mission archéologique de Mari 2/1). Paris: Geuthner.

Parrot, A. 1958. *Le palais: peintures murales* (Mission archéologique de Mari 2/2). Paris: Geuthner.

Parrot, A. 1968. Review of Heinrich and Seidl 1967. *Syria* 45: 155–7.

Pearce, L. E., and L. T. Doty. 2000. The activities of Anu-bēlšunu, Seleucid scribe. In *Assyriologica et Semitica: Festschrift für Joachim Oelsner* (Alter Orient und Altes Testament 252), ed. J. Marzahn and H. Neumann, 331–42. Münster: Ugarit-Verlag.

Pedersén, O. 1985–6. *Archives and libraries in the city of Assur: a study of the material from the German excavations*, 2 vols. Uppsala: Almqvist & Wiksel.

Pedersén, O. 1998. *Archives and libraries in the Ancient Near East, 1500–300 BC*. Bethesda: CDL.

Pedersén, O. 2005. *Archive und Bibliotheken in Babylon: die Tontafeln der Grabung Robert Koldeweys 1899–1917* (Abhandlungen der Deutschen Orient-Gesellschaft 25). Saarbrücken: SDV.

Peet, E. T., and L. Woolley. 1923. *The city of Ahkenaton*, vol. 3. London: Egypt Exploration Society.

Peltenburg, E. 1991. *The Burrell collection: Western Asiatic antiquities*. Edinburgh: Edinburgh University Press.

Pettinato, G. 1991. *Ebla: a new look at history*, trans. C. F. Richardson. Baltimore: Johns Hopkins University Press.

Pinches, T., and Sachs, A. 1955. *Late Babylonian astronomical and related texts*. Providence: Brown University Press.

Pingree, D., and E. Reiner. 1974–7. A Neo-Babylonian report on seasonal hours. *Archiv für Orientforschung* 25: 50–5.

Pohl, A. 1935. *Vorsargonische und sargonische Wirtschaftstexte* (Texte und Materialen der Frau Professor Hilprecht Collection 5). Leipzig: Hinrichs.

Pollock, S. 1999. *Ancient Mesopotamia: the Eden that never was*. Cambridge: Cambridge University Press.

Pomponio, F., and G. Visicato. 1994. *Early Dynastic administrative tablets of Šuruppak*. Napoli: Istituto Universitario Orientali.

Postgate, J. N. 1992. *Early Mesopotamia: society and economy at the dawn of history*. London: Routledge.

Postgate, J. N. 2005. New angles on early writing. *Cambridge Archaeological Journal* 15: 275–80.

Potts, D. 1998. *Mesopotamian civilization: the material foundations*. London: Athlone.

Potts, D. 1999. *The archaeology of Elam: formation and transformation of an ancient Iranian state*. Cambridge: Cambridge University Press.

Powell, A. B., and M. Frankenstein (eds.). 1997. *Ethnomathematics: challenging Eurocentrism in mathematics education*. Albany: SUNY Press.

Powell, M. A. 1971. Sumerian numeration and metrology. PhD diss., University of Minnesota.

Powell, M. A. 1976a. The antecedents of Old Babylonian place notation and the early history of Babylonian mathematics. *Historia Mathematica* 3: 417–39.

Powell, M. A. 1976b. Two notes on metrological mathematics in the Sargonic period. *Revue d'Assyriologie* 70: 97–102.

Powell, M. A. 1979. Ancient Mesopotamian weight metrology: methods, problems and perspectives. In *Studies in honor of Tom B. Jones* (Alter Orient und Altes Testament 203), ed. M. A. Powell and R. H. Sack, 71–109. Kevelaer/Neukirchener-Vluyn: Butzon & Bercker/Neukirchener Verlag.

Powell, M. A. 1987–90. Masse und Gewichte. In *Reallexikon der Assyriologie*, vol. 7, ed. D. O. Edzard, 457–517. Berlin: de Gruyter.

Powell, M. A. 1994. Review of Schmandt-Besserat 1992. *Journal of the American Oriental Society* 114: 96–7.

Powell, M. A. 1995. Metrology and mathematics in ancient Mesopotamia. In *Civilizations of the ancient Near East*, ed. J. Sasson, 1941–58. New York: Scribners.

Powell, M. A. 1999. *Wir müssen unsere Nische nutzen*: monies, motives, and methods in Babylonian economics. In *Trade and finance in ancient Mesopotamia* (MOS Studies 1), ed. J. G. Dercksen, 5–23. Leiden: Nederlands Historisch-Archaeologisch Instituut te Istanbul.

Proust, C. 2000. La multiplication babylonienne: la part non écrite du calcul. *Revue d'Histoire des Mathématiques* 6: 293–303.

Proust, C. 2002. Numération centésimàle de position à Mari. In *Florilegium Marianum 6: Recueil d'études à la mémoire d'André Parrot* (Mémoires de NABU 7), ed. D. Charpin and J.-M. Durand, 513–6. Paris: Société pour l'étude du Proche-orient ancien.

Proust, C. 2005. A propos d'un prisme du Louvre: aspects de l'enseignement des mathématiques en Mésopotamie. *SCIAMVS—Sources and Commentaries in Exact Sciences* 6: 3–32.

Quenet, P. 2005. The diffusion of the cuneiform writing system in northern Mesopotamia: the earliest archaeological evidence. *Iraq* 67: 31–40.

Quillien, J. 2003. Deux cadastres de l'époque d'Ur III. *Revue d'Histoire des Mathématiques* 9: 9–31.

Rawlinson, H. C. 1891. *A selection from the miscellaneous inscriptions of Assyria*, 2nd ed. London: Trustees of the British Museum.

Reade, J. 1986. Introduction. Rassam's Babylonian collection: the excavations and the archives. In *Tablets from Sippar,* vol. 1 (Catalogue of the Babylonian Tablets in the British Museum 6), by E. Leichty. London: British Museum Press.

Reade, J. 1993. Hormuzd Rassam and his discoveries. *Iraq* 40: 39–62.

Reschid, F. 1980. The tiles of king Marduk-shapik-zeri. *Sumer* 26: 124–49 (Arabic section).

Reschid, F., and C. Wilcke. 1975. Ein 'Grenzstein' aus dem ersten (?) Regierungsjahr des Königs Marduk-šāpik-zēri. *Zeitschrift für Assyriologie* 65: 34–62.

Richardson, S.F.C. 2002. The collapse of a complex state: a reappraisal of the end of the First Dynasty of Babylon, 1683–1597 BC. PhD diss., Columbia University.

Robins, G., and C. Shute. 1987. *The Rhind mathematical papyrus: an ancient Egyptian text*. London: British Museum Press.

Robson, E. 1995. Review of Nemet-Nejat 1993. *Bibliotheca Orientalis* 52: 424–32.

Robson, E. 1996. Building with bricks and mortar: quantity surveying in the Ur III and Old Babylonian periods. In *Houses and households in ancient Mesopotamia* (Comptes Rendus du 40ᵉ Rencontre Assyriologique Internationale, 1993), ed. K. R. Veenhof, 181–90. Leiden: Nederlands Historisch-Archaeologisch Instituut te Istanbul.

Robson, E. 1997. Three Old Babylonian methods for dealing with 'Pythagorean' triangles. *Journal of Cuneiform Studies* 49: 51–72.

Robson, E. 1999. *Mesopotamian mathematics, 2100–1600 BC: technical constants in bureaucracy and education* (Oxford Editions of Cuneiform Texts 14). Oxford: Clarendon.

Robson, E. 2000a. Mathematical cuneiform tablets in Philadelphia, I: problems and calculations. *SCIAMVS—Sources and Commentaries in Exact Sciences* 1: 11–48.

Robson, E. 2000b. The uses of mathematics in ancient Iraq, 6000–600 BC. In *Mathematics across cultures: the history of non-Western mathematics*, ed. H. Selin, 93–113. Dordrecht: Kluwer.

Robson, E. 2001a. Neither Sherlock Holmes nor Babylon: a reassessment of Plimpton 322. *Historia Mathematica* 28: 167–206.

Robson, E. 2001b. Technology in society: three textual case studies from Late Bronze Age Mesopotamia. In *The social context of technological change: Egypt and the Near East, 1650–1150 BCE*, ed. A. Shortland, 39–57. Oxford: Oxbow.

Robson, E. 2001c. The tablet house: a scribal school in Old Babylonian Nippur. *Revue d'Assyriologie* 95: 39–67.

Robson, E. 2002a. Guaranteed genuine Babylonian originals: the Plimpton collection and the early history of mathematical Assyriology. In *Mining the archives: Festschrift for C.B.F. Walker*, ed. C. Wunsch, 245–92. Dresden: ISLET.

Robson, E. 2002b. More than metrology: mathematics education in an Old Babylonian scribal school. In *Under one sky: mathematics and astronomy in the ancient Near East* (Alter Orient und Altes Testament 297), ed. J. M. Steele and A. Imhausen, 325–65. Münster: Ugarit-Verlag.

Robson, E. 2002c. Words and pictures: new light on Plimpton 322. *American Mathematical Monthly* 109: 105–20.

Robson, E. 2003. Tables and tabular formatting in Sumer, Babylonia, and Assyria, 2500–50 BCE. In *The history of mathematical tables from Sumer to spreadsheets*, ed. M. Campbell-Kelly, M. Croarken, R. G. Flood, and E. Robson, 18–47. Oxford: Oxford University Press.

Robson, E. 2003–4. Review of *Changing views on ancient Near Eastern mathematics*, ed. J. Høyrup and P. Damerow. *Archiv für Orientforschung* 50: 356–62.

Robson, E. 2004a. Accounting for change: the development of tabular bookkeeping in early Mesopotamia. In *Creating economic order: record-keeping, standardization, and the development of accounting in the ancient Near East* (International Scholars Conference on Ancient Near Eastern Economies 4), ed. M. Hudson and C. Wunsch, 107–44. Bethesda: CDL.

Robson, E. 2004b. Counting the days: scholarly conceptions and quantifications of time in Assyria and Babylonia, c. 750–250 BCE. In *Time and temporality in the ancient world*, ed. R. Rosen, 45–90. Philadelphia: University Museum Press.

Robson, E. 2004c. Mathematical cuneiform tablets in the Ashmolean Museum, Oxford. *SCIAMVS—Sources and Commentaries in Exact Sciences* 5: 3–65.

Robson, E. 2005a. Four Old Babylonian school tablets in the collection of the Catholic University of America. *Orientalia* 74: 389–98.

Robson, E. 2005b. Influence, ignorance, or indifference? Rethinking the relationship between Babylonian and Greek mathematics. *BSHM Bulletin* 4: 1–17.

Robson, E. 2006. Review of Glassner 2003. *American Journal of Archaeology* 110: 171–2.

Robson, E. 2007a. Gendered literacy and numeracy in the Sumerian literary corpus. In *Analysing literary Sumerian: corpus-based approaches*, ed. J. Ebeling and G. Cunningham, 215–49. London: Equinox.

Robson, E. 2007b. Mesopotamian mathematics. In *The mathematics of Egypt, Mesopotamia, China, India, and Islam: a source book*, ed. V. J. Katz, 57–186. Princeton: Princeton University Press.

Robson, E. 2007c. The clay tablet book in Sumer, Babylonia, and Assyria. In *The Blackwell companion to the history of the book*, ed. S. Eliot and J. Rose, 67–83. Oxford: Blackwell.

Robson, E. 2007d. The long career of a favorite figure: the *apsamikku* in Neo-Babylonian mathematics. In *From the banks of the Euphrates: studies in honor of Alice Louise Slotsky*, ed. M. Ross, 213–27. Winona Lake: Eisenbrauns.

Robson, E. 2007e. Mathematics, metrology, and literate numeracy. In *The Babylonian world*, ed. G. Leick, 414–27. London: Routledge.

Robson, E. Forthcoming 1. Clay mathematics: Euclid's Babylonian contemporaries. In *Euclid and his heritage*, ed. J. Carlson. Boston: Clay Mathematics Institute and the American Mathematical Society.

Robson, E. Forthcoming 2. Numeracy, literacy, and the state in early Mesopotamia. In *Literacy in state societies*, ed. K. Lomas. London: Institute of Classical Studies.

Robson, E. Forthcoming 3. Secrets de famille: prêtre et astronome à Uruk à l'époque hellénistique. In *Les lieux de savoir, I: Lieux et communautés*, ed. C. M. Jacob. Paris: Albin Michel.

Rochberg, F. 1993. The cultural locus of astronomy in Late Babylonia. In *Die Rolle der Astronomie in den Kulturen Mesopotamiens* (Grazer Morgenländische Studien 3), ed. H. D. Galter, 31–45. Graz: rm-Druck- & Vergesellschaft mbH.

Rochberg, F. 1998. *Babylonian horoscopes* (Transactions of the American Philosophical Society 88/1). Philadelphia: American Philosophical Society.

Rochberg, F. 2000. Scribes and scholars: the *ṭupšar Enūma Anu Enlil*. In *Assyriologica et Semitica: Festschrift für Joachim Oelsner* (Alter Orient und Altes Testament 252), ed. J. Marzahn and H. Neumann, 359–76. Münster: Ugarit-Verlag.

Rochberg, F. 2004. *The heavenly writing: divination, horoscopy, and astronomy in Mesopotamian culture*. Cambridge: Cambridge University Press.

Rochberg-Halton, F. 1983. Stellar distances in early Babylonian astronomy: a new perspective on HS 229. *Journal of Near Eastern Studies* 42: 209–17.

Röllig, W. 1980–3. Landkarten. In *Reallexikon der Assyriologie*, vol. 6, ed. D. O. Edzard, 464–7. Berlin: de Gruyter.

Römer, W.H.P. 2004. Zu einem Kudurru aus Nippur aus dem 16. Jahre Nebukadnezzars I. (etwa 1110 v. Chr.). *Ugarit-Forschungen* 36: 371–88.

Roth, M. T. 1989–90. The material composition of the Neo-Babylonian dowry. *Archiv für Orientforschung* 36–7: 1–55.

Roth, M. T. 1995. *Law collections from Mesopotamia and Asia Minor* (Writings from the Ancient World 6). Atlanta: Scholars Press.

Rouault, O., and M. G. Masetti-Rouault. 1993. *L'Eufrate e il tempo. Le civiltà del medio Eufrate e della Gezira siriana*. Milan: Electra.

Rubio, G. 2000. On the orthography of the Sumerian literary texts from the Ur III period. *Acta Sumerologica* 22: 203–26.

Rubio, G. 2003. Early Sumerian literature: enumerating the whole. In *De la tablilla a la inteligencia artificial: Homenaje al Prof. J. L. Cunchillos en su 65 aniversario*, ed. A. González Blanco, J. P. Vita, and J. Á. Zamora, 197–208. Zaragoza: Instituto de Estudios Islámicos y del Oriente Próximo.

Russell, J. M. 1999. *The writing on the wall: studes in the architectural context of Late Assyrian palace inscriptions* (Mesopotamian Civilizations 9). Winona Lake: Eisenbrauns.

Sachs, A. J. 1946. Notes on fractional expressions in Old Babylonian mathematical texts. *Journal of Near Eastern Studies* 5: 203–14.

Sachs, A. J. 1947a. Babylonian mathematical texts, I: reciprocals of regular sexagesimal numbers. *Journal of Cuneiform Studies* 1: 219–40.

Sachs, A. J. 1947b. Two Neo-Babylonian metrological tables from Nippur. *Journal of Cuneiform Studies* 1: 67–71.

Sachs, A. J. 1976. The latest datable cuneiform tablets. In *Kramer anniversary volume: cuneiform studies in honor of Samuel Noah Kramer* (Alter Orient und

Altest Testament 25), ed. B. Eichler, 379–98. Kevelaer/Neukirchen-Vluyn: But-zon & Bercker/Neukirchener Verlag.

Sachs, A. J., and H. Hunger. 1988–. *Astronomical diaries and related texts from Babylonia*, 5 vols. Vienna: Österreichischen Akademie der Wissenschaften.

Sachs, A. J., and O. Neugebauer. 1956. A procedure text concerning solar and lunar motion: BM 36712. *Journal of Cuneiform Studies* 10: 146–50.

Said, E. 1978. *Orientalism*. London: Routledge & Kegan Paul.

Sallaberger, W. 1999. Ur III-Zeit. In *Mesopotamien: Akkade-Zeit und Ur III-Zeit* (Orbis Biblicus et Orientalis 160/3), by W. Sallaberger and A. Westenholz, 121–414. Göttingen: Vandenhoek & Ruprecht.

Salvini, M. 1995. Sargon et l'Urartu. In *Khorsabad, le palais de Sargon II, roi d'Assyrie*, ed. A. Caubet, 133–57. Paris: La documentation Française.

Sassmannshausen, L. 1994. Ein ungewohnliches mittelbabylonisches Urkunden-fragment aus Nippur. *Baghdader Mitteilungen* 25: 447–58.

Sassmannshausen, L. 1999. The adaptation of the Kassites to the Babylonian civi-lization. In *Languages and cultures in contact: at the crossroads of civilizations in the Syro-Mesopotamian realm* (Orientalia Lovaniensia Analecta 96), ed. K. Van Lerberghe and G. Voet, 409–24. Leuven: Peeters.

Sassmannshausen, L. 2001. *Beiträge zur Verwaltung und Gesellschaft Babyloniens in der Kassitenzeit* (Baghdader Forschungen 21). Mainz am Rhein: von Zabern.

Scheil, V. 1900. *Textes élamites-sémitiques, première série* (Mémoires de la Déléga-tion en Perse 2). Paris: Leroux.

Scheil, V. 1902a. *Textes élamites-sémitiques, deuxième série* (Mémoires de la Délé-gation en Perse 4). Paris: Leroux.

Scheil, V. 1902b. *Une saison de fouilles à Sippar*. Cairo: Imprimerie de l'Institut français d'archéologie orientale.

Scheil, V. 1905. *Textes élamites-sémitiques, troisième série* (Mémoires de la Délé-gation en Perse 6). Paris: Leroux.

Scheil, V. 1915. Les tables 1 igi-x-gal bi etc. *Revue d'Assyriologie* 12: 195–8.

Scheil, V. 1916. Le texte mathématique 10201 du Musée de Philadelphie. *Revue d'Assyriologie* 13: 138–42.

Scheil, V. 1938. Tablettes susiennes. *Revue d'Assyriologie* 35: 92–103.

Schmandt-Besserat, D. 1992. *Before writing: from counting to cuneiform*, 2 vols. Austin: University of Texas Press.

Schmid, H. 1995. *Der Tempelturm Etemenanki in Babylon* (Baghdader Forschun-gen 17). Mainz: von Zabern.

Schmidt, J., et al. 1979. *XXIX. und XXX. Vorläufiger Bericht über die von dem Deutsche Archäologisches Institut aus Mitteln der Deutsche Forschungsgemein-schaft unternommenen Ausgrabungen in Uruk-Warka 1970/71 und 1971/72*. Berlin: Mann.

Schmidt, O. 1980. On Plimpton 322. Pythagorean numbers in Babylonian mathe-matics. *Centaurus* 24: 4–13.

Schneider, N. 1930. Die Geschäftsurkunden aus Drehem und Djoha in den Staatli-chen Museen (VAT) zu Berlin. *Orientalia*, 47–9.

Schroeder, O. 1916. *Kontrakte der Seleukidenzeit aus Warka* (Vorderasiatischer Schriftdenkmäler der Staatlichen Museen zu Berlin 15). Leipzig: Hinrichs.

Schroeder, O. 1920. *Keilschrifttexte aus Assur: verschiedenen Inhalts* (Ausgrabungen der Deutschen Orient-Gesellschaft in Assur. E, Inschriften 3). Leipzig: Hinrichs.

Seidl, U. 1989. *Die Babylonischen Kudurru-Reliefs: Symbole Mesopotamischer Gottheiten* (Orbis Biblicus et Orientalis 87). Freiburg: Universitätsverlag Freiburg.

Selin, H. (ed.). 2000. *Mathematics across cultures: the history of non-Western mathematics*. Dordrecht: Kluwer.

Shaffer, A. 2006. *Literary and religious texts*, vol. 3 (Ur Excavations. Texts 6/3). London: British Museum Press.

Sigrist, R. M. 1984a. *Les* sattukku *dans l'Ešumeša durant la période d'Isin et Larsa* (Bibliotheca Mesopotamica 11). Malibu: Undena.

Sigrist, R. M. 1984b. *Neo-Sumerian account texts in the Horn Archaeological Museum*, vol. 1. Berrien Springs: Andrews University Press.

Sigrist, R. M. 2000. *Texts from the Yale Babylonian Collection* (Sumerian Archival Texts 2–3). Bethesda: CDL.

Sigrist, M., H. H. Figulla, and C.B.F. Walker. 1996. *Catalogue of the Babylonian tablets in the British Museum*, vol. 2. London: British Museum Press.

Sigrist, M., R. Zadok, and C.B.F. Walker. 2006. *Catalogue of the Babylonian tablets in the British Museum*, vol. 3. London: British Museum Press.

Sjöberg, A. W. 1975. Examenstext A. *Zeitschrift für Assyriologie* 64: 137–76.

Slanski, K. E. 2000. Classification, historiography, and monumental authority: the Babylonian entitlement *narûs* (*kudurrus*). *Journal of Cuneiform Studies* 52: 95–114.

Slanski, K. E. 2003a. *The Babylonian entitlement* narûs (kudurrus): *a study in their form and function*. Boston: American Schools of Oriental Research.

Slanski, K. E. 2003b. Middle Babylonian period. In *A history of ancient Near Eastern law* (Handbuch der Orientalistik, I/72), ed. R. Westbrook, 485–520. Leiden: Brill.

Slanski, K. E. 2007. The Mesopotamian 'rod and ring': icon of righteous kingship and balance of power between palace and temple. In *Regime change in the ancient Near East and Egypt: from Sargon of Agade to Saddam Hussein* (Proceedings of the British Academy 136), ed. H. Crawford, 37–60. Oxford: British Academy.

Slotsky, A. L. 1997. *The bourse of Babylon: market quotations in the astronomical diaries of Babylonia*. Bethesda: CDL.

Smith, D. E. 1907. The mathematical tablets of Nippur. *Bulletin of the American Mathematical Society* 13: 392–8.

Smith, D. E. 1923–5. *History of mathematics*. Boston: Ginn & Co.

Sommerfeld, W. 1984. Die Mittelbabylonische Grenzsteinurkunde IM 5527. *Ugarit-Forschung* 16: 299–305.

Soubeyran, D. 1984. Textes mathématiques de Mari. *Revue d'Assyriologie* 78: 19–48.

Spar, I. 1988. *Tablets, cones, and bricks of the third and second millennia* BC (Cuneiform Texts in the Metropolitan Museum of Art 1). New York: Metropolitan Museum of Art.

Spycket, A. 2000. La baguette et l'anneau. Un symbole d'Iran et de Mésopotamie. In *Variatio delectat: Iran und der Westen. Gedenkschrift für Peter Calmeyer* (Alter Orient und Altes Testament 272), ed. R. Dittmann, B. Hrouda, U. Löw,

P. Matthiae, R. Mayer-Opficius, and S. Thürwächter, 651–66. Münster: Ugarit-Verlag.

Starr, I. 1990. *Queries to the Sungod: divination and politics in Sargonid Assyria* (State Archives of Assyria 4). Helsinki: Helsinki University Press.

Steele, F. R. 1951. Writing and history: the new tablets from Nippur. *University [of Pennsylvania] Museum Bulletin* 16/2: 21–7.

Steele, J. M. 2000. A 3405: an unusual astronomical text from Uruk. *Archive for History of Exact Sciences* 55: 103–55.

Steinkeller, P. 1989. *Sale documents of the Ur III period* (Freiburger Altorientalische Studien 17). Stuttgart: Steiner.

Steinkeller, P. 1991. The administration and economic organization of the Ur III state: the core and the periphery. In *The organization of power: aspects of bureaucracy in the ancient Near East* (Studies in Ancient Oriental Civilizations 46), 2nd ed., ed. M. Gibson and R. D. Biggs, 15–33. Chicago: University of Chicago Press.

Steinkeller, P. 2001. The Ur III period. In *Security for debt in ancient Near Eastern law* (Culture and History of the Ancient Near East 9), ed. R. Westbrook and R. Jasnow, 47–62. Leiden: Brill.

Steinkeller, P. 2003. Archival practices in Babylonia in the third millennium BC. In *Ancient archives and archival traditions: concepts of record-keeping in the ancient world*, ed. M. Brosius, 37–58. Oxford: Oxford University Press.

Steinkeller, P. 2004. The function of written documentation in the administrative praxis of early Babylonia. In *Creating economic order: record-keeping, standardization, and the development of accounting in the ancient Near East* (International Scholars Conference on Ancient Near Eastern Economies 4), ed. M. Hudson and C. Wunsch, 65–88. Bethesda: CDL.

Stephens, F. J. 1953. A surveyor's map of a field. *Journal of Cuneiform Studies* 7: 1–4.

Stève, M. J., H., Gasche, and L. De Meyer. 1980. La Susiane au II millennaire: à propos d'une interpretation des fouilles de Suse. *Iranica Antiqua* 15: 49–154.

Stol, M. 2004. Wirtschaft und Gesellschaft in Altbabylonischer Zeit. In *Mesopotamien: die altbabylonische Zeit* (Orbis Biblicus et Orientalis 106/4), by D. Charpin, D. O. Edzard, and M. Stol, 643–975. Fribourg and Göttingen: Academic Press and Vandenhoeck & Ruprecht.

Stone, E. C. 1987. *Nippur neighborhoods* (Studies in Ancient Oriental Civilization 44). Chicago: Oriental Institute of the University of Chicago.

Streck, M. 1916. *Aššurbanipal und die letzten assyrischen Könige bis zum Untergange Niniveh's*, 3 vols. Leipzig: Hinrichs.

Strommenger, E. 1980. *Habuba Kabira: eine Stadt vor 5000 Jahren. Ausgrabungen der Deutschen Orient-Gesellschaft am Euphrat in Habuba Kabira, Syrien*. Mainz: von Zabern.

Swerdlow, N. M. 1993. Otto E. Neugebauer (26 May 1899–19 February 1990). *Journal of the History of Astronomy* 24: 289–99.

Taisbak, M. 1999. Review of Netz 1999. *MAA Online* <http://www.maa.org/reviews/ netz.html>.

Tanret, M. 1986. Fragments de tablettes pour des fragments d'histoire. In *Fragmenta historiae Elamicae: mélanges offerts à M. J. Stève*, ed. L. De Meyer, H. Gasche, and F. Vallat, 139–50. Paris: Éditions Recherche sur les Civilisations.

Tanret, M. 2002. *Per aspera ad astra: l'apprentissage du cunéiforme à Sippar-Amnanum pendant la periode paléo-babylonienne tardive* (Mesopotamian History and Environment. Series III: Texts I/2). Gand: Université de Gand.

Tanret, M. 2003. Sur le calcul dans l'apprentissage scribale. *Akkadica* 124: 121–2.

Tanret, M. 2004. The works and the days: on scribal activity in Sippar Amnanum. *Revue d'Assyriologie* 98: 33–62.

Thompson, R. C. 1903. *Cuneiform texts from Babylonian tablets, &c. in the British Museum*, vol. 17. London: Trustees of the British Museum.

Thomsen, M.-L. 1984. *The Sumerian language: an introduction to its history and grammatical structure* (Mesopotamia 10). Copenhagen: Akademisk Vorlag.

Thureau-Dangin, F. 1897. Un cadastre chaldéen. *Revue d'Assyriologie* 4: 13–27.

Thureau-Dangin, F. 1903. *Recueil de tablettes chaldéennes*. Paris: Leroux.

Thureau-Dangin, F. 1912. La mesure du QA. *Revue d'Assyriologie* 9: 24–5.

Thureau-Dangin, F. 1918. Note métrologique. *Revue d'Assyriologie* 15: 59–60.

Thureau-Dangin, F. 1921. Numération et métrologie sumériens. *Revue d'Assyriologie* 18: 123–38.

Thureau-Dangin, F. 1922. *Tablettes d'Uruk à l'usage des prêtres du temple d'Anu au temps des Séleucides* (Textes cunéiformes du Louvre 6). Paris: Geuthner.

Thureau-Dangin, F. 1926. Notes Assyriologiques, XLIX. Le *še*, mesure linéaire. *Revue d'Assyriologie* 23: 33–4.

Thureau-Dangin, F. 1934. Une nouvelle tablette mathématique de Warka. *Revue d'Assyriologie* 31: 61–9.

Thureau-Dangin, F. 1935. La mesure des volumes d'après une tablette inédite du British Museum. *Revue d'Assyriologie* 32: 1–28.

Thureau-Dangin, F. 1938. *Textes mathématiques babyloniens* (Ex Oriente Lux 1). Leiden: Brill.

Tinney, S. 1999. On the curricular setting of Sumerian literature. *Iraq* 61: 159–72.

Toomer, G. J. 1988. Hipparchus and Babylonian astronomy. In *A scientific humanist: studies in memory of Abraham Sachs*, ed. E. Leichty, M. de J. Ellis, and P. Gerardi, 353–62. Philadelphia: University of Pennsylvania Museum of Archaeology and Anthropology.

Tripp, C. 2002. *A history of Iraq*, 2nd ed. Cambridge: Cambridge University Press.

Tybjerg, K. 2005. The point of Archimedes. *Early Science and Medicine* 10: 566–77.

Ulshöfer, A. 1995. *Die altassyrische Privaturkunden* (Freiburger Altorientalische Studien. Beihefte: Altassyrische Texte und Untersuchungen 4). Stuttgart: Steiner.

Unguru, S. 1975. On the need to rewrite the history of Greek mathematics. *Archive for the History of Exact Sciences* 15: 67–114.

Van De Mieroop, M. 1999. *Cuneiform texts and the writing of history*. London: Routledge.

Van De Mieroop, M. 1999–2000. An accountant's nightmare: the drafting of a year's summary. *Archiv für Orientforschung* 46–7: 111–29.

Van De Mieroop, M. 2004a. Accounting in early Mesopotamia: some remarks. In *Creating economic order: record-keeping, standardization, and the development of accounting in the ancient Near East* (International Scholars Conference on Ancient Near Eastern Economies 4), ed. M. Hudson and C. Wunsch, 47–64. Bethesda: CDL.

Van De Mieroop, 2004b. *A history of the ancient Near East, c. 3000–323 BC.* Oxford: Blackwell.

van der Meer, P. E. 1935. *Textes scolaires de Suse* (Mémoires de la Mission Archéologique de Perse 27). Paris: Leroux.

van der Meer, P. E. 1938. *Syllabaries A, B1 and B: with miscellaneous lexicographical texts from the Herbert Weld Collection* (Oxford Editions of Cuneiform Texts 4). Oxford: Clarendon.

van der Spek, R. 1985. The Babylonian temple during the Macedonian and Parthian domination. *Bibliotheca Orientalis* 42: 541–62.

van der Spek, R. 1987. The Babylonian city. In *Hellenism in the East: interaction of Greek and non-Greek civilizations from Syria to Central Asia after Alexander*, ed. A. Kuhrt and S. Sherwin-White, 57–74. Berkeley: University of California Press.

van der Spek, R. 2000. The *šatammus* of Esagila in the Seleucid and Arsacide periods. In *Assyriologica et Semitica: Festschrift für Joachim Oelsner* (Alter Orient und Altes Testament 252), ed. J. Marzahn and H. Neumann, 437–46. Münster: Ugarit-Verlag.

van der Spek, R. 2001. The theatre of Babylon in cuneiform. In *Veenhof anniversary volume. Studies presented to Klaas R. Veenhof on the occasion of his sixty-fifth birthday*, ed. W. van Soldt, 445–56. Leiden: Nederlands Instituut voor het Nabije Oosten.

van der Spek, R., and C. A. Mandemakers. 2003. Sense and nonsense in the statistical approach of Babylonian prices. *Bibliotheca Orientalis* 60: 521–37.

Van Dijk, J., and W. Mayer. 1980. *Texte aus dem Reš-Heiligtum in Uruk-Warka* (Baghdader Mitteilungen. Beiheft 2). Berlin: Mann.

van Soldt, W. H. 1995a. Babylonian lexical, religious and literary texts and scribal education at Ugarit. In *Ugarit: ein ostmediterranisches Kulturzentrum im Alten Orient*, ed. M. Dietrich and O. Loretz, 171–212. Münster: Ugarit-Verlag.

van Soldt, W. H. 1995b. *Solar omens of* Enūma Anu Enlil: *Tablets 23 (24)–29 (30)*. Istanbul: Nederlands Historisch-Archaeologisch Instituut te Istanbul.

van Soldt, W. H. 1999. The syllabic Akkadian texts. In *Handbook of Ugaritic studies* (Handbuch der Orientalistik 1/39), ed. W.G.E. Watson and N. Wyatt, 28–45. Leiden: Brill.

Vanstiphout, H.L.J. 1997. Sumerian canonical compositions. C. Individual focus. 6. School dialogues. In *The context of scripture, I: canonical compositions from the Biblical world*, ed. W. W. Hallo, 588–93. Leiden: Brill.

Vargyas, P. 2000. Silver and money in Achaemenid and Hellenistic Babylonia. In *Assyriologica et Semitica: Festschrift für Joachim Oelsner* (Alter Orient und Altes Testament 252), ed. J. Marzahn and H. Neumann, 513–21. Münster: Ugarit-Verlag.

Veenhof, K. R. 2003. Archives of Old Assyrian traders. In *Ancient archives and archival traditions: concepts of record-keeping in the ancient world*, ed. M. Brosius, 78–123. Oxford: Clarendon.

Veldhuis, N. 1997. Elementary education at Nippur: the lists of trees and wooden objects. PhD diss., University of Groningen. <http://socrates.berkeley.edu/~veldhuis/EEN/EEN.html>

Veldhuis, N. 1997–8. Review of Cavigneaux 1996. *Archiv für Orientforschungen* 44–5: 360–3.

Veldhuis, N. 2000a. Kassite exercises: literary and lexical extracts. *Journal of Cuneiform Studies* 52: 67–94.

Veldhuis, N. 2000b. Sumerian proverbs in their curricular context. *Journal of the American Oriental Society* 120: 383–99.

Veldhuis, N. 2004. *Religion, literature, and scholarship: the Sumerian composition «Nanše and the Birds»* (Cuneiform Monographs 22). Leiden: Brill.

Veldhuis, N. 2006. How did they learn cuneiform? Tribute/Word List C as an elementary exercise. In *Approaches to Sumerian literature: studies in honor of Stip (H.L.J. Vanstiphout)*, ed. P. Michalowski and N. Veldhuis, 181–200. Leiden: Brill.

Verbrugghe, G. P., and J. M. Wickersham. 1996. *Berossos and Manetho, introduced and translated: native traditions in ancient Mesopotamia and Egypt*. Ann Arbor: University of Michigan Press.

Visicato, G. 2000. *The power and the writing: the early scribes of Mesopotamia*. Bethesda: CDL.

von Weiher, E. 1983–98. *Spätbabylonische Texte aus Uruk*, vols. 2–5 (Ausgrabungen der Deutschen Forschungsgemeinschaft in Uruk-Warka. Endberichte 10–13). Berlin: Mann.

Waerden, B. L. van der. 1954. *Science awakening*, trans. A. Dresden. Groningen: Noordhoff.

Waerden, B. L. van der. 1983. *Geometry and algebra in ancient civilizations*. Berlin: Springer.

Waerzeggers, C. 2003–4. The Babylonian revolts against Xerxes and the 'end of archives'. *Archiv für Orientforschung* 50: 150–73.

Walker, C.B.F. 1981. *Cuneiform brick inscriptions in the British Museum, the Ashmolean Museum, Oxford, the City of Birmingham Museums and Art Gallery, the City of Bristol Museum and Art Gallery*. London: British Museum Publications.

Walker, C.B.F. 1990. Cuneiform. In *Reading the past: ancient writing from cuneiform to the alphabet*, 15–74. London: British Museum Publications.

Walker, C.B.F. 1998. Babylonian observations of Saturn during the reign of Kandalanu. In *Ancient astronomy and celestial divination*, ed. N. M. Swerdlow, 61–76. Cambridge, MA: MIT Press.

Walker, C.B.F., and D. Collon. 1980. Hormuzd Rassam's excavations for the British Museum at Sippar in 1881–1882. In *Sounding at Abu Habbah (Sippar)* (Tell ed-Der 3), ed. L. de Meyer, 93–114. Leuven: Peeters.

Wallenfels, R. 1993. *Apkallu*-sealings from Hellenistic Uruk. *Baghdader Mitteilungen* 24: 309–24.

Wallenfels, R. 1994. *Uruk: Hellenistic seal impressions in the Yale Babylonian Collection* (Ausgrabungen der Deutschen Forschungsgemeinschaft in Uruk-Warka. Endberichte 19). Berlin: Mann.

Wallenfels, R. 1998. *Seleucid archival texts in the Harvard Semitic Museum: text editions and catalogue raisonné of the seal impressions* (Cuneiform Monographs 12). Groningen: Styx.

Ward, W. H. 1910. *Seal cylinders of western Asia*. Washington, DC: Carnegie Institution.

Washburn, D. K., and D. W. Crowe. 1988. *Symmetries of culture: theory and practice of plane pattern analysis*. Seattle: University of Washington Press.

Washburn, D. K., and D. W. Crowe (eds.). 2004. *Symmetry comes of age: the role of pattern in culture*. Seattle: University of Washington Press.

Weidner, E. 1957–8. In aller Kürze. *Archiv für Orientforschung* 18: 393–4.

Weidner, E. 1967. *Gestirn-Darstellungen auf babylonischen Tontafeln* (Österreichische Akademie der Wissenschaften Philosophisch-Historische Klasse. Sitzungsberichte 254/2). Graz: Österreichischen Akademie der Wissenschaften.

Weisberg, D. 1991. *The Late Babylonian texts of the Oriental Institute collection* (Bibliotheca Mesopotamica 24). Malibu: Undena.

West, M. L. 1997. *The east face of Helicon: West Asiatic elements in early poetry and myth*. Oxford: Clarendon.

Westenholz, A. 1974–7. Old Akkadian school texts: some goals of Sargonic scribal education. *Archiv für Orientforschung* 25: 95–110.

Westenholz, A. 1999. The Old Akkadian period: history and culture. In *Mesopotamien: Akkade-Zeit und Ur III-Zeit* (Orbis Biblicus et Orientalis 160/3), by W. Sallaberger and A. Westenholz, 17–117. Göttingen: Vandenhoek & Ruprecht.

Westenholz, J. G. 1998. Thoughts on esoteric knowledge and secret lore. In *Intellectual life of the ancient Near East* (Comptes Rendus du 43ᵉ Rencontre Assyriologique Internationale), ed. J. Prosecký, 451–62. Prague: Academy of Sciences of the Czech Republic. Oriental Institute.

Whiting, R. M. 1984. More evidence for sexagesimal calculations in the third millennium BC. *Zeitschrift für Assyriologie und vorderasiatische Archäologie* 74: 59–66.

Wiesehöfer, J. 1996. *Ancient Persia from 550 BC to 650 AD*, trans. A. Azodi. London: Tauris.

Wilcke, C. 1987. Inschriften 1983–1984 (7.–8. Kampagne). In *Isin—Išān Bahrīyāt, III: die Ergebnisse der Ausgrabungen 1983–1984* (Bayerische Akademie der Wissenschaften Philosophisch-Historische Klasse. Abhandlungen 94), ed. B. Hrouda, 83–120. Munich: Bayerischen Akademie der Wissenschaften.

Wilcke, C. 1998–2001. Ninsun. In *Reallexikon der Assyriologie*, vol. 9, ed. D. O. Edzard, 503–4. Berlin: de Gruyter.

Winter, I. J. 1994. Light and radiance as aesthetic values in the art of ancient Mesopotamia (and some Indian parallels). In *Art: the integral vision*, ed. B. N. Saraswati et al., 123–32. New Delhi: DK Printworld.

Winter, I. J. 1999. Tree(s) on the mountain: landscape and territory on the victory stele of Naram-Sîn of Agade. In *Landscapes: territories, frontiers and horizons in the ancient Near East*, vol. 1 (History of the Ancient Near East Monographs III/1), ed. L. Milano, S. de Martino, F. M. Fales, and G. B. Lanfranchi, 63–72. Padova: Sargon.

Wunsch, C. 2000. *Das Egibi-Archiv, I: die Felder und Gärten* (Cuneiform Monographs 20). Groningen: Styx.

Yon, M. 2006. *The city of Ugarit at Tell Ras Shamra*. Winona Lake: Eisenbrauns.

Zettler, R. L. 1996. Written documents as excavated artifacts and the holistic interpretation of the Mesopotamian archaeological record. In *The study of the ancient Near East in the 21st century: the William Foxwell Albright Centennial*

Conference, ed. J. S. Cooper and G. M. Schwartz, 81–101. Winona Lake: Eisenbrauns.

Ziegler, N. 1999. *La population féminine des palais d'après les archives royales de Mari: le harem de Zimrî-Lîm* (Florilegium Marianum 4). Paris: Société pour l'étude du Proche-orient ancien.

Zimansky, P. 1993. Review of Schmandt-Besserat 1992. *Journal of Field Archaeology* 20: 513–7.

INDEX OF TABLETS

The page numbers of illustrations are shown in *italics*

SUBJECT INDEX

The page numbers of illustrations are shown in *italics*